Fusion

Fusion

Science, Politics, and the Invention of a New Energy Source

Joan Lisa Bromberg

The MIT Press
Cambridge, Massachusetts
London, England

This book was prepared under a contract with the US Department of Energy. The interpretations and viewpoints stated herein are, however, those of the author alone and do not represent the official position or policy of the US Government or of the Department of Energy.

Second printing, 1983

This book was set in Baskerville
by The MIT Press Computergraphics Department
and printed and bound by Halliday Lithograph
in the United States of America.

Library of Congress Cataloging in Publication Data

Bromberg, Joan Lisa.
 Fusion: science, politics, and the invention of a new energy source.

 Bibliography: p.
 Includes index.
 1. Controlled fusion. I. Title.
QC791.73.B76   1982   333.79′24   82–10039
ISBN 0-262-02180-3

MIT Press
0262521067
BROMBERG
FUSION

To Walter and David-Hahnon

Contents

Preface

Inspired, in part, by Richard G. Hewlett and Francis Duncan's admirable study *Nuclear Navy*, tracing the evolution of Rickover's submarine program, the leaders of the magnetic fusion energy program decided to commission a history of their own project. The idea was conceived by Robert L. Hirsch, in the midseventies, and put into effect by Edwin E. Kintner in 1977. Kintner arranged for the work to be funded by his division, the Office of Magnetic Fusion Energy (OMFE) within the Department of Energy, but supervised solely by Hewlett, the department's chief historian. This plan was faithfully carried out. From the start of my work until the day in June 1981 that I turned in my manuscript, the fusion energy leadership scrupulously refrained from pressing me to take any particular point of view or to include or omit any events.

In one important sense, this book is nonetheless a company history. The great bulk of the materials on which it is based comes from within the program. The chief sources were, first, the records of the Atomic Energy Commission, preserved in the archives of the Department of Energy's Historian's Office. Second, I used the files of the Office of Magnetic Fusion Energy, which include the files of its predecessors, the Controlled Thermonuclear Research Branch of the Division of Research (until 1971), the Division of Controlled Thermonuclear Research (1971–1976), and the Division of Magnetic Fusion Energy (1976–1977). In contrast to the material in the Historian's Office's archives, few of the OMFE files were marked for permanent retention, with the consequence that almost no items predating 1960 are included in this group of records. I also used a number of records collections and manuscript archives, including the central files of Oak Ridge National Laboratory, the records center of Los Alamos Scientific Laboratory, and the important collection of personal and project papers that Lyman Spitzer, Jr., donated to the archives at Princeton University.

Many of the program's personnel threw open to me their personal scientific files or the files of their research groups. My searches through this class of materials were decidedly unsystematic; nevertheless, some of the most valuable items of evidence turned up here. Finally, I talked to about 175 fusion scientists, engineers, and administrators. Our conversations ranged in formality from taped interviews to lunchtime chitchat, with, however, little correlation between the form of the interview and the value of its contents. These are the principal sources, in addition to the scientific literature, on which I constructed this history, and the range of views presented here, the controversies and criticisms, by and large reflect the opinions of people within, or very close to, the program.

I had no access to classified material. The chief result is that I did not attempt to include the history of laser fusion, which has had close ties to weapons research. I was also barred from examining the hearings on fusion conducted before 1958 by the Joint Committee on Atomic Energy, since these, unlike the rest of the materials generated by the magnetic fusion energy project, have never been declassified.

My readers will also want to know the boundaries I drew in point of laboratories covered and period treated. In order to get the work done within the allocated three years, I chose to concentrate on four programs on which the greatest part of the government money was spent, those at Lawrence Livermore National Laboratory, Los Alamos Scientific Laboratory, Oak Ridge National Laboratory, and the Princeton Plasma Physics Laboratory. Because of the same limitations of time, my history ends in September 1978 with the success of the Princeton Large Torus experiment (PLT) and the formulation of a new fusion strategy by the then director of the Department of Energy's Office of Energy Research, John M. Deutch. A great deal of interesting science was done outside the government laboratories. The period since September 1978 has been, likewise, as eventful as any in the program's history. I hope that my book will provide a useful starting point for historians who want to chronicle the recent years, or the nongovernment projects, as well as for those who want to deepen and revise the story I present.

Acknowledgments

I owe my greatest debt to Richard G. Hewlett, who guided this project throughout. He presided over my transition from an "internalist" to a historian of both science and its political context. He encouraged me in periods of difficulty and preserved me from my enemies. He read every chapter of the first three drafts and advised me on content, style, and evidence. I also want to thank the other members of the Historian's Office who gave me so much time, in particular Jack M. Holl, the current chief historian, Roger Anders, the archivist, and Betty Wise.

Robert B. Belfield worked for nine months as my research assistant. He concentrated on utility executives' attitudes and actions, but also made important contributions to chapters 3 and 6 and criticized one draft. I am also happy to acknowledge my debt to George K. Hess, Jr., whose insight into the ways of the federal bureaucracy made it possible to get this history project off the ground.

Essential to my research was the help afforded me by various archivists and record center managers. In particular, I am grateful to Lewis H. Strauss and Richard Pfau for allowing me access to the papers of Admiral Lewis L. Strauss.

The book could not have been written without the wholehearted cooperation that the men and women of the fusion community gave me. I should first like to express my gratitude to Edwin E. Kintner, director of the magnetic fusion energy program from 1976 through 1981. His interest kept the project alive, and his commitment to the integrity of history protected me to pursue the project as I saw fit. Literally hundreds of scientists, engineers, and administrators took the time to speak to me, show me through their experiments, and open access to their files. I should like to single out J. Rand McNally at Oak Ridge, Richard F. Post at Lawrence Livermore, Earl C. Tanner and Harold P. Furth at Princeton, Bruno Coppi at Massachusetts Institute of Technology, Harold Grad at New York University, Gareth E. Guest at General Atomic, Gloria B. Lubkin at *Physics Today*, and Lenore Ledman, Philip Stone, Rosalie Weller, and Franklin Valentine at the Washington office. I hope that those I have not mentioned will feel partially recompensed by the recognition given their contributions in the text of the book.

For reading large portions of the manuscript and alerting me to some of my errors, I heartily thank Harold P. Furth, Lyman Spitzer, Jr., and Robert L. Hirsch. The errors that remain are my responsibility.

Cynthia West made the typescript, assisted by Ann Aubin, Joan Goehler, Cindy Zentner, Flora Hollins, and Judie Stream.

Lyman Spitzer, Jr., and Melvin Gottlieb, 1959.

Demonstration model of Princeton's figure-8 stellarator, 1958.

James Tuck, 1958. Photo by Heka Davis, courtesy of *Physics Today*.

Los Alamos's Perhapsatron S-4, which was displayed at the second Geneva conference, 1958. James Phillips and Hugh Karr are shown.

Marshall Rosenbluth, 1979. Photo by Peter
Schweitzer.

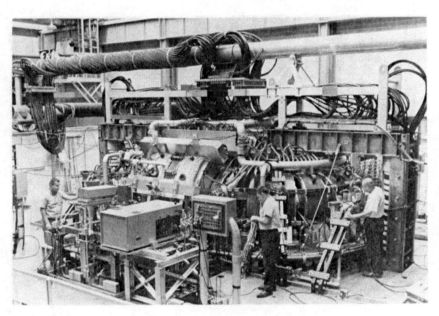

Princeton's Model C stellarator, 1961.

Arthur Ruark, 1969. Photo by Rosemary Gaffney, courtesy of *Physics Today*.

Livermore's ALICE, 1959. Richard Post (standing) and Charles Damm are shown.

Harold Furth, 1979. Photo by Peter Schweitzer.

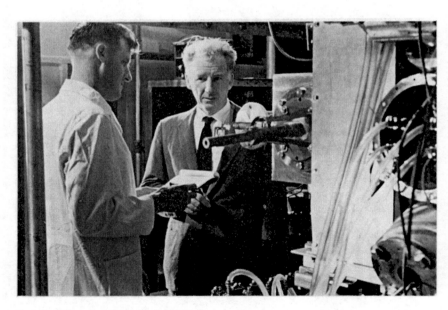

Ray Dandl and Arthur Snell, 1962.

Nicholas Christofilos, 1968. Photo by Gloria B. Lubkin.

Jerome Berkowitz and Harold Grad, 1970.

George Sawyer and Fred Ribe examining a component of Scyllac, 1974.

Los Alamos's Scyllac, 1974.

Herman Postma (left) showing visitors the Ormak conducting shell, 1973.

Oak Ridge's Ormak, 1971.

Tihiro Ohkawa and Robert Hirsch, 1974.

Lev Artsimovich at MIT, 1969. Courtesy of *Physics Today*.

T. Kenneth Fowler explaining a model of the 2XIIB machine, 1978.

John Deutch.

John Clarke and Edwin Kintner, 1980. Photo by Peter Schweitzer.

Magnet being built for the Mirror Feasibility Test Facility at Livermore, 1981.
Keith Thomassen, Ken Davis, and T. Kenneth Fowler are shown.

David Rose.

Fusion

1

Inventing a New Energy Source

Today, in major laboratories of the United States and other industrialized nations, with little public awareness or discussion,[1] a new nuclear energy technology is being readied for world use. The technology is that of magnetic fusion reactors, where the word "magnetic" refers to the technique of holding the fuel in place within the reactor core by means of strong magnetic fields.[2] The US government has supported a research program in fusion energy since 1951, and in the 30 years through 1980 it has expended more than $2 billion. Washington program administrators have estimated that the total cost, through the stage of commercialization, will approximate $15 billion.[3]

The final form or forms of this new energy source are not yet fixed; they will be determined in the next few decades by scientific results and governmental policies. The starting fuel will be some combination of light elements; it may range from the heavy isotopes of hydrogen to atoms, such as boron, that are several times heavier. Fusion reactors may eventually take the form of gigantic plants generating tens of thousands of megawatts, financed by consortia of utilities, or modest installations generating hundreds of megawatts. Depending on the technical possibilities and the choices made, the amount of radioactivity can vary from almost nothing to about one tenth of comparable breeder fission reactors.[4] If magnetic fusion becomes available for commercialization, the date will depend strongly upon the type of reactor selected; the most optimistic forecast points to the early part of the twenty-first century.

The public relations literature of fusion portrays each successive step in the program as a consequence of the technical developments that have preceded it. Yet it is a commonplace that the directions taken by the large scientific and technological research projects of the last few decades are heavily influenced by extrascientific pressures.[5]

And, in fact, the major decisions in fusion research have always emerged from a medley of technical, institutional, and political considerations. Resolutions have been required for issues such as what apparatus to build and what experiments to run on it, which projects to extend and which to terminate, how much latitude to allow scientists to wander from applied studies in pursuit of basic research, and how to divide resources between research on the engineering and the physics problems of fusion reactors. The decisions made on these and similar issues determine the shape that the emerging technology will have.[6] Collectively, they constitute the strategy of the US fusion research program. That strategy has changed significantly over the program's 30 years, but it has always been nourished by the double root of technical and external determinants.

To approach the history of the fusion program by focusing on its strategy is thus to bring into relief just this characteristic quality of the intermingling of science and politics. In selecting among the enormous output of scientific and technological results, the multitude of events, and the thousands of actors in order to capture in a few hundred pages something of the essence of the enterprise, I have needed a guide. My choice has been to write the story through the episodes that tell the most about program strategy and its evolution.

A history of strategy requires first of all an identification of the strategists. The determination of research direction has not been made by the same kinds of people over the three decades. Instead, there has been a steady shift of the decision-making function to higher and higher levels in the fusion hierarchy.

In the earliest years, 1951 and 1952, when there was as yet no hierarchy, the effective actors were the scientist-inventors who created or promoted the initial ideas for fusion reactors. In 1953 the Atomic Energy Commission (AEC), which financed the program, dictated its expansion. Money was poured in and full-time fusion teams were built, or enlarged, at several laboratories. The institutions of a fusion community began to form: A continuing series of interlaboratory conferences with written reports was established, a Washington administrator was assigned, and a coordinating council, the Steering Committee, was created. The group of policymakers at the laboratories grew larger. Augmenting the founders were the inventors of still other reactor ideas, an emerging cadre of senior scientists, and a handful of administrators whose background was in science but who were not active in fusion research.

Above the laboratory teams were additional strata: the Steering Committee, the Washington managers, and, for a time, the Atomic

Energy Commissioners themselves. Yet despite this superstructure, to a surprising extent the determination of strategy remained with the laboratory leaders. There were both scientific and administrative reasons. Scientifically, the state of research on the fusion fuel gas—the so-called plasma—was so rudimentary in the 1950s as to obstruct the formation of a consensus across the laboratories.[7] There was little theory. The majority of the techniques needed for an experimental exploration of hot plasmas were just being created. And when experimental results were forthcoming, the teams testing the various approaches found themselves in the classic situation of the blind men feeling the elephant: The range of parameters that a plasma can assume is so vast that each team was experiencing a different region of plasma phenomena.

On the management side, the Steering Committee was dominated by the laboratory leaders. The commissioners' intervention was significant from 1953 to 1958, but after Chairman Lewis L. Strauss retired in June 1958, the commission and the upper levels of AEC management lapsed into unconcern. In the years 1957–1965, when a technical basis for coordination was beginning to exist and centralized management of government research and development was beginning to be popular,[8] the Washington office was run by a physicist, Arthur E. Ruark, who had neither the taste for centralization nor the administrative skills for coordination.

The progressive centralization in governance began in 1966 with Amasa S. Bishop, Ruark's successor. Bishop put the major program decisions in the hands of a new Standing Committee, composed in equal parts of laboratory project heads and distinguished outside physicists. A second large step away from laboratory autonomy was taken by Robert L. Hirsch, who became director in 1972 of what was, by then, an independent division. Hirsch moved the function of formulating strategy more and more into the Washington office. The process culminated in 1977 when the new Department of Energy took over many of the programs that had been housed first in the Atomic Energy Commission and then in the short-lived Energy Research and Development Administration. At this point, some of the decision making for the now expensive program was shifted out of the Division of Magnetic Fusion Energy and up to higher levels of the department.

One constraint upon the strategists has been the needs of the institutions that grew to form the fusion program's infrastructure. The initiators—Lyman Spitzer, Jr., at Princeton University, James L. Tuck at Los Alamos Scientific Laboratory (LASL), and Herbert York and Richard F. Post at the Livermore branch of the University of California's

Radiation Laboratory—each originally assembled a small group of investigators around a particular scheme for a reactor. After the expansion that was mandated in 1953 when Strauss took office, the teams at Princeton, Los Alamos, and Livermore, now enlarged, took on the additional role of interest group. The code name Project Sherwood, given to the still secret program in late 1953, could now signify the institutional base as well as the research activities. In late 1957 the Thermonuclear Group at Oak Ridge National Laboratory (ORNL) graduated from an ancillary effort to the fourth major project as one consequence of the commissioners' decision to make a Sherwood exhibit the centerpiece of the 1958 Second International Conference on the Peaceful Uses of Atomic Energy. In the middle and late 1950s, more money was available than could easily be absorbed. Teammates in the four major laboratories were bound together by loyalty to a machine type (or "concept" in the jargon of the program), deprecation of the other concepts, and competition for the glory of being first.

Funding contracted sharply in the 1960s, and competition for money became a fact of life. Project leaders were henceforth unable to ignore the potential institutional impacts of major decisions. They struggled to promote reactor approaches or experiments in which they ardently believed, to maintain, as resources of value to the whole fusion enterprise, the scientific and engineering capabilities of their groups, and even to preserve jobs and ensure their groups' survival. The budgetary straitjacket began to loosen in 1971 when James R. Schlesinger moved from the Bureau of the Budget to the AEC chairmanship. It dissolved altogether in the following years; outlays zoomed from $33 million in fiscal year 1972 to nearly $300 million in fiscal year 1977. But by then experiments had become costly, and the interlaboratory contest for funding persisted.[9]

During the first epoch of fusion history, from 1951 to 1958, professional institutions, for example, the universities and learned societies with which the scientists were affiliated, had abnormally little influence because the program was secret. After declassification in 1958, however, they played a role in raising the quality of research. Another professional group that enters the story is the new subdiscipline of nuclear engineering. First spottily in the 1960s, and then in increasing numbers in the 1970s, nuclear engineers were drawn to apply their special perspectives to the subject of fusion reactors.[10] The effect was to illuminate, for the attention of the strategists, a set of problems that had been attended to only cursorily before.

The development of the Washington office as an articulated organization with full consciousness of its own distinct viewpoints and goals dates only from Robert Hirsch's time. There were nine employees, of whom five were technically trained, in the Division of Controlled Thermonuclear Research when Hirsch became acting director in 1972; three years later there were nearly 50 scientists and engineers in a total staff of 74. Hirsch and his large staff did more than centralize decision making. They tried to overcome the parochialism of the laboratories and forge a suprainstitutional loyalty to the goal of a magnetic fusion technology. They also brought into the program the peculiar capabilities and preoccupations of expert Washington managers: the ability to turn political circumstances to account in increasing the funding and priority of the program, a concern with planning, a concern with practical technical results as against scientific knowledge, and a stress on achieving milestones easily comprehensible to politicians and the public.[11]

It is a salient feature of fusion history that the political and social demands put to the program by both scientists and government leaders have continually changed. The unusual technical difficulty of controlling fusion power and the consequent fact that the program has stretched over decades has had this as an unavoidable correlative. In the 1950s, for example, AEC chairman Strauss jealously kept the program classified; he and the project scientists led a hard-driving scientific horse race, against both the USSR and Great Britain, to be the first nation to achieve fusion energy. In the 1960s, in contrast, the program leaders established close exchange of information with Soviet scientists that culminated in joint participation in international projects.

As one would expect, the recent attention to energy has had a crucial impact on fusion. Certainly, in 1951 thoughtful people realized that fossil fuels would some day run out. But the kind of concern and action that translates itself into political pressures only appeared in the 1970s. Increasing awareness of environmental degradation as a political issue, culminating around 1970, was also important. Environmentalists' objections were directed not only against the light-water fission reactor, but against the breeder program. The breeder is fusion's true competitor because the estimated world resources of uranium-238, on which the breeder could run, are about the same as the world supply of lithium, the limiting substance for the most easily realized fusion reactor, which would be fueled on a 50-50 mix of hydrogen-2 (deuterium) and hydrogen-3 (tritium).[12] Hence in their opposition to fission breeders, environmentalists turned to fusion. The nuclear engineers had hardly begun to delineate the range of radioactivity to be expected from various types of fusion plants in 1970, and it was easy for the press to oversell fusion as "virtually pollution-free." As utility executives grew sensitive to the popular dismay with fission, they also became more supportive of fusion.

We may put all this into some order through a periodization by decades of the history of fusion's external determinants. In the 1950s the most prominent external factor was the international competition for prestige. In the 1960s the breeder program was ascendant within the Atomic Energy Commission, and there was very little outside interest in fusion. Social demands put to the program then were really guesses as to what might benefit the public, advanced by the fusion community itself in its worried search for adequate funding. Overwhelmingly important for the 1970s were the environmental and energy issues.

We can now survey briefly how the fusion strategists evaluated the technical possibilities for creating a power engine in each epoch of the program's history, and what, in light of these possibilities and of the extratechnical constraints, they thought needed to be done. The first scientific phase was one of paper calculations. These showed that a fusion fuel would need to be kept at a temperature above 50 million degrees centigrade in order to allow for a net positive energy output. At such temperatures, the atoms comprising the fuel would completely decompose, or "ionize," into their constituent positive nuclei and negative electrons. It was easy to show that this hot plasma must be isolated from the "cool" (roughly 1,000 degrees centigrade) walls of the confining chamber. Spitzer, Tuck, and Post each offered a different design by which magnetic fields could be used to do the job. Spitzer and Post made use of the assumption that if the field configuration were adequate to contain an individual positively or negatively charged particle, it would be adequate to contain the collection of such particles that is the plasma. In each case the scheme for confinement had to be coupled with a plan for raising the fuel to the requisite temperature.

The computations needed to be verified by actual experiments. The supposition that a plasma could be treated as a collection of single particles was merely a tentative working hypothesis. Relevant empirical data were almost nonexistent since plasmas of the necessary temperatures had never been examined in laboratories. Extrapolation from results on the cool and only slightly ionized plasmas that so far had been studied was risky. The ensuing experimental phase was marked by formidable technical tasks. Complicated apparatus for producing the plasmas had to be devised and constructed, and high-field electromagnets in convoluted geometries had to be built and supported against their own mechanical stresses. Qualitative results showed that each of the trio of initial magnetic field configurations — the "stellarator" at Princeton, the "pinch" at Los Alamos, and the "mirror" at Livermore—showed at least some degree of confinement. Ideas for improvements were also at hand, as were still other, novel confinement schemes.

Quantitatively, however, the properties of the plasmas were not well understood. As well as inventing the confinement and heating methods, the fusion scientists had to invent the array of instruments needed to diagnose the state of the plasma by measuring its temperature, density, and other parameters. Moreover, the behavior of the hydrogen plasmas that the scientists wanted to test was being masked by a flood of impurities that boiled off the walls of the containment vessel. A more sophisticated level of vacuum technology was needed.

Faced with uncertainties about the adequacy of their simple, single-particle theory and about the performance of their devices, the scientists of Project Sherwood had two alternatives. They could laboriously build up a sure scientific and technical base before turning to the problem of creating a working technology or they could attempt, by trial-and-error empiricism, to invent their way to a reactor. A number of factors influenced their choice. They were in a mood—general among physicists in the dozen years following World War II—of enthusiasm over their capacities for problem solving. Secrecy prevented them from forming a realistic picture of the state that the arts of fusion science and technology had reached on a world scale; for all they knew, the Soviets were already mass-producing desk-sized reactors! There was a profusion of ideas, albeit of varying quality, and plenty of money for translating them into hardware. Strauss and his fellow commissioners were pressing for quick results. Finally, there was the goad of competition, both among the laboratories and internationally. As a net result, the Sherwood scientists opted for optimism. Whatever the differences of particular tactics among the laboratories, and these were large, the overarching strategy was founded upon "the belief that the controlled-fusion problem could be solved by short-term technological pressure, just as the problem of uncontrolled fusion has been."[13] And, at first, the route chosen appeared to be justified. By the end of 1957 some of the pinch machines, in particular the British ZETA pinch, seemed to have reached plasma conditions which could allow them to be scaled up to make reactors.

Against this background, 1958 unrolled as a year of trauma. The success of ZETA and its relatives proved an illusion, the product of insufficiently thorough experimentation. Fusion was declassified, and the crumbling walls of secrecy left revealed the fact that the large and able Soviet program had not advanced any farther than had the Americans and the British. Impurities came slowly under control, and it was possible to see that plasma was confined far less well than had been thought.

Already, during the Age of Optimism, there had been an alarm. In 1954 Sherwood Steering Committee Chairman Edward Teller had advanced qualitative arguments indicating that the independent-particle model was too simple. A more appropriate theory would treat the plasma as an ideal fluid, albeit a fluid whose properties were enormously complicated by its ability to generate, and respond to, electric and magnetic fields. Two disquieting conclusions followed. First, not one of the magnetic field configurations invented up to that time could work because a plasma, conceived as a fluid, would be unstable in each of them. Second, the theoretical basis of fusion physics was sketchier than it had been assumed to be because much less was known about magnetohydrodynamics (MHD), the science of electrically active fluids, than was known about the motion of single charged particles. The problem that Teller had highlighted was met, however. The pinch and stellarator teams modified their magnetic configurations. The mirror team had reasons to judge that MHD was not relevant to the mirror. The theorists of the Sherwood program applied themselves to, and greatly developed, magnetohydrodynamic theory.[14]

After 1958, however, it was discovered that Sherwood plasmas often behaved in ways that even the complex theories of ideal MHD could not describe. What made this serious was not merely the fact that the plasma was inaccessible to the theoretical methods at hand, but that the plasma was behaving badly. Small local perturbations were apparently ballooning up in a phenomenon called "microinstability" to distinguish it from the instabilities that could be treated by MHD theory. Microinstabilities were dumping plasma particles onto the walls, preventing densities from rising to the required levels and limiting the applicability of the heating methods that had been devised. To top it off, it was found that even mirror plasmas could be magnetohydrodynamically unstable. Many of these discoveries were made possible by the vigorous and imaginative development of refined diagnostic techniques within the fusion laboratories.

The effects of the shocks of the watershed year of 1958 and the subsequent uncovering of microinstabilities were reinforced by the rising professional standards that followed declassification. The consequence was a call for "greater recognition that physics rather than technology is the present avenue to success."[15] The fusion scientists turned to research into the physics of plasmas and, above all, to the experimental and theoretical investigation of instability.

Science and politics combined to bring the second epoch of US fusion history to an end around 1968. The tokamak, a variant of the pinch that Soviet scientists had been continually improving since the mid-1950s, broke into prominence in that year. The Russians had brought the tokamak to a point where it seemed almost free of the microinstabilities that were dumping the plasma out of the stellarator. They had also coaxed it into sustaining a fairly high temperature. The tokamak was less suitable than the stellarator as an experimental apparatus for the elucidation of plasma behavior. But its plasma was closer to a reactor plasma and hence a more interesting object for scientific study. In addition, one could hope that, having done so well, the tokamak could be made to do even better, whether or not its theory could be mastered. In 1969 Amasa Bishop and the Standing Committee decided to bring the tokamak into the US program; it was de facto also a decision to terminate the stellarator.

The success of the tokamak was the principal scientific input to the strategy that Robert Hirsch began to formulate about 1970, while he was still a member of the Washington staff under Roy W. Gould, Bishop's successor. Hirsch was convinced that in the tokamak magnetic fusion had finally found at least one scheme that could, assuredly, be made into a reactor. He concluded that the time was ripe to reorient the program from its mode of basic research to a mode in which an orderly series of steps could lead to the final, practical product.

There were several components to the Hirsch strategy. One was the formulation of long-range plans. Up to that time, the single articulated goal of the program had been the demonstration of "scientific feasibility." By scientific feasibility, the fusion community meant that the parameters of density, temperature, and length of time during which the plasma remains confined would be brought up to the values characteristic of the plasma of a working reactor. Hirsch prodded the leadership into planning up through the step of a demonstration reactor plant. Once he became director, moreover, he laid out a brisk pace for the several steps, with the demonstration plant projected first for the year 2000 and ultimately, by 1973, for the year 1995. As a concomitant, Hirsch and his staff wanted more attention given to the development of the technological components of a reactor, items such as the high-field superconducting magnets that had come to be identified as essential for a number of reactor types. The Washington staff, similarly, wanted the small, supporting, plasma investigations to be more tightly tailored to the needs of the larger machines that would form the program's backbone.

Another element in the strategy was a new attention to the guidance furnished by the nuclear engineers. A systematic program of paper studies of fusion electric generating plants had got under way in the late 1960s. These studies had the potential of illuminating the extent to which the different confinement schemes would be feasible from the viewpoint of engineering. Hirsch elevated engineering feasibility into one of the touchstones by which to judge which major approaches should be retained and which discarded. He worked to trim the program down to a handful of approaches: the tokamak on the one hand, and one or two practical and vigorous competitors on the other.

Hirsch brought about a major tactical change when he induced the fusion community to alter the nature of the demonstration of scientific feasibility. Hitherto the feasibility experiment had been planned to run on a deuterium fuel in order to avoid the complexity of working with radioactive apparatus. Hirsch persuaded the scientists to build into the tokamak feasibility experiment—the so-called Tokamak Fusion Test Reactor (TFTR)—the capacity to operate with a deuterium-tritium plasma. This was important to Hirsch because it would make possible a greater amount of research on reactor technology and because it would force the physicists to concentrate on studies of the burning plasmas that would be encountered in actual reactors. Not least, a deuterium-tritium TFTR could be politically persuasive because it would produce measurable power. Political factors were always weighty for Hirsch because he was persuaded that an adequate budget—one in the hundreds, rather than in the tens, of millions—was the sine qua non for fusion power. Funding of this magnitude had finally become possible in the early 1970s, through the conjunction of the program's scientific successes and the rise of environmental and energy issues. Hirsch believed that the fusion leaders had to offer the politicians a quid pro quo; they had to set and meet concrete milestones on the road to the reactor goal, of a type that would be comprehended by the politicians as constituting progress.

Milestones like better plasma parameters can be reached, however, by several paths. One path is via apparatus designed on the basis of a secure understanding of the physical phenomena. In the 1970s, that was still the slow method; despite 20 years of effort by capable men and women in all the leading scientific nations to master the complex phenomena involved, plasma theory had achieved only the most limited ability to guide fusion energy research.[16] Brute force methods constitute a second, often quicker path. In some machines, and the tokamak is one of them, better plasmas can be achieved simply by going to larger models. This can be understood in a crude way by reflecting, for example, that a parameter such as confinement time is highly likely to be increased if the walls are pushed farther away from the plasma. Pressure to achieve milestones on schedule can lead to pressure for large experiments.

There is a built-in circularity in this situation. Large experiments are expensive, expensive experiments are highly visible, and visibility generates still greater pressure for producing the parameters that have been promised. Thus large machines tend to squeeze out the fundamental research that might have offered an alternative, cheaper pathway. First of all, the big experiment tends to eat up the resources of smaller projects, which are sacrificed to ensure that the giant project will not fail. Second, the large machine itself is too tightly programmed toward its goal and too important to be diverted to fundamental investigations. The large-machine problem had confronted the program since the late 1960s; with the new emphasis that the Washington office put on milestones and with the commitment to the TFTR, it became central.

A second difficulty with Hirsch's strategy was that its elements had the potential for contending among themselves. There was, first of all, no a priori assurance that the eminence given to the tokamak concept was compatible with the use of engineering feasibility as a touchstone. Hirsch had formed his faith in the tokamak's capacity to serve as a reactor before the engineering studies on the tokamak had been completed. Second, Hirsch formulated the strategy to promote fusion reactors as an energy source, but the budget that Hirsch needed to carry it out was so large as to make it inevitable that his strategic plans would be reviewed at higher levels, by administrators with no prior commitment to fusion energy. In order to give fusion a secure place in the energy economy, Hirsch believed that a demonstration fusion reactor must be operating by the start of the twenty-first century. This was important politically because it was hard to get members of Congress, whose political lives stretch over a handful of decades at best, to fund a project whose payoff will be delayed until long after they have left office. It was important in view of fusion's rivalry with the breeder because a well-established breeder economy could forestall the need for fusion. Hirsch's superiors, however, whose careers were far less intimately linked with fusion's success, did not have the same motivation to abide by the timing of the Division of Magnetic Fusion Energy's master plan.

By 1976 the requisite engineering results were forthcoming, and it was clear that there were formidable technological obstacles to the success of the tokamak. Then, in 1977, incoming President Jimmy Carter reorganized energy research and development within a newly formed Department of Energy. One of the first jobs undertaken by the leaders of the new department was a review of the program whose budget had just burgeoned so astonishingly. The review committee concluded that the program—at that point under the direction of Edwin E. Kintner—was well run and its budget was appropriate. But the schedule was drastically changed and the preeminent position of the tokamak was put in doubt. John M. Deutch, director of the Department of Energy's Office of Energy Research, laid down a plan calling for a demonstration plant around 2015 and full commercialization in 2050, when the nation's need for power plants fueled by virtually inexhaustible substances should have matured. The more leisurely timing, in Deutch's view, would provide the opportunity to explore alternative confinement schemes that might have better engineering properties than the tokamak.

Whether even Deutch's less ambitious plan can be realized is by no means certain. It is now overwhelmingly probable that the TFTR, and similar foreign experiments, will demonstrate fusion's scientific feasibility during the 1980s. But neither the tokamak nor any other confinement scheme presently being tested has yet shown conclusively that it can meet the criteria of reactor engineers. Only the future will decide whether magnetic fusion energy will be technologically, as well as scientifically, feasible.

2

The Initiators

Lyman Spitzer and the stellarator

The proximate cause of the United States fusion program was Argentine dictator Juan Peron. Peron, estranged from his own scientists and intellectuals, and enamored of things German, had set up an island laboratory in the late forties for Ronald Richter, a mediocre German scientist who had come to him with a scheme for fusion power. On March 25, 1951, Peron's report on Richter's results received banner headlines in US newspapers. Argentina, proclaimed Peron, had surveyed the efforts that scientists in other nations were making to design uranium fission reactors and had elected to run the risk of taking the entirely different direction of designing fusion reactors. On February 16, "there was held with complete success the first tests which, with the use of this new method, produced controlled liberation of atomic energy."

No details of their process were released, by either Peron or Richter, and elsewhere in the issue, prominent American and European scientists declared their skepticism. "Only a superficial knowledge of the facts will establish at once that such a claim involves . . . several factors, each known to be impossible under the immutable laws of nature, as far as present knowledge goes," explained the *New York Times* science reporter two days later.[1]

At Princeton University Professor Lyman Spitzer, Jr., was preparing for a skiing vacation in Aspen, Colorado, when a telephone call from his father alerted him to the reports in the *Times*. Spitzer hurried out to get a paper. The articles he read could scarcely have fallen upon a more prepared mind. For one thing, although an astrophysicist by profession, Spitzer had always been intrigued by technology. As a student, he had tried his hand at designing both a particle accelerator and an intercity transportation system. Later, he had been one of the first astronomers to interest himself in the possibilities that space satellites offered for celestial observations.[2] Second, the core of a fusion reactor is a plasma—a sensibly neutral gas formed of electrons and positive ions, with larger or smaller admixtures of neutral atoms and other particles. Cosmic plasmas and their behavior in the vast magnetic fields of space were then of great interest to astrophysicists. Spitzer himself had been doing considerable work on the interstellar medium, a cool, tenuous material known to be as much a variety of plasma as the hot, dense interiors of stars.[3]

Finally, Spitzer had recently been introduced to the problems of hydrogen fusion. The explosion of the first Soviet atomic bomb in August 1949 had triggered a controversy in government circles that had ended with a decision to undertake a crash program to develop a working hydrogen bomb.[4] One of the people drawn into the hydrogen bomb project was John A. Wheeler, a Princeton physicist with whom Spitzer had been in close scientific contact. Wheeler had gone to the Los Alamos Scientific Laboratory in New Mexico to work on the weapon in the spring of 1950, and he had proposed that a salient of the Los Alamos group be established at Princeton so that work could continue when he returned there in the summer of 1951.[5] Spitzer had agreed to join this group as a part-time theorist and had suggested Project Matterhorn as the code name for the Princeton bomb work. In December 1950 he had stopped at Los Alamos on his way east from California to be briefed on the thermonuclear weapons program.[6]

Given Spitzer's background, it is understandable that the Argentine claims inspired him to ask how one might go about inventing a fusion reactor. In the relaxed atmosphere of the ski slopes, he began to make calculations. In a plasma containing hydrogen nuclei, energy would be released whenever two nuclei coalesced. For every collision that led to a fusion, however, there would be hundreds of collisions that led merely to a deflection because of the strong repulsive force two positively charged nuclei exert on each other. The plasma had somehow to be confined so that the particles could be made to encounter each other repeatedly. Simply encircling the plasma with material walls would not work. It was essential that the nuclei have high energies in order to increase the ratio of fusions to deflections, and if plasma at high energy hit walls at about 1,000 degrees centigrade, the plasma would lose its energy to the walls. "To keep the ions from hitting the wall, some type of force is required that will act at a distance. Gravitational forces are too small. Electrical forces act oppositely on positive ions and electrons and cannot simultaneously confine both. . . . A magnetic field seems to offer the only promise. In the presence of a strong magnetic field, a charged particle simply circles about the . . . lines of force."[7]

Spitzer had here arrived at one of the central ideas of fusion energy—the plasma must somehow be confined by magnetic fields. A second central idea is the concept of the power balance. The power released by fusions in each volume of plasma must exceed the power lost. Spitzer took the dominant source of loss to be *Bremsstrahlung* radiation, the energy radiated by electrons that are decelerated by encounters with ions. He calculated that with the most favorable fuel mix, and at particle energies corresponding to a temperature of 100 million degrees centigrade, the fusion energy would exceed the *Bremsstrahlung* loss by a factor of nearly 100. This ratio of power gained to power lost would come to be denoted Q. Instead of fixing a temperature and deducing Q as Spitzer did, it is also possible to set $Q = 1$ for the "break-even" condition of as much power created as lost. In this case, one derives the "ideal ignition temperature," the minimum allowable reactor temperature for a given fuel mix. Early calculations were to give an ideal ignition temperature of 50 million degrees centigrade or more, hotter than the interior of the sun and larger by many factors of 10 than any temperature theretofore known in the laboratory.

Spitzer's first problem, which was common to the originator of every early fusion scheme, was to choose a magnetic field pattern that would successfully confine the plasma and to design a configuration of conductors such that electric currents through them would produce the requisite fields. He proposed that a magnetic field be created by currents in coils placed so as to encircle the tube that would contain the plasma.

Spitzer first worked up the details of an ideal device in which the containing tube is an infinite cylinder and the electromagnetic field coils generate a straight and perfectly homogeneous field. Real tubes, however, are finite, and plasma placed within them will flow out the ends. "The simplest means to [correct] this is to bend the tube into a circle and join the ends together, forming a torus." This seemingly simple solution, however, gives rise to profound difficulties. In a torus, the field created by the external coils is neither straight nor homogeneous, but curved and stronger at the inside of the doughnut than at the outside. The effects of the curvature and the inhomogeneity are additive: together they give rise to a force that sends the positive ions down toward the bottom of the containing vessel and the electrons up toward the top (see figure 2.1). This separation of positive and negative charges produces an electric field that interacts with the magnetic field already present to push both ions and electrons, without

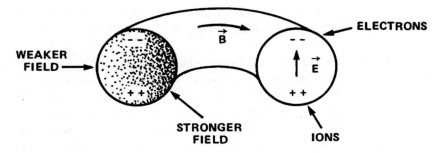

Figure 2.1
The electric ($\vec{E}$) and magnetic ($\vec{B}$) fields produced by coils encircling a torus.

regard to sign, to the outside wall. Spitzer's calculations showed that all the plasma in the tube would strike the outer wall before the particles had circled the tube once. This was as far as Spitzer's thinking had progressed by the time he left Aspen on April 8.[8]

Some days later, back in Princeton, Spitzer discovered a solution. His idea was to close the tube in a figure-8. "This . . . consists essentially of two tori, with magnetic fields in opposite directions, linked together to form a pretzel. It is readily seen that the drift velocity of an ion going around one end of the pretzel [due to field curvature and inhomogeneity] will be in the opposite direction from the drift velocity for the same ion going around the other end. . . . Thus . . . the drifts will cancel out in one circuit of the tube" (see figure 2.2).

Spitzer envisioned two types of fuel for his reactor. The fusion of deuterium with tritium nuclei[9] produces energetic neutrons and helium nuclei:

$$_1D^2 + {}_1T^3 \rightarrow {}_2He^4 + {}_0n^1 \text{ (+ kinetic energy of the resultant particles)}.$$

Deuterons fused with deuterons produce neutrons, protons, tritium and the isotope helium-3:

$$_1D^2 + {}_1D^2 = \begin{cases} {}_2He^3 + {}_0n^1 \text{ (+ energy)} \\ \text{or} \\ {}_1T^3 + {}_1p^1 \text{ (+ energy)}. \end{cases}$$

The deuterium-deuterium reaction is the more difficult to achieve. A larger containment vessel and stronger magnetic field coils would be needed. On the other hand, the deuterium-tritium reaction introduces the complication that tritium does not occur naturally. Spitzer proposed

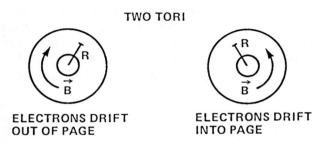

TWO TORI

ELECTRONS DRIFT
OUT OF PAGE

ELECTRONS DRIFT
INTO PAGE

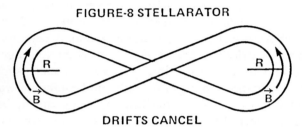

FIGURE-8 STELLARATOR

DRIFTS CANCEL

Figure 2.2
The principle of a figure-8 stellarator. R is the radius and $\vec{B}$ the magnetic field.

making use of the fact that neutrons can react with one of the isotopes of lithium to produce tritium:

$$_3Li^6 + n = {}_2He^4 + {}_1T^3 \; (+ \; \text{energy}).$$

Outside the pretzel-shaped tube "would be a shell of water-cooled lithium in which the neutrons would be absorbed and tritium generated." Finally, Spitzer anticipated that the gas could be heated to the requisite 100 million degrees centigrade by inducing a current in the plasma itself.[10]

At a meeting in late May, Spitzer, together with Wheeler and another Princeton colleague, selected the name "stellarator" for this invention that was to harness for humanity a mode of energy production until that time found only in the stars.[11] On the eve of his involvement in the crash program for the hydrogen bomb, therefore, Lyman Spitzer was diverted instead into the investigation of hydrogen reactors. It would be too simple to characterize this as a turn from weapons research to research on civilian applications. In 1951 a fusion reactor appeared to have important military potential. Fusion reactions between the two heavy hydrogen isotopes, deuterium and tritium, produce large amounts of neutrons that can be used to generate heavy fissile materials for use in either fission reactors or atomic bombs. Fusion reactions between two deuterium nuclei can also produce tritium, and at that time, with America on the verge of a deuterium-tritium hydrogen bomb, "any device which might produce considerable quantities of tritium is . . . of great military interest."[12]

The stellarator versus the pinch

Although Spitzer did not know it in April 1951, wartime work on the atomic bomb at Los Alamos had led quite naturally to discussions of fusion energy. Edward Teller, the Hungarian immigrant physicist, had directed studies on the hydrogen fusion bomb, or "Super," side by side with the main laboratory effort on the atomic bomb. A persistent accompaniment to this activity had been speculations about hydrogen fusion reactors. As the war drew to a close in 1945, this speculation had become more systematic. A basis in scientific understanding was laid by a masterful series of lectures that Enrico Fermi, the Italian Nobel laureate and architect of the first American fission reactor, had given on the physics of the Super.

Thereafter, several of the sessions of Edward Teller's "wild ideas" seminar were devoted to schemes for fusion reactors. Robert R. Wilson sketched a spherical copper reactor vessel within which the plasma is held together by magnetic fields and heated by radiofrequency waves and made a preliminary experiment together with Robert Cornog searching for thermonuclear reactions. James Leslie Tuck, a member of the British team that had been sent during the war to aid the Manhattan Engineering District Project, collaborated with mathematician Stanislaus Ulam on a theoretical study of the possibility of allowing jets of deuterium atoms at very high velocities to collide, and Tuck had followed this up with experiments aimed at achieving jets of the requisite energies. Participants remember accelerator physicists Luis W. Alvarez, Edwin M. MacMillan, and Donald W. Kerst, Princeton mathematician and polymath John von Neumann, and theorists Rolf Landshoff and Emil Konopinski as among those who made contributions. These activities had all come to an end by late 1946, as one after another most of the men who were involved left the isolated and beautiful New Mexican mesa and returned to civilian careers.[13]

James Tuck was one of those who left Los Alamos in 1946, going back to England to work at the Clarendon Laboratory at Oxford University. He soon found people in the United Kingdom who, quite independently of the group at Los Alamos, had also been thinking

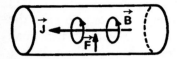

Figure 2.3
The linear pinch. The current $\vec{J}$ down the length of the plasma (shaded area) creates a circular magnetic field $\vec{B}$. The interaction of the field and the current then gives rise to an inward force $\vec{F}$ on the plasma.

about fusion reactors. One of these was Sir George P. Thomson, professor of physics at the Imperial College at London University. Thomson and a colleague, Moses Blackman, had already arrived at a solution to the problem of isolating a hot plasma from the walls of its vessel by a magnetic field. The pattern of magnetic force lines selected by Thomson and Blackman, and the way in which this field was to be generated by electric currents, was quite unlike the stellarator Spitzer was to invent in 1951. Thomson and Blackman elected to use the "pinch" effect.

The pinch effect is a phenomenon that had already been studied in investigations of the passage of electric currents through conducting gases. When a current of sufficient strength is discharged through the gas, the magnetic forces that form in a circular pattern around the current react back upon the gas to pinch it into a thin filament (see figure 2.3). A plasma is a conducting gas since it contains charged particles. The pinch effect clearly pointed to a way of keeping a hot plasma from the walls of its container.

G. P. Thomson's idea for a fusion reactor centered around exciting the pinch effect in an endless plasma within a toroidal vessel. He and Blackman had been sanguine enough to file for a patent in the spring of 1946. In hindsight, the Thomson-Blackman patent shows the same blend of high ingenuity and ingenuous optimism that was later to characterize Spitzer's first stellarator proposal. A continuous current was to be maintained in a deuterium plasma by radio waves fed into the torus through waveguides. The radio waves would also heat the plasma to the temperatures at which fusion reactions could begin. Neutrons from the fusion reactions would leave the vessel and deposit their energy outside, but charged particle products of the reaction were expected to remain within the vessel and, by giving up their energy to the plasma, maintain it at the hundreds of millions of degrees centigrade necessary. The process would thus be self-sustaining, a state of affairs that was later to become known as "reaching ignition."

Early in 1947, Britons Alan A. Ware and Stanley W. Cousins returned from their war service to start doctoral research. Thomson gave the two of them a review of his ideas and set them to work on two small experiments, one on a linear and the other on a toroidal pinch.[14]

At Oxford, Tuck also encountered the young Australian, Peter Clive Thonemann. Thonemann had brought with him from his home country original ideas for a pinch fusion energy device, although when Tuck met him, he was occupied in preparing a doctoral thesis in nuclear physics. Tuck himself would have liked to continue with controlled fusion in Oxford. There were, however, more pressing matters. Lord Cherwell, whose wartime aide Tuck had been and who had arranged for Tuck's return, wanted help in getting the Clarendon Laboratory back on its feet. Tuck spent his first 6 months helping to put a new electron synchrotron into operation.[15]

In late 1947 Tuck and Thonemann both found an opportunity to turn to fusion. A series of meetings was organized under the aegis of the UK Atomic Energy Research Establishment to provide an arena for the discussion of fusion research by those interested, to coordinate their efforts, and to arrange for funds. The tone of the meetings was noticeably cautious. The participants, who included Thomson, Blackman, Thonemann, and Tuck among others, agreed among themselves that the calculations underlying the Thomson-Blackman patent had been very approximate. They concluded that before there could be any thought of inventing, they would need to probe theoretically and experimentally whether the self-field created by the current in the pinch effect was truly adequate to isolate the plasma from the walls. Tuck, in addition, believed that getting sufficient energy into the plasma to achieve thermonuclear temperatures would not be easy. The greatest power, given the technology of the day, could be obtained by discharging a bank of condensers through the deuterium gas, but Tuck did not foresee raising the gas temperature to more than some tens of millions of degrees centigrade by that method—enough to give sufficient fusions to yield detectable numbers of neutrons, but not enough to lead to a self-sustaining plasma "burn." He brought forth the idea that a more realizable, intermediate aim might be to use hydrogen fusion for a highly pulsed, very intense source of neutrons.[16]

As a result of the meetings, Tuck received money for a pinch experiment and started assembling apparatus. Before he reached the point of actual experiment, however, another project enticed him. Edward Teller was recruiting scientists to come to Los Alamos and work on the hydrogen bomb. Tuck had hugely enjoyed his Los Alamos years and the stimulation and culture provided by the superb group of physicists assembled there. When Teller invited him to join the bomb work, therefore, the British physicist laid down his fusion research and returned to America.[17]

In late April or early May 1951, Kenneth W. Ford, a Princeton graduate student, spent a short time at Los Alamos, and Tuck learned of Spitzer's idea for a stellarator. His reactions were conditioned by both his own prior work on the pinch and by the 6 years he had already given to thermonuclear power. Tuck had misgivings about the physics of Spitzer's invention. He believed that loss of heat by thermal conduction across the plasma would be so serious in a steady-state device like the stellarator that it would be impossible to keep it going. It was for this reason that he was excited about the highly impulsive pinch with its exceptionally contracted plasma filament. In addition, Tuck mistrusted Spitzer's optimism—an optimism expressed by the very name, stellarator, that Spitzer had chosen, and by the fact that the Princetonian was planning to take out a patent. Tuck and his British colleagues had already arrived at a far more measured view of fusion energy's problems. Tuck therefore judged Spitzer's project as praiseworthy but "incredibly ambitious." "A self-sustaining thermonuclear reaction is so extremely difficult that it should not be thought of until problems of ionization, conduction and the effects of magnetic fields have been worked out experimentally and a detectable thermonuclear reaction has been created, all on a small scale."[18]

Spitzer, for his part, thought that the formula from which Tuck was deducing the quantity of heat conduction was unjustified. Furthermore, a pinch would be a fast-pulsed device, whereas Spitzer judged that, from an engineering viewpoint, only a steady-state device would be practical.[19] On May 11, Tuck and Spitzer met in person at a Washington meeting called to discuss the possibilities and implications of fusion power. The scientific evidence on the stellarator and the pinch was inconclusive; each man assessed it differently, and each concluded that his own approach was more promising and that the other had severe shortcomings. It was an example, at the very outset of the program, of the fierce partisanship that was to mark the first decade of research.[20]

First steps

In July 1951 Paul W. McDaniel, deputy director of the Atomic Energy Commission's Division of Research, informed the five commissioners that the division was awarding Spitzer $50,000 so that he and his colleagues could do theoretical studies related to the stellarator.[21]

The innocent phrase "theoretical studies related to the stellarator" masks some complexities, and it will be worth our while to pause and attend to them. Plasma theory is less concerned with the discovery of new fundamental laws than with the application of known laws to inordinately complex phenomena. The complexity of the phenomena dictates that in every application an approximation, or "model," appropriate to the individual features of the case under consideration be made. What Spitzer needed to do at this point was to decide which model or models should be singled out for investigation and elaboration.

Two different models were clearly of possible interest. In one, the plasma particles were assumed to interact only at those rare moments when they came very close.[22] At all other times, the motion of an individual particle would be affected solely by the large fields induced by the stellarator coils. The entire plasma could be regarded simply as a summation of individual ions and electrons moving in obedience to the fields and to the laws of two-body collisions. In the other model, called the "cooperative" or "collective" model, electromagnetic interactions among particles were presumed to produce effects large enough to be taken into account. Alternatively, the cooperative model was invoked when the motions of the particles were presumed to have enough regularity to affect phenomena. Collective motions, for example, might set waves of one or another kind propagating through the machine. Or the plasma particles might act collectively to set up local, fluctuating electric or magnetic fields, little eddies of force that would react back on the particles around them. It is a point of importance that the knowledge then available as a basis for developing each of these models was at a very different level. A great deal was known about the behavior of individual charged particles moving in complex magnetic and electric fields. Similarly, there had been work done, by Spitzer and his pupils among others, on two-body interactions between charged particles.[23] On the other hand, the study of cooperative phenomena such as waves in plasmas and small-scale turbulence in fields was in its infancy. Preliminary studies of some classes of waves had been made by astrophysicists involved in the new field of cosmic electrodynamics, but little was known about turbulence.[24] For this reason in 1951 the very question whether there was an adequate background of physical theory to permit the design of fusion reactors had to be considered unsettled.

Experiments had been done on plasmas, but they did not allow an unequivocal choice between the individual particle model and the collective model. One of the most important series of experiments had been carried out during World War II by a joint British-American team working at the University of California in Berkeley on discharges induced in gases of uranium compounds. They had found that turbulence was indeed a major factor. The catch was that the Berkeley team was working with plasmas of utterly different properties than those that would be needed in a stellarator. Their plasmas were only weakly ionized and so were composed largely of neutral atoms and molecules. Purified fusion plasmas would be completely stripped—nothing but bare nuclei and electrons. The Berkeley plasmas were "cold," with temperatures of a few thousands of degrees centigrade. Fusion plasmas would be 100,000 times hotter. It was, at bottom, a matter of scientific intuition whether one thought that collective effects would also exhibit themselves under fusion conditions. Tuck was inclined to discount independent-particle calculations, given the Berkeley experiments and other results pointing in the same direction. Spitzer, on the contrary, was more suspicious of the experiments. "Frankly, I am very skeptical of most previous work along this line. The experiments all involve much lower temperatures than I have assumed, and the presence of neutral atoms and molecules probably plays an important role."[25]

Spitzer used his AEC contract to finance research in both independent-particle and collective models, enlisting theoretical scientists from several Princeton departments to give part time to the calculations. Detailed computations were made of the types of magnetic fields that would be set up in a stellarator and the degree to which individual charges would be confined by such fields. This would indicate how well or badly an entire plasma would be confined in the absence of collective effects. As for research into the scarcely plowed field of collective effects, it had occurred to Spitzer that if a kink were somehow to occur in the filament of plasma in a pinch device, then cooperative phenomena might cause the entire filament to become unstable and thrash into the sides of the encasing tube (see figure 2.4). Spitzer persuaded his friend and colleague, the astrophysicist Martin Schwarzschild, to collaborate with Martin Kruskal, a young Princeton mathematician, in a theoretical exploration of this question. They found that Spitzer's conjecture was confirmed by more rigorous calculations.[26] They also calculated, however—and this was particularly important from Princeton's point of view—that such a "kink instability" would not show itself in the sort of stellarator that Spitzer was proposing. Spitzer also worked with a colleague to examine in detail the processes by which a gas is ionized into a plasma and then heated, for the problem of heating had only been touched upon cursorily in his original proposal.[27]

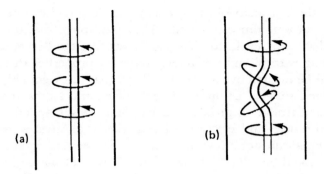

Figure 2.4
Kink instability in a pinch. Part (a) shows a pinch plasma with its associated magnetic field. In (b) a kink has formed, distorting the accompanying field. [From A. S. Bishop, *Project Sherwood* (Reading, MA: Addison-Wesley, 1958), p. 27]

By the fall of 1951, enough theoretical results had been obtained to make Princeton eager to start an experimental program. Above all, Spitzer thought it vital to determine empirically whether cooperative effects were present. Accordingly, the Division of Research convened a meeting to review the Princeton work in November 1951. One is struck by the inevitable sobering of expectations that occurred whenever scientists began to consider in detail a scheme for fusion energy. Like the members of the British Atomic Energy Research Establishment committee who had considered the Thomson-Blackman pinch reactor in 1947 and 1948, the Atomic Energy Commission panelists concluded that "a major developmental and engineering design study is not warranted in view of the fundamental questions remaining to be answered." Princeton's theoretical results deserved an experimental program, but one on "the fundamental research aspects."[28] In January 1952, therefore, Spitzer and his colleagues sent a formal proposal to the Division of Research for an experimental program "to explore some of the basic physical problems associated with the Stellarator."[29]

Spitzer felt that the crucial first step was to reach 1 million degrees centigrade. At these temperatures "ionization of hydrogen will be nearly complete and the plasma may be regarded as a classical assembly of free charged particles. The phenomena . . . should follow simple scaling laws, and observations at one million degrees should make possible accurate prediction at one hundred million degrees. In conventional plasmas, with electron temperatures of about ten thousand degrees and with neutral particles predominating . . . extrapolation to higher temperatures is more hazardous."[30]

The first experiment Spitzer contemplated was a "Model A" stellarator the size of a table top in which the goal would be to use ohmic currents to create a fully ionized plasma and heat its electrons to 1 million degrees centigrade. Such a device would be too small to confine the plasma for times long enough to allow the electrons to bring the plasma ions to their own temperature. The further step, therefore, would be the "Model B," a larger device in which Spitzer hoped that the plasma could be contained for the approximately 30 milliseconds (3/100 of a second) he computed it would take for the ions to reach 1 million degrees centigrade. At this stage it would be crucial that no instability or other collective effect would intervene to disrupt the plasma before the 30 milliseconds had passed. The final step would require a still larger machine, the "Model C." The magnetic field in this last device would be capable of being raised a hundredfold, from the experimental value of 200 gauss to a final value of 20,000 gauss (2 tesla). The rising field would cause the plasma to compress and hence would heat it by a factor of 100 to thermonuclear temperatures. If all went well, the Model C would be a reactor prototype.[31]

During all of 1951, James Tuck was deeply immersed in work supporting the Super. He was heading a team seeking more precise experimental determinations of the probabilities ("cross sections") for fusion reactions between deuterium and tritium nuclei. He had also found time, however, to lecture on controlled thermonuclear research. His enthusiasm was a powerful stimulus to interest. There were, moreover, others at Los Alamos who had independently developed a curiosity about fusion power. As the work on cross sections reached completion toward the end of the year, Tuck and his team members began to think of using part of their time for fusion energy research.[32]

In late 1951 or early 1952 Tuck approached Norris E. Bradbury, director of Los Alamos Scientific Laboratory, for permission to initiate experimental studies. At that time the laboratory was still overwhelmingly occupied with weapons research and Bradbury still conceived of its mission as predominantly that of developing better bombs. He was, however, interested in and hopeful about fusion power. Furthermore, there were possibilities in this line of investigation for military applications. Bradbury agreed to turn over $50,000 in discretionary funds to the project.[33]

Tuck's group[34] decided to build a toroidal pinch. Tuck christened it, with Ulam's help, the "Perhapsatron," in a direct allusion to the contrast he perceived between his appreciation of the state of the art and Spitzer's. A pinch device was a natural choice. The Los Alamos men were concerned, first of all, not to overlap the stellarator project.

Second, Tuck was enthusiastic about the pinch's possibilities. It seemed to him to possess particularly excellent properties for confinement. It also had a wonderful simplicity in that the plasma current provided its own magnetic field. Finally, it was economic because the field was strongest precisely where it was needed, namely, at the surface of the plasma column. Tuck stressed the essential difference between bomb research and energy research; fusion power generators would have to be competitive in cost, and magnets were likely to be among the most expensive components of the reactor.[35] The Perhapsatron was well suited to Los Alamos's capabilities. It did not make unreasonable demands on the laboratory's modest energy supplies. Moreover, diagnostic techniques that had been developed at Los Alamos to measure the rapidly evolving phenomena of bomb explosions were there to be applied to the roughly 10-millisecond sequence of signals the team expected from their pinch. It is true that Schwarzschild and Kruskal had been showing that pinches were inherently unstable. This theoretical prediction did not, however, bother Tuck unduly. The approximations that Schwarzschild and Kruskal had made did not correspond exactly to the conditions that were expected in the Perhapsatron, and there was a possibility that the difference would be decisive.[36] Tuck's first goals were modest. He wanted to see whether or not a pinch could actually be formed, and to ascertain its stability. Over the longer term, he had in mind his old aim: the production of a plasma hot enough to produce detectable yields of fusion neutrons.[37]

Richard Post and the mirror

Another, radically different, configuration of magnetic field lines was still to be proposed in 1952, this time in California. In the early months of 1952 scientists and commission officials were debating the future of a complex of laboratories on the outskirts of Livermore, a small-sized town in a cattle- and wine-producing valley southeast of the San Francisco Bay area. Formerly a naval training camp, the site had been acquired by the University of California at Berkeley in 1950 at the instigation of Ernest O. Lawrence, the director of the University's Radiation Laboratory. Now the buildings housed the MTA, a linear accelerator that had been designed by Lawrence's laboratory as a prototype for a machine that could produce neutrons for the manufacture of fissile materials. Lawrence expected that the full-scale production model would soon be built in Weldon Springs, Missouri. Meanwhile, Lawrence was backing Edward Teller in his agitation for a second weapons laboratory to supplement Los Alamos. The new laboratory was intended to be heavily committed to the development of thermonuclear weapons. Such a laboratory, Lawrence thought, might suitably be placed at Livermore, since the MTA prototype was soon to be phased out there.[38]

It was natural for Lawrence to involve Herbert F. York, a young Berkeley physicist, in the discussions, for York was the one of Lawrence's students who had been most closely connected with the work on the hydrogen bomb. Lawrence's intention was that York be chosen head of the Livermore weapons laboratory, should it be possible to bring it into existence. In the early months of 1952 York traveled to several major laboratories and universities to confer about the ideas of Teller and Lawrence. These trips brought him to both Princeton and Los Alamos, and he took the opportunity to talk with Spitzer and Tuck about controlled fusion research.

York was very interested in fusion research. Although the overwhelming emphasis at the Livermore installation was to be on military work, York wanted to broaden the laboratory somewhat beyond that. A few additional projects would add variety and ultimately strengthen the laboratory's reputation; both these effects would in turn aid recruiting. Thermonuclear fusion was an obvious choice. It rested upon scientific and technological expertise that was similar to that for hydrogen bombs. Weapons scientists and controlled fusion scientists would be able to get some benefit from talking together, and personnel could move easily from one type of project to the other.[39]

In intervals of time free from the main organizing tasks, York surveyed the nascent American fusion program. Two distinct approaches to confinement were then known, as we have seen. One was the pinch, which confines the plasma by means of the interaction of the current in the plasma with its self-produced magnetic field. The other, the stellarator, made use of fields created by external coils surrounding the plasma tube. In the latter case there was the problem of geometry: If the tube were a cylinder, plasma would be lost out the ends; if it were a torus, plasma would be lost to the sides. Spitzer had suggested escaping the dilemma by forming the tube into a pretzel, thus decreasing the sideways loss. York now adopted a strategy of his teacher, Lawrence. It could be fruitful to do a thing differently merely for the sake of being different. He decided Livermore would use externally produced fields, unlike Los Alamos. Unlike Princeton, it would use a straight cylinder.[40] The advantage of York's idea was that the magnetic field coils would be much simpler than those for the pretzel-shaped stellarator. Hence they would be less costly and create fewer difficulties of design, construction, and maintenance. The disadvantage was that the field would be "open" instead of "closed." That is, instead of magnetic force lines that closed on themselves, and thereby remained wholly within the plasma vessel, as in both the stellarator and pinch, York's force lines would extend out the ends of a cylindrical vessel. The charged particles, which wind their way around the lines like vines around a bean pole, would therefore flow out along the magnetic field. It would be necessary for the people who became involved in the project to find some way of preventing the plasma's particles from pouring out the ends of the tube. As one possibility, York proposed using cavities filled with radiofrequency electromagnetic waves to dam the ends. What he had in mind was the force that electromagnetic waves exert upon matter due to their radiation pressure, a force that had been known since the late nineteenth century.[41]

In the academic year 1951–1952 York gave three lectures on the problems of fusion at the Berkeley Radiation Laboratory. One of his listeners was Richard F. Post, who had just joined the laboratory to work with McMillan on the synchrotron. Post had recently completed his doctor's degree at Stanford University with studies that had introduced him to both microwaves and the problems of plasmas. In addition, he had a pronounced inclination toward research problems that combined scientific interest with social utility. Post was immediately captured by the fusion problem. He wrote a long memorandum to York that explored the radiofrequency dam and added other suggestions. York, in turn, invited Post to join the group of Berkeley scientists that was preparing to move to Livermore: Post would set up an experimental thermonuclear fusion group there. Meanwhile, Post, with the aid of others, began theoretical studies and small experiments at Berkeley.[42]

One of Post's earliest results led away from York's suggestions.[43] Calculations showed that radiation pressure dams could not be made strong enough to hold in a plasma of reasonable density and temperature unless an exorbitant amount of power were fed into them. Post and his associates turned to the use of these fields as a supplementary device to augment some other method, and the method chosen was based upon the behavior of cosmic rays in the earth's magnetic field. If charged particles spiral along a magnetic field that is not constant, but is stronger at the ends than at the center, some of them will be reflected. This kind of field is called a magnetic mirror (figure 2.5). The condition for reflection is that $v_\perp$, the component of the particle's velocity which is perpendicular to the direction of the field, be sufficiently large with respect to $v_\parallel$, the component of velocity parallel to the field. The Livermore group proposed to enhance mirrors

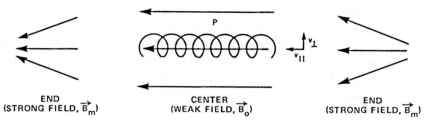

Figure 2.5
A magnetic mirror field. Particle P spirals to the left along a magnetic field line. Its velocity is composed of a component $v_\parallel$ parallel to the field $\vec{B}_0$ and a component $v_\perp$ perpendicular to $\vec{B}_0$. The particle will be reflected at the left "mirror" if $v_\perp^2/(v_\perp^2 + v_\parallel^2) \geq B_0/B_m$.

with radiofrequency fields, used now not as dams, but as a way of increasing the perpendicular velocity component with respect to the component parallel to the field.

Thus began a slide that in about a year was to carry the Livermore Laboratory group from its original focus on radiofrequency fields for confinement, heating, and measurement to a program centered on the mirror. For the time being, however, in early 1952, radio-fields were still Livermore's primary interest and Richard F. Post was jocularly introducing himself to his new colleagues in the fusion field as "radiofrequency Post."[44]

The Atomic Energy Commission

In March 1952 the Division of Research took Princeton's new proposal for an experimental program to the commissioners. The five men that the division leaders faced were Gordon E. Dean, a lawyer with lengthy government experience who had been promoted to chairman on the retirement of the first Atomic Energy Commission chairman, David E. Lilienthal; Henry D. Smyth, author of the "Smyth report" on the atomic bomb project and head of the Princeton Physics Department; Thomas E. Murray, former industrialist and engineer; T. Keith Glennan, who had come to the agency from the presidency of Case Institute of Technology; and Eugene M. Zuckert, like Dean a lawyer and government administrator.[45]

There was solid support for controlled thermonuclear research within both the division and the commission. The new director of the Division of Research, Thomas H. Johnson, had taken an immediate and unusually strong interest in the Princeton work when he had come from

Brookhaven National Laboratory in October 1951 to take up his post. Johnson was a cosmic ray specialist, and as such was very knowledgeable in plasma physics. Smyth chaired Princeton's Committee on Projects and Inventions, which, among its other responsibilities, oversaw Matterhorn. Smyth favored a general expansion of AEC-funded research activities; in the particular case of fusion power, his backing was also based upon an optimistic estimate of its potentialities. Murray was forcefully and consistently championing federal development of nuclear power, and his support of fusion was a corollary of this overarching position. The commissioners had satisfied themselves the previous spring that there was probably little behind the Argentine claims. But they knew that the British were working in the field. They had an eye on US scientific prestige and an obligation to place the AEC in the forefront of nuclear research related to national security and the nation's technological capacity. And commission and division leaders alike "all felt that [Spitzer's] suggestion simply had to be tested in practice . . . that *if* the process worked as Spitzer proposed trying, then the real economic benefits were great indeed."[46]

Johnson asked the commissioners to approve Princeton's proposal and by so doing to endorse a general research program in controlled thermonuclear fusion of which the Princeton project should be a particular instance. On an "educated guess," he estimated that "it would cost the Commission something on the order of $1,000,000 over the period of three and a half to four years to answer the question 'Can a plasma at high temperature be confined by a magnetic field? If not, why not?'" Without much debate, the commissioners gave their concurrence.[47]

On another issue, however, Johnson got a more cautious response. The AEC fusion program was born classified. Not only the details of the program, but the very fact that the AEC was sponsoring work on thermonuclear reactors, and the names of the sites at which the research was being conducted, were secret. Johnson emphatically recommended that it be declassified. In this, he had the strong support of the director of classification, as well as the concurrence of the General Advisory Committee, the Committee of Senior Responsible Reviewers, the Division of Military Application, and the group of scientists which had convened in Princeton the previous November. Johnson's argument was that an open program could be pursued far more effectively and successfully. It would, for example, be substantially easier to recruit into it excellent scientists. Not stated, but perhaps also in Johnson's mind, was the fact that there was opposition among the Princeton faculty to expanding a classified program on their campus.

Here, the commissioners balked. Dean suggested instead that a second secret conference, on the pattern of that of November, could serve both as a forum for a further discussion of declassification and as a means of easing recruitment problems by stimulating interest among persons who were already cleared, but were not in the program.[48]

The meeting that Gordon Dean thus initiated was held in Denver on June 28, 1952, with about 80 persons in attendance, the majority from outside the program. The occasion marked the birth of the American fusion community. Information was drawn together as the scattered results so far achieved, most of which were theoretical calculations, were presented. Spitzer and Tuck gave major reviews, and smaller ones were presented by Post, William P. Allis from the plasma physics group at MIT's Research Laboratory for Electronics, and Winston H. Bostick of Tufts University, who had been working on plasmas for the Air Force. Social ties were formed, as the scientists had the chance to meet, often for the first time.[49]

Only one among the participants spoke with any vehemence for continued secrecy for the program, and his argument was that once the field was thrown open, it might be difficult to classify new developments. Otherwise, "most scientists felt that the best hope for future progress lies in declassification."[50] Nevertheless, the commissioners decided to keep controlled thermonuclear research secret for the time being. Their prudence was scarcely surprising, given the attacks the commission had sustained over its 5-year history for allegedly loose security.[51]

Aside from their active interest in the issue of classification, the Dean commission did not intervene in the conduct of the fusion program. The commissioners were encouraging and anxious to be informed, but the supervision was left to the Division of Research. As for the division's attitude, it was described by Johnson when in August 1953 he told the commission that he believed the project should be allowed to find its own momentum, one dictated by the judgment of the project leaders.[52] In fact, the overall pace of the program was leisurely. Spitzer, Tuck, and Post, for example, were each devoting only part-time to it. As Johnson wrote in 1953, "The work on the thermonuclear reactions has . . . been done in a relaxed manner with no feeling of urgency such as might be associated with specific military or economic objectives."[53]

The US program for controlled thermonuclear research did not originate in Washington but was a grass-roots enthusiasm initiated by working scientists connected with the national laboratories. Spitzer first sketched out the principles of the stellarator and then approached the Division of Research for funding. Tuck was stimulated by Spitzer's work and by interest in Washington to ask his laboratory director for some of Los Alamos Scientific Laboratory's discretionary monies. York was recruiting, and sought to make his new weapons laboratory more attractive.

But why did the scientists propose fusion research? For one thing, like their colleagues in fission energy research, the fusion scientists saw a social need. Although utilities were scarcely clamoring for nuclear power in the early fifties, it was recognized by thoughtful people that fossil fuels were bound to be exhausted. Fusion fuels were potentially inexhaustible. Tritium, it is true, is an artificial isotope, and must be manufactured from lithium, which is about as abundant as uranium-238.[54] Reactors fueled by deuterium-tritium would thus be limited by lithium supplies. Machines run on pure deuterium, however, if they could be made, would not have this problem. Deuterium comes from heavy water, a naturally occurring substance which is abundant enough to provide fuel for thousands of millions of years. Second, there was the magnitude of the challenge. Fusion was clearly more difficult than fission energy, and its scientific and technological problems fascinated the professionals. Tuck called it "the most intriguing [problem] in technical nuclear physics."[55] Greater difficulty also suggested the possibility of greater recognition; creating this technology would be an achievement of Nobel prizewinning caliber. Third, developments in a number of scientific fields, among them space physics and astrophysics, cosmic ray research, particle accelerators, and microwave tubes, were exciting an interest in the research questions that underlay fusion.

The commissioners were also interested in new ways of providing electrical power. But they had still other reasons for embracing the scientists' proposals. They were concerned about maintaining American scientific prestige vis-à-vis other nations. They believed that they had a mandate to keep up with potential new nuclear technologies so as to ensure a strong base for both military and economic strength. They were aware that should fusion reactors become commercially important, a pioneering government program would provide US reactor manufacturers with important competitive advantages.

A most pervasive stimulus for the fusion program was American weapons research. It is no accident that the program began in a period marked by an intensification of the cold war with Russia, by a hot war with North Korea, and by a crash program to create a thermonuclear bomb. Controlled thermonuclear research was compatible with hydrogen weapons work, and was so judged by York at Livermore and by Bradbury at Los Alamos. Spitzer had been about to participate in the Matterhorn project for the Super when he got started on stellarators. Tuck and his team were doing measurements for the same bomb. At Washington, the potential use of fusion reactors for the production of fissionable (or "fusion-able") material for warheads was a possibility that the commissioners felt constrained to take into account. So were the chances of other, as yet undetermined, military applications. American fusion research spread in conjunction with the spread of military atomic research. Starting at Los Alamos, where it was a natural companion to speculation about a hydrogen explosive, it moved to Princeton as Wheeler brought Matterhorn to the campus, and then to Livermore when the weapons laboratory there was organized.

One striking feature of the program's first years is the astonishingly short period that the director of the Research Division estimated for the first, or feasibility, state of fusion research. Thomas Johnson estimated that it would take 3–4 years and $1 million to discover whether it is feasible to contain a hot plasma. Today it is clear that it will have required more than 30 years to demonstrate feasibility, while more than $2 billion had been spent on the US fusion program by the end of 1980. What lies behind a mistake of this order, 10-fold too optimistic in time and 1,000-fold too optimistic in dollars?

Clearly, there was room for wide extremes of optimism or skepticism in 1952. Almost nothing was known about the behavior of plasmas under the conditions that they would have to assume in reactors. Would such plasmas be quiescent or turbulent? How much and what kind of theory would be needed? If it was hard to evaluate the extent to which a scientific base existed, then the situation with respect to technology was perhaps even more wide open. Magnetic fields of the strength that the fusion scientists needed had never been created on so large a scale. Plasmas of millions of degrees centigrade had never been made in any laboratory, and no one really knew how many months or years would be needed to devise ways to produce them, if indeed they could be made at all.[56]

Spitzer and his colleagues at Princeton, Post and his coworkers at Livermore, the Los Alamos team, York, Bradbury all chose the stance of optimism. It is worth considering two of the social factors that may have influenced their judgment. First, there was the postwar exuberance over the possibilities of technology; anything can be done, many felt, if only the decision is made to make it a matter of national priority. Second, there was no one in 1952 who commanded all the bits and pieces of scientific and engineering knowledge that underlie the invention of fusion generators. To put the same thing differently, neither the disciplines of fusion physics and fusion engineering nor their sociological counterpart, a fusion community, existed in the early fifties. There were astrophysicists like Spitzer who knew about the rare, cool plasmas of interstellar space and the hot plasmas of the interior of stars. There were gas dynamicists like G. P. Thomson who had familiarity with laboratory experiments on weakly ionized, cool discharges. There were accelerator specialists like Tuck and Wilson who understood the careful design of magnetic fields and the potentialities of power supplies. There were cosmic ray scientists like Thomas Johnson who had studied the behavior of charged particles in complex magnetic fields. There were weapons physicists like those at Los Alamos who could contribute some rudimentary methods for measuring rapid plasma phenomena. But no one had an overview, and in its absence there seems to have been an entirely understandable tendency for the fusion scientist to underestimate the difficulties that were likely to emerge in that part of the enterprise that was outside his or her particular expertise. It is important also that there was no one at all in controlled thermonuclear research who possessed the perspective of the nuclear engineer. This specialty simply did not exist in 1952, when the technique of building fission reactors was in its barest beginnings.

The strength of the optimists' faction must not be overestimated. Most of the scientists who attended the meeting in Denver in June 1952 represented the other, pessimistic, end of the spectrum.[57] An example was Edwin M. McMillan, not himself working as a fusion physicist, but rather one of the larger "nuclear community." He used to point out that to confine a plasma with a magnetic field was to place it in a distinctly unnatural situation, somewhat like the situation that would result from trying to push all the water in one's bath to one side of the tub and keep it there. McMillan thought the plasma would avail itself of every permissible mechanism, including instability and turbulence, to escape.[58]

Interestingly, Johnson himself considered fusion power a long shot. Yet it was Johnson who, basing himself upon Princeton's figures, had given the commissioners the figure of 3–4 years. Here we need to observe that Spitzer and his associates had mentioned only two of the possible alternatives in their proposal. Either the reactor plasma would be quiescent, in which case the simple theory, which predicted confinement times of a second or more for a plasma of 100 million degrees centigrade, would be valid. In this case "it seems likely that a Stellarator can be made to function . . . as envisaged." Or the experiments would reveal turbulence that would "take particles to the wall in a small fraction of a second." In this second case, there was a clear implication in both Princeton's proposal and Johnson's report that a declaration of nonfeasibility would follow and the program for fusion reactors would be dropped.[59] In the event, as we shall see, neither of these scenarios came to be followed.

Project Sherwood

"A quantum jump"

In mid-1953 Dwight David Eisenhower appointed Lewis L. Strauss, Republican and a financial advisor to the Rockefeller interests, to replace Gordon E. Dean as chairman of the Atomic Energy Commission. Strauss had a passionate amateur interest in science and technology. He had been strongly drawn to physics as an adolescent. Subsequently, he had formed friendships with leaders of science like Ernest Lawrence and New York University mathematician Richard Courant. As a New York investment banker, he had financed a number of scientists and inventors, and during his wartime stint as a rear admiral in the naval reserves he had involved himself with weapons innovations.[1]

Strauss had already served a term on the commission, from the AEC's inception in 1946 until the spring of 1950. His tenure had spanned the debate that led up to Truman's decision to accelerate research on the hydrogen bomb, and he had been a leading advocate of the crash program.[2] The debate had kindled in Strauss an abiding interest in the peaceful use of hydrogen fusion energy for power. He believed it would be a major contribution to human needs since he saw it as clean, safe, and, because of the inexhaustibility of its fuel, able to serve as the "ultimate" power source. Moreover, he believed that the achievement of fusion energy would help demonstrate the superiority of the capitalist system. Strauss was a committed cold warrior, and he shared the conviction that the Western world stood locked in competition with the Communist nations before the eyes of Third World countries. An element of this competition, as he saw it, was the struggle to be first in applying nuclear technology to peaceful uses.[3]

Then, too, Strauss had a strong streak of contrariness. The scientists on the AEC General Advisory Committee who had opposed accelerating the hydrogen bomb program in 1949, in Strauss's memory, had also claimed that no practical peaceful application would emerge from it. This alleged judgment on the part of men who were also against the hydrogen bomb had only served to strengthen Strauss's own support for a fusion energy program.[4] When he returned to the commission as chairman, the controlled thermonuclear research (CTR) program was high on his list of priorities. In July 1953 the realization of a fusion reactor did not appear to be impossibly far in the future. It looked as if the basic lines of the device might be laid down in a matter of 5 or 10 years, and thus during Strauss's tenure. Strauss badly wanted to be, himself, the man to bring this benefaction to mankind. In the biblical terms familiar to him as a student of Jewish tradition, he expressed his determination to the program's scientists: In my time, this shall come to pass.[5]

When Strauss inquired into the status of the program, he found that about $1 million had been spent from the time that the first contract for $50,000 had been awarded to Princeton's Project Matterhorn in 1951. Of this, about $500,000 had gone to Princeton, about $300,000 to the University of California programs at Livermore and Berkeley, and the remaining $200,000 to Los Alamos. The number of persons in the field had gone from 8 in 1952 to 30 in 1953. Many of the scientists employed, however, were devoting only part of their time to controlled thermonuclear research.[6] Their loyalties were thus somewhat divided.

In Washington, funding was being provided under three separate divisions. The Division of Military Application funded the Los Alamos program, the Division of Reactor Development funded the University of California program, and the Division of Research funded Princeton. Thomas H. Johnson, director of research, was responsible for all the scientific activities.[7] Although very interested in fusion, Johnson had many other things to worry about, and no one else in the Research Division was following the program. With neither a full-time Washington coordinator, a common budget, nor a sizable group of scientists relying on the research for salary, there was very little institutional infrastructure.[8] In Strauss's view, the program, as he found it, was "in the doldrums."[9]

Strauss's personal interest in thermonuclear power had the effect of mobilizing the goodwill that the program already enjoyed among the other commissioners. At this time, Murray, Smyth, and Zuckert were still serving on the commission, while Glennan had been replaced by financial administrator Joseph Campbell. Commissioner Zuckert, in August 1953, reminded Strauss of the memorandum that Strauss himself had composed in 1949, proposing " 'a quantum jump' in the hydrogen bomb development effort." "I suggest," Zuckert continued, "that the Commission examine immediately the implications of a similar 'quantum jump' in our approach to controlled thermonuclear reactions."[10] The commissioners now convened a meeting of fusion project leaders and AEC scientific advisors to brief them on the status of the program and to suggest how "the effort [could] be intensified."[11] Present to give a picture of the project were Post, Spitzer, Tuck, and Teller. Teller (in his alter ego of advisor), John von Neumann, I. I. Rabi (then chairman of the AEC's General Advisory Committee), and Ernest O. Lawrence were there to address the second issue of intensification.

The advice that the commissioners got did not add up to the recommendation for "substantial expansion" that Strauss wanted to hear. "Dr. Rabi emphasized that considerable basic research remains to be done before the projects can be profitably expanded to any great degree . . . . Dr. von Neumann said . . . expenditures . . . might well be increased. However . . . the primary question should be whether or to what extent the present projects could profitably be expanded . . . ." Teller thought "it would be unprofitable now to attempt to predict the time scale . . . or attempt to establish a controlling overall organization." Johnson himself advocated that nothing be done beyond supporting the projects at the levels requested by the laboratory leaders.[12] Nor did Brigadier General Kenneth E. Fields, director of the Division of Military Application, want the program expanded. Both Los Alamos and Livermore were heavily involved in preparations for a set of thermonuclear weapons tests known as the Castle series. Fields did not want either Norris E. Bradbury, head of Los Alamos Scientific Laboratory, or Herbert York of Livermore to divert more of their laboratories' resources to controlled fusion at that point. Balanced against the advice of their scientific consultants and the wishes of their Division of Military Application, however, was the importance that the commissioners themselves attached to the thermonuclear enterprise. In the fall of 1953 they directed Johnson to prepare a plan for a new, more effective organization and a budget up to three times as large as that on which the CTR program was then operating.[13]

While promoting the expansion of fusion, Strauss also increased the secrecy surrounding the project. The general trend of administration policy in 1953 and 1954 was actually toward lessening the extent to which nuclear technology was classified. Eisenhower's Atoms for Peace policy, first announced at the end of 1953, demanded a measure of declassification, and the revised Atomic Energy Act of 1954 reflected this demand by incorporating an amendment permitting an exchange of classified information between the United States and Britain.[14] Strauss, however, had a passion for secrecy born of a concern to keep information with any conceivable military application away from the Soviets.[15] Hitherto a good part of the CTR program had been classified "confidential." A proposal the commissioners received in the summer of 1953 provided Strauss with a peg upon which to hang a tighter classification status.

The proposal had been forwarded to the commission by Robert R. Wilson, then professor of physics at Cornell University. His scheme was to generate a shock wave in a cylindrical plasma by means of a magnetic field, causing the plasma to implode. Thermonuclear reactions in the transiently dense, high-temperature plasma converging on the axis would supply enough energy to make the ensuing explosion more powerful than the implosion had been. The exploding plasma would drive the magnetic field outward over a coaxial cylindrical solenoid, inducing a reverse current stronger than the one that had powered the implosion. This thermonuclear "reciprocating internal combustion engine" caught the imagination of the commissioners. If it worked, it would have the major advantage that thermonuclear energy would be converted directly to electrical energy, instead of passing through the intermediate stage of heat. The commissioners asked Johnson to make sure that Wilson's proposal was aggressively pursued. But Strauss also required that it be classified "secret with strict compartmentalization." This meant that even cleared Sherwood personnel could not learn about it unless they had a demonstrated "need to know." Strauss went so far as to insist that certification of "need to know" be done at the top, in the office of the general manager. The inference that the Research Division read from this was that *all* of the program's research should be handled as "secret with strict compartmentalization," and Johnson immediately took steps to classify it in this way.[16]

Strauss's step, however, distressed the commission staff. Johnson and Director of Classification James G. Beckerley, who both vigorously supported declassification, had until then been making headway. Although the commission as a whole had wanted to maintain secrecy in 1952, the AEC General Advisory Committee had decided to support declassification in May 1952, and Commissioner Smyth was moving toward the same position. Strauss's instructions were a setback. Retrenching, Johnson and Beckerley reoriented their strategy and began to press for a lifting of compartmentalization. Their broader goal, of complete openness for controlled thermonuclear research, would necessarily be deferred now that the new chairman was in office.[17]

An institutional base

To coordinate the enlarged controlled thermonuclear effort that the commissioners wanted, Johnson decided to assign one of the Research Division staff members to work full-time on the program. He chose Amasa S. Bishop. Bishop was then 33 and had just joined the division, in the summer of 1953, as a member of the physics branch. Behind him lay a PhD in high-energy physics from the University of California and 3 years of postdoctoral work at the Federal Institute of Technology in Zurich, Switzerland. Energetic and eager, Bishop was well suited to the role of expediter and troubleshooter for the accelerating program.[18]

A second element in Johnson's plan for coordination was the establishment of a Steering Committee, composed of one member from each project. Johnson recommended Edward Teller to serve as chairman and to represent Livermore, and Tuck and Spitzer to represent Los Alamos and Princeton, respectively. Shortly after the committee started meeting, in March 1954, Johnson added William N. Brobeck, Lawrence's chief engineer at the Berkeley Radiation Laboratory, to strengthen the group's engineering expertise. Formally the Steering Committee was a body of equals, but in fact Teller had a predominant role. He was committee chairman, an intimate of Lewis Strauss, and also senior to the other scientists by 10 or 15 years. Above all, he was, on the one hand, profoundly knowledgeable about fusion, and on the other, notoriously forceful and persuasive.[19]

Official responsibility for managing the program remained with Johnson. Yet the Steering Committee had, of course, a much greater and more detailed knowledge of the state of fusion work. Steering Committee advice therefore tended to become Research Division policy.[20] Much of the committee's policy, moreover, took the form of a ratification, with superadded suggestions, of the laboratories' policies. This was natural. First of all, the science underlying controlled fusion efforts was fragmentary. Teller, Spitzer, Brobeck, and Tuck were by no means always able to agree on the scientific situation, and consequently on the path to be followed. Second, each of the committee members not only brought an expert's advice, but also represented an institution actively engaged in fusion research. Given a situation in which the formation of a consensus was difficult, in which each man had his own self-interest, and in which each enjoyed the respect of the others for his scientific ability, the outcome was bound to be that the laboratories mainly followed their own, autonomously determined directions.

The Steering Committee meetings—three or four each year—also became the means by which the so-called Sherwood conferences became institutionalized. Each committee meeting was preceded by one or two days of talks that were attended by most of the CTR scientists and some outsiders. These conferences were, in effect, a continuation of the series that Johnson had begun with the organization of the Denver meeting of June 1952 and a follow-up gathering in Berkeley in April 1953. But now the occasions for reviewing work were multiplied from one to three or more per year, and as a result the very scheduling of Steering Committee meetings had the effect of increasing the sense of urgency in CTR research. In one dramatic episode, last-minute results were telephoned in in the very middle of a presentation; in the more usual cases, teams at the projects struggled hard to get their work completed by the deadlines imposed by the conference schedule.[21]

One task to which Johnson and Bishop immediately bent their energies was that of intensifying the existing programs. They wanted acceleration and expansion, increased coordination, and full- rather than part-time participation from the program scientists.[22] By and large, reaction in the field was favorable. James Tuck was an exception. He had no personal inclination to increase his administrative responsibilities. Nor did he have a love for elaborate equipment; in the context of the fifties, he appeared as something of a string-and-sealing-wax experimentalist. Within the Steering Committee, Tuck began to argue against the rapid rise in budgets that Strauss and the other commissioners were mandating.[23] But Los Alamos Director Norris E. Bradbury had begun to support an expansion of his laboratory's CTR activities. Bradbury had begun to feel that research on weaponry would soon be winding down and that the laboratory needed to find other missions. "Although . . . five or ten years of very hard [weapons] work can be foreseen, the future beyond that point looks somewhat unrewarding . . . the Laboratory must have an intelligible, realistic, and exciting future for its staff if they are to remain and work hard on weapons *now*."[24] One element Bradbury saw for that future was the controlled fusion field.

At the Livermore and Berkeley branches of the Radiation Laboratory, Ernest Lawrence and Herbert York had some worries about drawing too heavily on the resources of their laboratories, but on the whole they supported the acceleration. "We were in the process of getting the Castle tests going. We didn't want a major interference. But we were young and enthusiastic about everything—and Lawrence was middle-aged and enthusiastic about everything."[25] Post was eager to move forward. He and his team had a "table-top" mirror machine under construction, and he had already asked for funds to start a pilot model reactor. The expansion in the CTR budget would make this possible.[26]

At Oak Ridge National Laboratory, a small group under Assistant Research Director Elwood D. Shipley had been conducting theoretical studies on controlled fusion problems since 1952, using discretionary funds from ORNL Director Clarence E. Larson. In addition, some small experiments were being funded at Oak Ridge by the Research Division. Oak Ridge Research Director Alvin M. Weinberg did not see the point of encouraging fusion. "His main thesis . . . appeared to be 'why bother with [controlled thermonuclear research], where the possibilities are both remote and questionable, when we already know how to produce fissionable power which is economically competitive to other sources?' " Shipley, however, actively promoted the program.[27]

With the impulse from Washington, and support in the field, more formal organizations were instituted and the number of personnel spurted. Hitherto, at California, Richard Post had directed the mirror experiments at Livermore while pinch work went on under William Baker at Berkeley. Now, in recognition that the laboratory program was taking on "the status of an independent research and development effort," the administration of the two programs was amalgamated. Chester M. Van Atta, who was helping to oversee Lawrence's moribund Materials Testing Accelerator project, was given the additional job of coordinating the University of California's controlled thermonuclear research.[28]

By March 1954 Bishop reported roughly 16 workers at Princeton and 24 at California. Together with the 5 he counted for Los Alamos, this made a total for the whole program of about 45, as compared with 30 in the previous year. By 1955 the total had more than doubled, to 110, and it doubled again in 1956.[29] The major fusion laboratories thus began to sport rapidly increasing teams. More and more members of the groups were devoting their full time to the project, with a concomitantly greater emotional commitment. More equipment was simultaneously being installed. Increasing commitment, increasing numbers of personnel, and increasing amounts of equipment all reflecting the acceleration mandated at the top of the Atomic Energy Commission produced a buildup of institutional momentum in the laboratories at the program's base.

While strengthening the existing groups, Johnson also took steps to bring new organizations into the program. The most important was

New York University's newly formed Institute of Mathematical Sciences. From the viewpoint of the fusion program, the institute offered the advantage, not only of its computer, but of its staff's expertise in analysis and applied mathematics.[30] Teller was sent to give a recruiting speech in the institute's new building in a dilapidated street of factories and apartment buildings in lower Manhattan. He got a particularly enthusiastic response from Harold Grad, a former student of Courant and at that time an assistant professor in his early thirties. Grad had worked in kinetic theory of gases, fluid dynamics, and electromagnetism, and he was intrigued by the chance to draw these studies together in the much more difficult subject of magnetohydrodynamics, the interaction of charged gases and fluids with electromagnetic fields. As an applied mathematician, rather than a physicist, he had a new approach to bring to CTR. A physicist tends to isolate physical problems and to bring to bear upon them whatever mathematical approximations (or models) are available that can lead to a solution. The applied mathematician, in contrast, fastens upon one of the models and explores its ramifications and consequences, whether these effects are presently manifest in laboratory experience or not. The mathematical approach carries the risk of solving nonproblems, but has the benefit of extending and deepening the mathematical resources available for the research.[31]

An NYU group formed under Grad's leadership. NYU would not have any experiments of its own, but rather would keep the whole US program under its survey and contribute where it could. This assignment gave the group an unusual opportunity to develop a perspective over the whole enterprise, a role that would be strengthened by Grad's personal qualities. Physically smaller than average, he had far more than the average measure of contentiousness. He was to fight over the scientific priority of NYU work, of which he was jealous; he was to be equally jealous not to compromise the scientific and mathematical facts as he saw them, and therefore equally combative in defense of scientific honesty. Grad was one of the more profound scientists recruited into the program, and also one of the least susceptible to the tendency to minimize difficulties that sometimes infects participants in highly politicized research and development programs. He was to become a gadfly, and a conscience, to the project.[32]

The program expansion was underwritten by a continually increasing budget. As early as October 1953 Johnson took steps to transfer more money into controlled thermonuclear research. Table 3.1 shows operating expenditures for the four largest laboratories. The fifth laboratory, New York University, was given about $50,000 in fiscal year 1954 and $100,000 in each of the two following fiscal years.[33]

Table 3.1
Expenditures in millions of dollars*

Fiscal year	Los Alamos	Livermore	Oak Ridge	Princeton	Washington	Operating total
1951–53[b]	0.2	0.3	0.0	0.5	0.0	1.00
1954	0.3	0.8	0.1	0.6	0.0	1.74
1955	1.0	2.2	0.4	1.0	0.1	4.72
1956	1.1	3.2	0.6	1.5	0.3	6.73
1957	1.8	4.5	0.9	2.6	1.0	10.74

*Other than construction and equipment funds.
[b]Rounded off.

Strauss was determined that the invention of a fusion reactor was not to be crimped by lack of money. He had an admiration bordering on awe for scientists, and more than the ordinary faith in their capability, given the resources, to deliver.[34] He protected the research program from congressional cuts and then went still further. "I remember," Spitzer recalled, "Strauss making a number of spectacular announcements to various controlled fusion groups. He said 'Well, now, what would happen if we offered a million dollar prize to the first person or first group that develops controlled fusion?' ... Then he would come down and say ... 'let's assume for the moment that money is no object. ... How can this program be pushed most rapidly, most effectively, most energetically?' ... That's the way he thought of his program, that it should be pushed without regard to the cost."[35] In the event, an embarrassment of riches resulted. By spring of 1956, after an intensive survey of the program, Carnegie Institute of Technology Professor Edward Creutz reported to the Steering Committee, "Surely there is no shortage of money allocated for this work; in fact, one gets the feeling in visiting the various sites that the number of dollars available per good idea is rather uncomfortably large. There is certainly a feeling of some pressure to spend the money made available."[36]

At the same time that the controlled fusion program expanded, it also finally received a code name. Sometime toward the end of 1953, Johnson called McDaniel into his office to discuss possible sources of money for the Los Alamos group. McDaniel recalled to Johnson that the division was closing out some wartime research in the Hood building at Massachusetts Institute of Technology. "Tom said to me," McDaniel later wrote, " 'Since we are robbing Hood to pay for Friar Tuck I would say that we are in Sherwood Forest. Let's call the project "Project Sherwood." ' "[37] The name Project Sherwood was symbolic of the fact that CTR was no longer a part-time endeavor funded from a miscellany of sources, but a program that was developing a structure of institutions and therefore a sharp identity of its own.[38]

The Model C

To accelerate Sherwood, Johnson decided "to focus efforts on those projects which are the most advanced and look the most promising and to carry them as rapidly as feasible from the stage of small experiments to the stage of scale model reactors gradually working up in size to a prototype power or neutron producer."[39] The stellarator work at Princeton certainly qualified as one of the most advanced and most promising projects in late 1953. The first stellarator, the table-top Model A, had been constructed and operated and had proved that Spitzer's innovation of a figure-8 geometry did indeed enhance plasma confinement over what was possible in a simple torus. While experiments continued on the Model A, a second machine, the Model B, was in process of construction. Model B was not much larger than A in its physical dimensions, but it was designed for a considerably larger magnetic field strength, 50,000 gauss, as against A's 1,400 gauss. The Princeton team hoped it would give temperatures up to 1 million degrees absolute, and densities of 3×10^{13} particles per cubic centimeter. If its design succeeded, the Princeton scientists would be able to carry out detailed checks on the theoretical predictions of heating and confinement phenomena, at a temperature and density which they believed would allow their results to cast light on practical reactor conditions.[40] Johnson therefore asked Spitzer to accelerate the work at Matterhorn by making plans for the next machine, the Model C, which would be a prototype stellarator reactor.

Spitzer, in turn, took two immediate steps. First of all, he arranged that the Model B, then under construction, should be provided with an auxiliary method of heating. In 1951 Spitzer had suggested that stellarator plasmas be heated by passing through them a strong electric current: the method came to be called "ohmic" heating because it is effective by virtue of the plasma's electrical resistance, and resistance is measured in units of ohms. By 1953, however, Princeton theorists had concluded that ohmic heating could not yield a temperature above 1 million degrees centigrade. To raise it still further, Spitzer proposed "magnetic pumping." Magnetic pumping imposes an extra time-varying magnetic field that alternately compresses and releases the plasma, thereby heating it. Magnetic pumping could only take effect after prior ohmic heating had created a completely ionized plasma of more than 100,000 degrees centigrade.

The introduction of magnetic pumping at this time meant going over to a concurrent rather than a sequential ordering of research; the work was being started before the results upon which it should naturally build were available. Spitzer underlined to Johnson the extent to which the commissioners were converting Sherwood into a crash program. "There is, of course, an appreciable probability that turbulence, oscillations, and other cooperative phenomena will so impair the confinement in a magnetic field that the Model B Stellarator will be incapable of producing a completely ionized plasma at 10^5 degrees or more. We are pushing ahead with this supplementary heating method, in advance of any observations with Model B, in conformity with the AEC request that we start immediately on work leading to a larger model, in advance of tests with Model B."[41]

The second step Spitzer took immediately was to initiate a paper study of a full-scale stellarator reactor. Spitzer reasoned that the Model A and Model B had been designed to explore the basic physical phenomena of fusion. Model C, on the other hand, would be a scale model of a reactor. It made sense to determine in advance what the features of a full-sized reactor might be and to use this information in designing the Model C. Spitzer also thought that such a study would be a useful way of bringing to the fore technological and economic problems. As logic demanded, the full-scale device to be studied was named the Model D.[42]

Spitzer was anxious to bring industrial scientists into the Model D study. A commercial reactor would necessarily be produced by industry; Spitzer believed that involving industry in the initial design would allow it to begin to acquire knowledge in the new field. Industry, for its part, was eager to participate. Some large contracts were in prospect—most immediately, the contracts for the Model C facility. Having their own scientists and engineers at Matterhorn would put company managers in a better position to bid.[43] It appeared, moreover, that a full-scale reactor would follow closely on the heels of the prototype Model C. With Sherwood classified secret, there was no way for industry to learn about fusion energy other than to lend scientific and technical personnel to the AEC projects.[44]

Spitzer secured the services of two men from General Electric and two from Westinghouse for the Model D study. From General Electric came Dutch-born Willem Westendorp, a specialist in electromagnets and a crack engineer with a strong intuitive feel for the problems he tackled, and plasma specialist Lewi Tonks, author of prewar theoretical studies of the pinch. From Westinghouse came Don J. Grove, who would ultimately remain at Princeton, although always on the Westinghouse payroll, and serve as a permanent liaison between Princeton and his parent company. Grove had just finished a doctoral thesis on the cyclotron, a machine that, like fusion devices, involved the application of the classical theory of electricity and magnetism. At Westinghouse he had also developed into an expert on vacuum systems. Vacuum technology would be important because the highly rarefied fusion plasmas would have to be kept free of contamination by heavier elements. The second man from Westinghouse was Woodrow E. Johnson, a member of the atomic power group; Johnson's expertise was in problems of heat transfer. Spitzer himself served as the group head.[45]

The Model D study, "The Problems of a Stellarator as a Useful Power Source," was completed in June 1954. A remarkable document of nearly 300 typed pages, it broke entirely new ground. It dealt with questions like the machine's detailed geometry, the structure of a lithium-bearing "blanket" that could function both to generate new supplies of tritium as the initial charge was used up and to transfer heat to the boilers of a steam turbine, and the problem of removing spent fuel from the plasma chamber and injecting fresh fuel. One favorable result of the calculations was that a much higher value of the parameter β (beta) seemed possible than had previously been assumed. β is the ratio between the pressure within the plasma and the magnetic pressure, proportional to the square of the magnetic field strength, at the plasma surface. The output of thermonuclear energy increases with plasma pressure, while the necessary input of magnetic field energy increases with the square of the field strength, so that β gives a rough indication of the fusion reactor's economic feasibility. The Model D team computed that βs ranging from 25% in the curved sections to 75% in the straight sections would be possible.[46]

The machine as a whole was uncomfortably large—over 500 feet in length. The output was also large. With a magnet of strength 75,000 gauss, which could be produced with existing technology, close to 5,000 megawatts of electricity would be generated, at a capital cost of about $209 per kilowatt.[47] This was about 10 times the power output of large power plants in operation in 1954. Princeton scientists were not overly concerned with the size and cost of the Model D. They reasoned that fusion reactors were still in the future; by the time they were ready for commercial use, it was likely that 5,000 megawatts would be an acceptable unit size.[48]

With the Model D study concluded, the question was raised anew whether to go ahead with the first step in the design of the Model C prototype reactor. This would be the step of the "conceptual design," which would produce a broad frame upon which the engineers could then elaborate detailed plans for the various components. The members of the Steering Committee talked it over at their second meeting, in late June 1954. It was obvious to everyone, and certainly to Spitzer and the Model D study group, that the features that the group had chosen for the Model D reactor were tentative and that the study itself was probably more a mechanism for locating problems than an actual blueprint for a reactor. This fact meant that whatever guidance the engineering study could offer in the design of the Model C was also tentative. Teller was anxious to see Sherwood move forward quickly, and he argued that the conceptual design would only cost $400,000, a gamble worth taking toward accelerating the program. Johnson, however, wondered whether it made sense to begin the Model C design when so many of its features were likely to be made obsolete by new concepts.[49]

The uncertainty felt by the Steering Committee did not, however, extend to the commissioners. On July 1 Johnson telephoned Spitzer from Washington. He wanted the Princeton scientist to know that "in spite of a certain negative feeling on the part of the Steering Committee at the recent Los Alamos meeting, the Commission is 100% back of you in the Model C design study and will supply all possible support in forwarding this work, which it regards as of great importance."[50] Shortly thereafter, the work started on the Model C design, with an initial group of six engineers and physicists.[51]

The concept of a pilot, intermediate-scale stellarator was not, of course, a creation of Strauss or the other commissioners. At least by late 1951, Lyman Spitzer had begun to envision it as an orderly step, following the table-top A and the experimental B. There is no evidence, however, that before late 1953 Spitzer had any intention of starting on the Model C before the Model B produced results. It was entirely the consequence of the commissioners' desire for a "quantum jump" in effort that design work on the Model C was inaugurated in the fall of 1954. At that time the Model B had just begun to run, but was not yet operating in a satisfactory manner.

The road for the C was to be uncommonly rocky in the years that followed. In the end, when it came to be assembled, it would be the single biggest piece of hardware emerging from the crash program for Sherwood that Strauss instituted in 1953.

The fusion program was at a level of about $500,000 and 30 scientist-years in July 1953, when Lewis Strauss took over as AEC chairman; 2 years later, in fiscal year 1955, it had reached a level of close to $5 million and more than 100 scientist-years. Controlled thermonuclear research had been a relaxed, minor program that had sought its own pace; now it was entering a period of active recruiting of personnel and institutions. It had been characterized by the prudent building of one result upon another; now, as it began to embark upon concurrent, rather than sequential research, CTR was assuming the aspect of a crash program.

The change had two important implications for program strategy. First, it increased the attractiveness of trial-and-error methods. This tendency toward empirical testing of proposed magnetic configurations had already been present. Theory was rudimentary; moreover, the handful of theorists that the program possessed would not have been adequate to analyze all the ideas that were emerging. The commission's orders for a speedup further exaggerated the trait of empiricism. The overabundance of funds and the pressure to find ways to spend them meant that equipment for almost any plausible confinement scheme would be funded.

Second, as the institutional structure of Sherwood took shape, fed by the infusion of funds and personnel, a new factor necessarily entered into the determination of strategy. The more people, equipment, and experience a project team acquired, the more it would seek to maintain or expand its strength, and to keep going the particular lines of research in which it had developed expertise. Strategic decisions in CTR would henceforth begin to take into account the institutional needs of the laboratories.

4

The Problem of Instability

Stability theory and development strategy

When James Tuck and his five colleagues initiated experiments on the toroidal pinch in September 1952, theoretical calculations had already indicated that a pinched filament, once formed, should be unstable. Tuck's immediate goals were to learn whether a pinch could be formed and to test its stability experimentally. At first, the team could find no sign whatsoever of pinching, although discharge currents as high as 40,000 amperes were obtained—currents large enough to cause marked constriction and even thermonuclear temperatures. Encouraged by interest shown at the Berkeley Sherwood conference in April 1953, the six men decided to repeat their experiments. Their equipment consisted of cameras arranged to photograph the light emitted by the plasma, through one or more slits in the torus, as the luminosity varied with time. This time they found the pinch. Their error had been in being too leisurely; the pinches they observed were not forming and coming apart in *milli*seconds, as they had expected; rather, the photographs showed that the instability set in a few *micro*seconds after the initial constriction. The process—pinch formation and unstable disruption—was highly reproducible. Tuck pronounced the Perhapsatron "VUI" (Very Unstable Indeed). It was a textbook case of a theoretical prediction followed by an experimental confirmation, and Tuck and his group promptly reconsidered their strategy. Tuck foresaw two possible courses. One was to work toward a highly pulsed reactor in which the plasma imploded sufficiently rapidly to give a substantial spurt of energy, and the other was to explore methods of stabilizing the pinch.[1]

The other two reactor conceptions, the stellarator and the mirror, did not seem to share the pinch's severe instability problems. This is not to suggest that the Princeton and Livermore teams were performing careful and systematic investigations of the stability of the plasmas in their respective machines. The art of fusion research had not yet reached the point at which such methodical experimentation was possible. Most of the effort at both laboratories was necessarily going into finding techniques for creating plasmas, setting up magnetic fields of the required configurations, and heating. It was also essential to invent methods for measuring plasma properties ("diagnostics"), for "the primitive state of experimental knowledge in the field of Sherwood is exceeded only by that of the diagnostic methods."[2]

Some attempts to test both mirror and stellarator plasmas for stability were nonetheless made. Princeton's Martin Kruskal probed stellarator stability theoretically in 1953. He concluded that instability could be caused by the plasma current that Spitzer was counting on for the initial heating of the plasma, if it rose too high. In late 1954 the Model B was still not in sufficiently satisfactory shape to allow experimental confirmation of Kruskal's prediction. From the measurements that could be carried out, no other plasma instabilities were evident. Spitzer expected that the current limitation implied by the Kruskal calculation would not be a serious constraint on the heating process.[3]

The Livermore mirror team looked for plasma instabilities by comparing experimental values of the confinement time with theoretical values calculated on the assumption that no collective effects of any kind, including instabilities, were occurring. The team inferred an experimental confinement time from data on densities, which it measured in rough fashion through the use of interfering microwaves; when the density fell to near zero, the confinement was over. The scientists concluded that the experimental and theoretical results were reasonably close, and they inferred that the plasmas were stable. The results held only for very low plasma densities, and thus very low β values, of no interest for an actual reactor. The Livermore researchers were acutely aware of this limitation.[4]

In view of such evidence as they had in late 1954, Princeton and Livermore were both building their development strategies on the assumption that the stellarator and mirror were stable. The path to a reactor seemed straightforward, and their expectations were high. The Princeton team, under Spitzer's scientific and administrative leadership, had by this time decided not to test magnetic pumping on the Model B. Instead, Princeton had initiated an additional model, the B-2, specifically dedicated to this form of heating. The B-2 was under construction, and it was hoped that fully thermonuclear temperatures of 100 million degrees centigrade would be realized. Model B itself

B itself had been rechristened B-1. A new B-1′, a modification of the B-1, was begun late in 1954. There had been trouble with the B-1 coils. The heavy, rapidly rising currents used in the experimental runs had pushed them out of alignment. Model B-1′ was to incorporate a solution to this problem. In the conceptual design stage at the same time was the Model C, to be used "partly as a research tool on a larger scale, and partly as a prototype for the fullscale power producing stellarator, which we have been calling Model D."[5]

Livermore's immediate goal was to raise the value of β to levels of economic interest. It was crucial for the Livermore scientists to ascertain whether the plasma stability they seemed to have would persist for reactor-level β values. Their plan for raising β was to go to a larger machine. A larger machine would also have the effect of increasing containment time; Post's mirror group was aiming at 0.1 second confinement. A technique for producing and injecting hot plasmas was also high on the Livermore list of needed inventions. Some ingenious team members had invented a plasma source made up of a stack of titanium washers on which deuterium had been absorbed, separated by disks of mica. A spark sent down the stack released a plasma into the chamber that was usable for experiments but too cold for reactors. Finally, the group was seeking to push up temperatures. After that, "If the immediate goals are reached, a reappraisal of the mirror machine must be made to see how it might be converted from an experimental tool to a potentially economically attractive device. . . . The experiments now being done are the beginnings of a technology of the manipulation and control of hot plasmas. . . . Once the crucial question of stability and long containment at useful values of β is established, it is probable that these 'components' can be arranged and rearranged in a great variety of ways to produce working machines. . . . Although the experimental effort which must yet be made is large, at this point the prospects appear hopeful indeed."[6]

The premise of stellarator and mirror stability was seriously shaken in October 1954. The occasion was a Sherwood conference—this time held in the Princeton gun club, which shared the grounds of Princeton's Forrestal campus with Project Matterhorn and other research projects. It was Edward Teller who did the shaking. In his address opening the meeting, he presented a new preliminary calculation. He had examined the energy of a system composed of an ideal plasma and a surrounding magnetic field, and had derived from this consideration a criterion indicating whether a perturbation of the plasma arising from some unspecified cause would decay or intensify. His conclusion was that "one can show that the stability criterion is not fulfilled in simple

LINES OF MAGNETIC FORCE

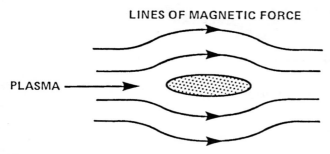

PLASMA

Figure 4.1
An unstable force configuration.

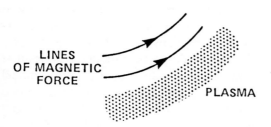

LINES
OF MAGNETIC
FORCE

PLASMA

Figure 4.2
A stable force configuration.

models of the magnetic mirror."[7] Teller suggested that the instability could be traced to the fact that the lines of magnetic force curve away from the plasma in the mirror geometry (figure 4.1). "Intuitively, magnetic lines of force are like elastic bands, and if you try to confine a high pressure gas by elastic bands, this intuitively is not a good business to do. The elastic bands are apt to snap in and the gas is apt to flow around them."[8] Conversely, in configurations in which the magnetic lines are convex toward the plasma, stability should, intuitively, obtain (figure 4.2). The stellarator ideally represented a neutral case. In fact, however, the discrete magnetic coils actually used created small, periodically recurring concave bumps in the force lines (figure 4.3). The question for Spitzer, as he returned to the Matterhorn buildings from the gun club, and for Post, as he rode back to California on the train, was this: What influence should Teller's highly idealized calculation have on the plans that they had formulated for their programs?[9]

The question that Post and Spitzer were considering had two parts. First of all, was Teller right? Would his approach, pursued in more detail and with more rigor, truly lead to the result of instability?

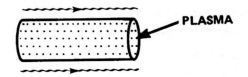

LINES OF MAGNETIC FORCE

Figure 4.3
A rippled force field.

Second, if Teller was proved right, would that fact have any relevance for mirrors and stellarators? To understand the second part of the question, it is necessary to expand a distinction that was introduced in chapter 2. Plasmas can be described by a variety of models, corresponding to a variety of sets of assumptions about the kind of physical system that a given plasma is. The single-particle model assumes that the plasma ions and electrons move independently, the collective model that they significantly influence each other. Here it needs to be added that the collective model is really a multiplicity of models. One particularly simple set of assumptions—the ideal magnetohydrodynamic (MHD) model—pictures the plasma as a simple fluid, with a pressure that is the same in all directions and an ideally infinite electrical conductivity. A contrasting model employs an anisotropic fluid in which the pressure in the direction of the magnetic force differs from the pressure perpendicular to the force. Still another set of assumptions includes the postulate of finite conductivity. And so on. Teller's remarks at the gun club were predicated upon the ideal MHD model. Even if his theoretical prediction of instability was corroborated by more detailed calculations, it would still be necessary to determine whether ideal MHD theory was a realistic approximation to the actual plasma of a mirror or stellarator.

Four months later, in February 1955, at the Sherwood conference in Livermore, a special evening symposium was convened on the topic of stability. The main paper was presented by a young Princeton theorist, Edward A. Frieman. Like Teller, Frieman framed his analysis in terms of energy; it was a step in the direction of making actual plasmas more amenable to mathematical treatment because the energy approach is capable of application to more complex geometric configurations than the "normal mode" approach, used until then, which treats the changing shape of the plasma surface. Frieman assumed a simple, isotropic pressure and an ideal conductivity for the plasma. He postulated that there was an absolutely sharp boundary between the plasma and the surrounding magnetic field, with no material particles wandering away into the region of vacuum. Even with such highly idealized assumptions, the analytical tools were not available to treat realistic geometries. Instead of a finite mirror with its little spindle of plasma suspended in the field, Frieman considered an infinite chain of mirrors.[10] At the end of his computations, he succeeded in proving that for at least one initial perturbation, which was, unfortunately, uninteresting, Teller was correct. Teller summed up Frieman's proof in the discussion: "What you have proved is essentially that this thing is unstable merely for the special case where there are no particles in the region outside the plasma, and then it is unstable only with respect to very fine perturbations for which, practically, there might be the difficulty that the concepts of magnetohydrodynamics no longer apply. . . . But at any rate in principle we have seen a little bit what the thing is, and that is some progress."[11]

The tasks that remained for Sherwood theorists were clearly numerous. Frieman had taken an important step toward the development of an energy principle for ideal MHD models, but much more work was needed to reach a complete expression of this principle. In addition, ways had to be found to apply the principle to realistic geometry. Even were that accomplished, though, the relation between the ideal MHD model and actual plasmas was completely unclear and required elucidation. The properties of more complex models also needed to be explored. Beyond that, as Tufts University physicist Winston H. Bostick pointed out at the symposium, lay yet another assignment. A model, whether it be MHD or some other, is always embodied in a set of equations. At the point of actually solving these equations and writing down a result, theorists of the midfifties were inevitably forced to introduce yet another distorting assumption. This was the so-called linear approximation—the approximation that the displacement of the unstable plasma from the stable equilibrium positions remained small compared with the overall dimensions of the plasma. To assume otherwise would have required new analytical methods or an unmanageable labor on the small desk calculators then available for numerical computations. Even the few electronic computers that were coming into use at special centers like New York University were not adequate for a nonlinear plasma theory. As Bostick expressed it, "In the present state of the art the instabilities can be handled theoretically only on the basis of small amplitude disturbances." And he added, "Nature, however, does not stop at small amplitude disturbances."[12] Somehow, a nonlinear approach, and the means of calculating with it, had to be developed.

If the tasks that the Teller instability set for the theorists were immense, the issues it posed for the strategists were vexing. This was the first appearance of a question that was to become a recurring, and a characteristic, feature of fusion research. Hopes that reactors could be constructed on the basis of single-particle theory had been seriously called into question. Plasma theories of cooperative behavior were starkly tentative and preliminary. How much weight, then, should theory be given in framing a strategy of advance toward a fusion reactor?

Responses to the threat of instability

The several Sherwood projects were bound to respond differently to the threat posed by the Teller instability simply because the ideal MHD model from which the instability was deduced had a different degree of correspondence with the plasmas formed by different machines. But over and above that, the state of stability theory made it possible for sensible men and women to take distinct attitudes toward the new situation. There was room here for the influence of politics and, even more, for the idiosyncrasies, or "styles," of the laboratories to assert themselves. In fact, the three major laboratories did respond differently. The mirror group at Livermore made the judgment that the prediction of instability was probably not applicable to its experiments. The Los Alamos group, in communication with and stimulated by the applied mathematicians at New York University, worked on the design of new devices that could comply with the Teller criterion for stability. Princeton for a time followed the plan it had charted while simultaneously exploring instability theory. By the end of 1956, however, the theoretical results were to lead it to alter its program strategy significantly.

For Post at Livermore, the paramount consideration was that the experimental results from the mirror machines were indicating stability. Confinement times were being measured that were up to 1,000 times as long as the interval that Frieman and the theoretical group at Princeton were deducing as the time it would take for the Teller or "flute" instability to disrupt the plasma.[13] As a consequence Post, and many other leading fusion scientists, did not frame the question as, Why has our apparatus not yet shown us the instability? Rather, Post asked, Why is the MHD result not applicable to the mirror plasma? "The fact that plasmas [are] demonstrably being confined, for periods perhaps a thousand times as long as the theoretical instability times, led to the realization that the model used in the early calculations did not correspond closely enough to the real situation."[14]

As the theory of mirrors was further elaborated, considerations did emerge that could explain the discrepancy between the MHD results and the mirror situation. Mirror plasmas, of all the various plasmas then under investigation, seemed to be the most unlikely to satisfy the condition of the MHD theory that the pressure is the same in all directions. We have seen that a particle is reflected by a magnetic mirror if $v_\perp$, its velocity perpendicular to the axis of the containing vessel, is sufficiently large with respect to $v_\parallel$, its longitudinal velocity. The obvious corollary is that it is *not* reflected if $v_\parallel$ predominates. There is therefore a preferential loss of particles with high longitudinal velocities from a mirror plasma. Put otherwise, the contained particles have velocities skewed in the direction of high $v_\perp$. Fluid pressure is intimately related to the distribution of particle velocities; a skewed velocity distribution causes an anisotropic pressure. University of Chicago physicist Marvin L. Goldberger, working at a summer 1955 study group at Los Alamos, recalculated the rate of growth of the flute instability using an anisotropic pressure and arrived at numbers that corresponded more closely with actual confinement data for one of the mirrors than the numbers computed from the MHD model.[15]

Yet another deviation of MHD theory from mirror reality is the postulate embodied in the theory that the plasma is perfectly continuous. Plasmas have a structure, or graininess, due to the fact that their particles revolve continually around the magnetic lines of force—the electrons in small circles, due to their small masses, and the ions in correspondingly larger circles. The crucial factor is the scale of the structure with respect to the overall dimensions of the plasma. Confined mirror plasmas were spindle-shaped blobs typically about 20 centimeters long and 2–4 centimeters thick. The size of an electron or ion was evanescent in comparison. The size of an ion's circle of gyration, on the contrary, was not. The "gyroradius" for ions reached one fifth of the plasma radius.[16]

In sum, sound scientific reasons were at hand by the fall of 1955 for judging that predictions deduced from the ideal theory should not be applied to mirrors. Post, moreover, was of a sanguine cast of mind. While he did not think that fusion would be easy, when his groups' experiments went well he was not disposed by temperament to imagine that they were really going badly. The spirit at Livermore was also different. At Ernest Lawrence's Radiation Laboratories there was a prevalent attitude that invention and experiment were more useful than theory. Lawrence himself had counseled the program in 1953 that a certain amount of "Edison-type experimentation" could profitably be undertaken in the immediate future to provide quick solutions to problems that would otherwise require difficult basic research.[17] Lawrence also liked to move forward on a project even before the way ahead could be completely seen, with the confidence that problems would be solved when they were met. Post clearly shared this disposition; his attitude was made the more reasonable by the circumstance that the various mirror experiments being carried out at Livermore at this time were relatively inexpensive.

Post's decisions about the mirror program thus were grounded in a combination of the results that theorists were bringing him and his own style as a scientific leader. "Although some months ago," he concluded in October 1955, "the calculations of Frieman and others at Princeton and Tuck, Longmire and others at Los Alamos seemed to indicate severe instability problems for the stellarator and the mirror machine, it is now becoming clear that in the case of the mirror machine at least these calculations do not apply in detail." He prepared, for the time being, to make the "unproved but reasonable assumption" that stability obtained at low β values, and was probably possible even at high β.[18] The Livermore mirror strategy remained, one year after Teller's talk, what it had been before: to push forward to higher temperatures and densities and longer confinement times within the classical geometry of the simple mirror magnetic field.

James Tuck at Los Alamos was at the opposite extreme from Richard Post. Post was deeply committed to the mirror concept and determined to show that, in one form or other, it could make an important contribution to fusion energy. Tuck was willing, and even eager, to investigate totally new confinement schemes. Tuck's flexibility was at times an irritant both to the managers at Washington, who were mindful of the investments already made in the Los Alamos pinches, and to the scientists under him. Tuck, the scientists joked among themselves, was the type of leader who would charge up a hill yelling "Follow me!"; but when they were halfway up, they would suddenly encounter him heading downhill and calling to them "Why are you going in that direction?"[19] Tuck himself felt that flexibility and exploration were the most effective ways to proceed: "What are the best possible things to be done? . . . Accentuate the theoretical and exploratory approaches rather than attempt large installations at this time."[20]

Already in the summer of 1954 Tuck and Los Alamos theorists had been examining calculations for a new idea called the "picket fence." Current-carrying wires laid out in parallel and supplied with sinusoidally

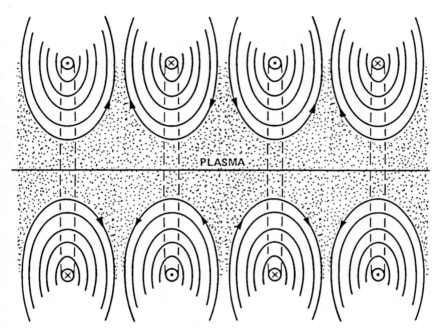

Figure 4.4
The picket-fence concept. Current-carrying coils are so arranged that the resulting magnetic field lines bend convexly away from the plasma at all points, thus
giving a stable configuration. Key: ⊗, conductor with current into paper; ⊙,
conductor with current out of paper. [From A. S. Bishop, *Project Sherwood* (Reading, MA: Addison-Wesley, 1958), p. 89]

varying currents would create the magnetic field whose configuration
suggested the name.[21] (In figure 4.4, the wires are perpendicular to
the plane of the page.)

Tuck had conjectured that the configuration would use magnetic
energy inexpensively. By the time of the October 1954 meeting, the
Los Alamos group had found that the picket fence did not offer this
advantage. But Tuck saw that it had precisely the property Teller was
pointing out as the criterion for stability: The force lines curved with
their convex face to the plasma.

The picket fence was a special case of a "cusp" configuration, so
named because of the shape of the curve in which opposing lines of
magnetic force met (see figure 4.5). After the meeting at the Princeton
gun club, Harold Grad and his associates at New York University began
a methodical exploration of cusp geometries, including a cusp-based
reactor concept, which they fleetingly called "Cuspidor." It was a
natural path for the NYU group. Grad had selected the MHD model
for special study when he first entered the program, in early 1954.

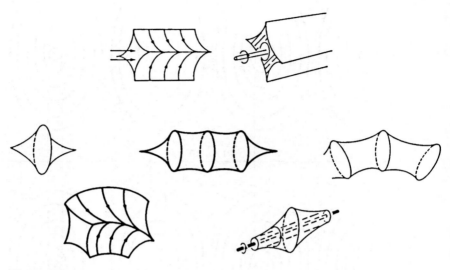

Figure 4.5
Some cusp geometries. [From J. Berkowitz et al., "Cusped Geometries," Geneva Conference (1958), vol. 31, p. 173]

He had therefore been thinking about MHD stability criteria even before the October meeting.[22]

The great advantage of the picket fence, and other cusp geometries, was that they were the only available type of magnetic configuration that seemed to have MHD stability. The disadvantage, perhaps equally great, was that they leaked; particles could squeeze out between the force lines where these came together at the cusps. In a simple, and perhaps overly pessimistic picture, such a machine would lose one seventh to one tenth of its particles in the time it took an ion to bounce from one side of the plasma volume to the opposite side.[23] Tuck at Los Alamos played with solutions involving multiple picket fences ("thickets") and "moving," that is, time-dependent, fences. Grad and his colleagues began a more accurate computation of particle losses, with the aim of learning whether a more refined picture gave a less grim result.[24]

In contrast to the other projects, Princeton was at the exact focus of the commissioners' interest. The Model C stellarator that was being planned would be the most costly installation in the program—in 1955, the Division of Research estimated its price at $10 million.[25] Model C would be the first machine in Sherwood to be more than an experiment; it would have some of the features of a pilot plant. Major decisions on the Model C were being brought directly before the commission.

To a far greater extent than at Los Alamos and Livermore, Princeton strategy was the concern of people other than the Princeton leadership. The Division of Research, the Steering Committee, and the commission itself were all actors.

In February 1955 the Steering Committee discussed the Model C at its Berkeley meeting. The Model C work was moving along with a nice momentum. The conceptual design was due to be completed in June, and Spitzer wanted authorization to go on to the next stage of a detailed engineering design. Westinghouse was willing to supply the personnel; a decision now would enable the engineering design to start by July.

Other members of the Steering Committee were not enthusiastic. There were two problems. First, the Model Bs were not going as well or as quickly as had been hoped. Confinement in the B-1 was poor. The B-2, it was expected at that time, would not be ready to operate until summer or early fall. The committee members were worried that the Model C would deflect attention from work on the Bs. They also felt that it would be prudent to wait and use the experimental results from the B machines in the design of the larger model. Second, Teller and the others were concerned about "the rather convincing theoretical evidence for instability." Spitzer "expressed a relative lack of concern at most of these arguments. He stated that in view of the relatively low value of β in model B-2 (approximately 3%) he would be exceedingly surprised if it had an instability. . . . On the other hand, Model C (with its β values of 25% and 75%) could serve as a means of studying and (hopefully) eliminating such instabilities, should they develop." He felt that he understood how to improve the B-1 confinement times and that the general shape of the Model C design was well enough conceived so that results from the B-2 would not perturb it.

The Steering Committee members were now in a quandary. The majority of them were inclined to put a brake on the Model C. But there was a strong tradition on the committee of allowing the leaders of the various projects to determine their own directions. The committee members also respected and trusted Spitzer and the Princeton scientists.[26] Finally, they were well aware of Strauss's desire to accelerate the fusion program. Their conclusion was a model of ambivalence:

After lengthy discussion it was agreed that the Committee would be foolish to recommend unequivocally that Spitzer not proceed with the design of Model C. It was recommended, however, that Spitzer give careful consideration to the concern of the other members of the

Committee, and, specifically, that he do everything in his power to speed the completion of the B-2. In addition, it was recognized that considerations other than those of a purely scientific nature are involved in the final decision and that it is neither within the competence nor the responsibility of this Committee to evaluate the political expediency of proceeding rapidly. The Committee concluded its discussion of this matter by recommending that the initiation of the Model C design be delayed unless the political arguments are felt by the Commission to be sufficiently compelling.[27]

The results of the committee's deliberations were epitomized in a staff paper submitted to the commission in March, which, however, used them as the grounds for a diametrically opposed conclusion. The director of research and the general manager both strongly endorsed proceeding to the engineering design of the Model C. Johnson identified the issue before the commission unequivocally as one of concurrent versus sequential program planning. "In a conservative approach, one would await the outcome of experiments with Model B-2 before deciding to construct Model C. . . . The possibility that experimental results from Model B-2 will appreciably alter the design of Model C must be weighed against the seriousness of delaying the Stellarator program by nine months or more." Introducing the issue of instability that had been raised at the Steering Committee, the Division of Research gave its emphasis to Spitzer's position that it would probably be better to test stability on the larger Model C than on the B models. Like Spitzer, in making this argument the research staff sidled from the C as prototype to the C as experimental machine. "It is clear that if the instability problem is to be of major importance, it is desirable to have an experimental tool capable of producing and studying such instabilities. It is likely, however, that the plasma pressures in Model B-2 will not be sufficient to cause instability, and hence a nine-month delay would serve no useful purpose in this regard. Model C, on the other hand, will develop sufficient plasma pressures to demonstrate the importance of this question and can serve as a tool to study and resolve the difficulty if it exists." The report concluded, "In view . . . of the importance of this program to the AEC, it is deemed justifiable to proceed with the detailed design of Model C as soon as possible, deferring the decision regarding its construction to a later date."[28]

By this time, two new members, both with scientific backgrounds, had been appointed to the commission. Henry Smyth had ceded his place to Willard F. Libby, a radiochemist. Zuckert had also retired, and in his stead was John von Neumann, who had participated in the discussions of fusion at Los Alamos immediately after World War II.

Campbell had left and not yet been replaced. The commission thus consisted of Strauss, Murray, Libby, and von Neumann. In April, the commissioners accepted the recommendation to start the engineering work. Approval for the detailed design would not entail approval for actual construction, for which a separate decision was being planned.[29]

All this time, the Princeton theoretical group was making an intensive effort to clear up the question of the validity of the Teller instability, dropping for the time work on other problems, such as plasma heating.[30] In June 1955, at the Los Alamos Sherwood conference, Frieman presented Princeton's results. The group had used ideal MHD theory and had assumed, for simplicity, an infinite cylinder, instead of the actual contorted tube of the stellarator. The magnetic field was assumed to be slightly rippled, as it would be in a real stellarator. Under those conditions, the results definitely yielded instability.

Of the three major fusion laboratories, Princeton was the one that most guided itself by the predictions of theory. "The main difference between the Princeton Laboratory and the other [major fusion laboratories] for many years was the great emphasis that has been placed on theory, the strong theoretical program that we had and the heavy reliance that we placed on theory in designing the next few stages of our experimental program."[31] It was an approach that emanated from Spitzer himself, who in this period exercised the definitive scientific and administrative leadership.

At the Steering Committee meeting, Spitzer now advocated a delay of at least several weeks in the initiation of engineering design studies for the Model C to allow for more calculations. The other members of the committee were more conservative still; they strongly urged that the engineering design be indefinitely postponed, pending experimental stability studies on the B models as well as further theoretical work. This time the Division of Research concurred. Spitzer informed the Princeton oversight committee, "The difficulties encountered both theoretically and experimentally now make it seem necessary and desirable to expand the project with a number of models at the B size, rather than as had been expected during the Spring, to jump to a C size model on the basis of only two or three B models."[32] Spitzer took some of the men who had worked on the Model C conceptual design and put them to the task of planning a B-3, which was to be engineered with a thoroughness and care beyond what had hitherto been brought to bear on the B machines.[33]

In October 1955 none of these moves yet added up to a drastic change in outlook. Spitzer still viewed the path ahead as relatively short.[34] Nevertheless, the moves carried within them the seeds of a more profound change. On the one hand, problems with the B models were to continue through 1955 and 1956. On the other, theoretical studies of mechanisms for stabilization of the plasma, done in the same years, were to reveal an essential limitation, namely, a critical value of β above which stabilizing magnetic fields would cease to be effective. The β values of 25% in the curved sections and 75% in the straight sections, taken from the Model D study and incorporated in the conceptual design of Model C, would therefore come to look like pie in the sky. This was important since β was a figure of merit for the economic feasibility of the fusion machine as a reactor. These two factors together were to give Spitzer and his scientists a very different sense of the time scale on which fusion energy could be reached.[35]

The starkest manifestation of the change in Princeton's expectation of the pace of advance was in the thinking about the Model C. That model had originally been intended to function as a pilot plant for the full-sized Model D. It now lost that role entirely. "The original conceptual design for Model C," Spitzer wrote in January 1957, "proposed a large machine which would serve partly as a research facility, partly as a prototype or pilot plant for a full-scale power producing reactor. The present philosophy is entirely different. The Model C Stellarator is planned entirely as a research facility, without any regard for problems of a prototype." Indeed, a series of C machines were now envisaged.[36] Princeton's strategy was gradually changing. From a simple progression through four levels of machines—A, B, pilot plant C, full-scale D—the laboratory's plan becoming one of prolonged experimentation on a series of devices of varying size, with no jump to a power plant in clear sight.

Teller's talk at the Princeton gun club has been portrayed as a crisis and a turning point in the controlled thermonuclear program.[37] It was, indeed, a crisis insofar as the talk and its aftermath sent a wave of pessimism through the Princeton group. It was a turning point in a number of ways. Princeton changed its strategy, putting aside for the time the rapid and straightforward progression to a demonstration reactor that it had envisaged and taking up a program oriented toward experimental research. A theoretical exploration of the new cusp magnetic geometries was undertaken at Los Alamos and New York University. This work was not translated into machines in the fifties; but in the early sixties it served as the basis for the important set of devices called "multipoles." Finally, Teller's talk stimulated theory. First, it gave rise to an important and intensive development of ideal MHD theory. Second, the desire to explain the mirror machine's seeming lack of a flute instability gave impetus to theories uniting the particle aspects of plasmas like the finite gyroradius and the particle velocity distribution, with the fluid aspects.

One must not be overly simplistic, however, about the role of the Teller instability. Theoretical predictions about stability were not compelling but admitted a wide latitude of response. No one pointed this out more forcefully than Teller himself. Recollecting that 100 years ago most theoretical physicists would have predicted that the onset of fluid turbulence would have occurred a hundred to a million times more quickly than it actually does, Teller told the Sherwood conferees in February 1955 that this "remark . . . carries a warning against excluding any machine for the superficial reason of alleged instability."[38]

The inability to use theory in a straightforward way to determine fusion strategy has always puzzled outsiders. In particular, it has puzzled members of Congress, who have accused fusion scientists of mere, unreflective, "tin-bending."[39] At the root of this puzzlement are two aspects of fusion theory. One is that the theory was necessarily approximate. Nonlinear problems could not be treated, and even linear problems required the use of approximations whose appropriateness needed verification. The computer was not to become fast and cheap enough to use for the purpose of simulating experiments until the late sixties. Trial and error methods, already supported on the twin legs of optimism and the pressure for quick results, were further supported by the fact that there was often no other way to proceed than by building an apparatus and trying it out. The other aspect was that the theory was not sophisticated. In fusion science, the attempt to create the technology and the attempt to discover the science underlying the technology have proceeded in parallel to an unusual extent. Harold Grad has pointed up this special characteristic: "A reasonable analogy for the fusion energy problem would be a mission-oriented program to land a man on the moon where, as part of the project, one must at the same time discover (and exploit) Newton's laws of motion, Maxwell's equations, electronics, and solid state physics. In addition, this is expected to follow a program plan which synchronizes and coordinates delivery of condenser banks with the planned date of the discovery of the magnetic monopole."[40]

5

A Sherwood Spectacular

The gloom that followed Teller's talk at the Princeton gun club began to dissipate by late 1955. For one thing, the mirror machines were showing good confinement, in violation of theoretical predictions that their plasma should be unstable, and new schemes such as the cusp and picket fence, which would not be susceptible to MHD instabilities at all, were being explored on paper. For another, there was good news from Oak Ridge. Two scientists there had collaborated on a theoretical and experimental study of the diffusion of plasma particles across magnetic field lines. Some particle diffusion would be inevitable; the key question was the rate at which the particles would be lost. The research on electrical discharges in uranium-bearing gases subject to magnetic fields, carried on at Berkeley during the war, had yielded a diffusion rate proportional to the temperature and inversely proportional to the field strength ($D \propto T/B$). This finding had come to be known as Bohm diffusion, after David Bohm, an American theorist working with the Berkeley team. The Oak Ridge scientists were testing whether this proportionality would also hold in the hotter plasmas of controlled thermonuclear research. Their results, evaluated as "the most important development in Project Sherwood during [fiscal year] 1955," was that the correct relation between the diffusion and the magnetic field was not Bohm's $D \propto 1/B$ but rather the much more favorable $D \propto 1/B^2$.[1]

The wave of pessimism that had swept over the stellarator workers was also receding as a path to MHD stability began to open. The magnetic fields in Spitzer's original stellarator had been "longitudinal," that is, parallel to the axis of the tube. By October 1955 the Princeton theoretical group had been able to use their energy-stability equation to predict that the addition of a small auxiliary field, directed perpendicular to the longitudinal field and therefore lying in the plane of the cross section of the tube, would result in a stable plasma. Then, in early 1956, Spitzer suggested that the auxiliary field be produced by supplementing the coils for the longitudinal field with another set wound around the tube in a helix. As an unexpected bonus, the tortuous figure-8 shape would be rendered superfluous because the helical windings were competent both to secure MHD stability and to prevent toroidal drift. The Model C could now be redesigned as an oval "racetrack," and a simpler shape always promised easier experimentation.[2]

The sense of excitement was not limited to the United States. The first International Conference on the Peaceful Uses of Atomic Energy was opened in Geneva in August 1955 by Homi J. Bhabha, chief of India's Atomic Energy Commission. Bhabha touched on fusion energy and prophesied its success: "The technical problems are formidable, but . . . I venture to predict that a method will be found for liberating fusion energy in a controlled manner within the next two decades."[3]

At the time of Bhabha's remarks, the very existence of Sherwood was still secret. His speech in fact forced the first step toward the program's declassification. Acting upon a prior arrangement among the commissioners, Chairman Strauss revealed US involvement in controlled thermonuclear research on the following day. All other aspects of the program remained secret, and Americans at Geneva were not free to comment upon Bhabha's speech. The British, however, did speak out. Members of the UK fusion team, now concentrated at the Atomic Energy Research Establishment at Harwell, pronounced themselves "quite optimistic." "There is no question that the problem can be solved," one of them said to *New York Times* reporter William L. Lawrence. "The difficulties in the way are mainly technological. Just give us a little more time."[4]

Thermonuclear neutrons and the pinch

Nowhere was the mood of optimism more evident in the United States than among the pinch teams. A new recruit to the Sherwood program had just contributed a theory of the dynamics of pinching that pointed the way to the achievement of temperatures high enough to produce thermonuclear neutrons. Marshall N. Rosenbluth was a burly young man whose theoretical investigations would eventually earn him the nickname of "the pope of plasma theory." In 1954 he was a recent arrival at Los Alamos Scientific Laboratory, having shortly before taken his doctorate at the University of Chicago under Edward Teller. He had been in the process of resolving to abandon weapons research when he attended a seminar given by Jim Tuck. The physics of the pinch captivated him. Up to that time, fusion scientists had gone no farther than to acquire an empirical feel for the dynamics of the process of pinching. Rosenbluth decided to work out a theory for the pinching process. With his wife and Richard Garwin, he developed the M, or motor, theory; it predicted that the heating effect of the compression should increase significantly if the strength of the electric field producing the discharge current could be raised. Thus high temperatures could be reached if high electric fields could be applied.[5]

Both the Los Alamos group and the pinch team under electrical engineer William R. Baker at Berkeley decided to follow up the predictions of the M theory. High electric fields were easier to get from linear pinch tubes than from toruses like the Perhapsatron. Both Los Alamos and Berkeley therefore worked with linear machines, and by the fall of 1955 they both thought they had the results they were after—bursts of millions of neutrons were emerging in synchrony with the plasma constriction. Their measurements showed that the neutrons were being produced within the body of the pinched plasma, and not in some phenomenon occurring at the walls or ends of the tubes. If, as the scientists conjectured, these neutrons were coming from true thermonuclear reactions, the plasma was reaching temperatures of millions or tens of millions of degrees centigrade.[6]

Tuck was uncharacteristically sanguine. "In my opinion, the last twelve months have been by far the greatest thrust forward yet made in Sherwood." His first goal had been a plasma hot enough to produce substantial numbers of thermonuclear reactions, as evidenced by a large yield of thermonuclear neutrons. That goal seemed to be at hand, and Tuck now began to speculate about how the pinches could be scaled up into reactors. He reasoned as follows. The studies of 1954 and 1955 on MHD instability had condemned both the mirror and the stellarator. ("Clearly as reactors, the static Stellarators and Mirror Machines retire to join the other victim of instability—the Perhapsatron.") There were two possible ways out. One was to proceed with the exploration of configurations like the cusp, which used static fields of a sort not liable to the Teller instability. The other was to build upon fast, pulsed devices like the Los Alamos linear pinch Columbus. The idea was that one could form a pinch so quickly, and with such a great in-rushing velocity, that significant numbers of fusion reactions would occur in the few millionths of a second that would transpire before the plasma filament had thrashed itself into extinction.[7]

Tuck did some back-of-the-envelope calculations on the size of a Columbus type of device that could be a reactor. Very large tube bores would be needed. Energy of the order of a ton of TNT fed into the plasma should suffice for break-even; in working reactors, the output per pulse would be equivalent to several tons of TNT.[8] Los Alamos now began to contemplate "a steady progression to a Columbus of enormous size, using very high voltage, very low inductance [and] special condensers."[9]

One barometer of the state of optimism was opinion on declassification. Like some of the AEC staff—notably Division Director Thomas Johnson, Branch Chief Amasa Bishop, and Classification Director James Beckerley and his successor, Charles Marshall—the scientists in the field had favored revealing the existence of Sherwood and the sites at which it was conducted. Such a disclosure was certainly in their professional interest. As long as the sites were secret, they could not publish scientific papers that in themselves were not secret but that were inspired by Sherwood problems because to do so would enable outsiders to guess where CTR work was going on.[10] But Johnson, Bishop, and Beckerley consistently championed complete declassification, while the people in the field wavered. They wanted open research, but they also wanted glory. They would have preferred to declassify simultaneously with achieving a successful thermonuclear reaction, if this did not mean waiting too long. Whenever a breakthrough appeared imminent, therefore, there was sentiment for postponing declassification.[11]

Coincident with the first triumphant reports of success, however, came the first doubts about the alleged high pinch temperatures. Livermore physicist Stirling A. Colgate had been inspired by the M theory to start his own pinch research. At the October 1955 Sherwood meeting, Stirling Colgate now punctuated the general exuberance of the pinch experimentalists with vigorous objections. The crux of his opposition was that neutrons were being produced even under conditions in which M theory predicted temperatures too low for thermonuclear reactions. Debate at the meeting was sharp. After it, on Colgate's suggestion, Baker's group at Berkeley compared the energy of the neutrons that emerged from the discharge tube in the direction of the plasma current with the energy of those emerging in the opposite direction. The measurements, made with nuclear emulsion plates, showed a clear asymmetry. True thermonuclear neutrons would have shown complete symmetry. For reasons not understood, a few privileged ions were being accelerated in one direction along the discharge, while the bulk of the plasma had not heated up enough to give out any substantial power. The fast pinch had not brought scientists to the goal of thermonuclear reactions.[12]

A confirmation of the seductiveness, and of the subsequent disillusionment, of the fast pinch now came from an unexpected source. While the Sherwood experimenters were reaching their conclusions isolated from the American physics community by the AEC's security system, the Russian scientist Ivor V. Kurchatov took the same story to the world. He had accompanied Soviet leaders Khrushchev and Bulganin on a trip to Britain and, in the course of a visit to Harwell Laboratory in April 1956, had requested permission to give a talk. To the astonishment of his British hosts, whose fusion work was also classified at this time, his public lecture was entitled "On the Possibility of Producing Thermonuclear Reactions in a Gas Discharge." It was a lucid and elegantly illustrated presentation of the Soviet experience with the fast pinch. Already in 1952 they had found the bursts of neutrons; by 1956 they too had realized the neutrons' spurious nature.[13]

The failure of the fast pinch, however, made only a dimple on the wave of American enthusiasm for the pinch concept in 1956. The reason was, again, a new theoretical prediction. Two ideas had already been advanced, starting in 1953, for making the pinch more stable. One was to add a longitudinal magnetic force, produced by external coils wound around the pinch tube, to the circular magnetic field produced by the pinch's own internal current. Another was to encase the confinement vessel in a metallic conducting shell. On Colgate's suggestion, Rosenbluth now amalgamated these ideas within a rigorous mathematical theory. He showed that with both a longitudinal magnetic "backbone" and a wrap-around conductor, the plasma should be stable for suitably chosen, albeit restrictive, values of the compression and other parameters. This new theory unexpectedly opened a vista that had been cut off since 1953, namely, the possibility of a completely stable pinch. The idea of running the pinch as a rapidly pulsed reactor delivering almost unmanageable bursts of energy suddenly seemed obsolete. By June 1956 both the Los Alamos and California groups had preliminary confirmations of the new theory of pinch stabilization to present at a Sherwood meeting in the Tennessee mountains at Gatlinburg.[14]

It was now the stabilized pinch that appeared to be the confinement scheme that everyone was seeking, the one that would scale up to a successful reactor. Tuck wrote Commissioner Libby that "it seems a particularly unsuitable time to take a hasty action like complete declassification just now, since we surely ought to wait and see whether the very new ideas on stabilization [of the pinch] constitute a breakthrough."[15]

If the pinch could be made stable, the crucial physics problem would have been solved; the rest of the road to a reactor would lead through relatively routine tasks. Tuck told Teller, "In my view, once a Stellarator or Pinch is found to be stable, the rest is mere slugging at detail."[16] Edward Creutz, a professor at the Carnegie Institute of Technology, whom the Steering Committee had engaged to make an independent survey of Sherwood, reported back to them, "It seems probable that the problem of controlled thermonuclear reactions is becoming increasingly one of technology, even though there are, of course, many basic scientific problems not yet solved."[17]

The first of these "technological" tasks would be to reach the old goal of raising the plasma temperature to the point of getting a large yield of thermonuclear neutrons. At the Gatlinburg conference, Tuck sketched out a possible experiment for a linear stabilized pinch producing 10^{10} neutrons per microsecond per unit centimeter at the reactor-level temperature of 100 million degrees centigrade. Colgate presented a design for a toroidal pinch that would generate enough power to compensate for the power loss to the magnets. Enthusiasm was the order of the day among the American pinch scientists.[18]

The second Geneva conference: Choosing Sherwood

The first Geneva conference had had a distinguished US exhibit with a swimming pool reactor as centerpiece. The second conference was scheduled to be held in 1958. The Eisenhower administration viewed the conference as an instrument to further American prestige by demonstrating preeminence in nuclear technology. For this conference also, therefore, the commission wanted an "impressive central exhibit."[19]

An Office of Special Projects had been established to arrange the American participation in the second conference, and in September 1956 its executive director, Edward R. Gardner, sent a list of possibilities for the Geneva "spectacular" to the commissioners. One item was "experimental models used in the controlled thermonuclear fusion research program." Other suggestions were drawn from six other areas of Atomic Energy Commission activity. Reviewing the list, the commissioners were especially enthusiastic about the idea of an Army Package Power Reactor, connected so as to provide all the power for the conference proceedings. Their list also included an exhibit on radioisotopes and one on the industrial uses of nuclear power; it did not include fusion. Invitations were sent out to several large manufacturers for reactor proposals. The responses, however, indicated that a suitable reactor would cost about $5 million, and the AEC had no authority to spend this amount without special congressional authorization.[20] In November 1956, therefore, the commissioners faced the need to make a new choice for the major conference exhibit. The fusion energy program now, once again, came under consideration.

Sherwood could stand as a candidate for a public exhibit because by the fall of 1956 the question in the minds of the commissioners was no longer whether it was to be declassified, but rather how and when. First of all, the direct military justification for concealing the program had evaporated by 1956. Strauss had defended secrecy on the ground that a thermonuclear reactor would produce neutrons that could be used to manufacture fissionable materials for warheads.[21] New exploration, however, had revealed more and more uranium in the ground. Fission reactors had also been used successfully to make nuclear materials, yet fission reactor technology was being declassified. As James G. Beckerley, former AEC director of classification, had tartly remarked to the Joint Committee on Atomic Energy early in 1956, "It seems extremely strange . . . [to be] reluctant to recommend release of information on a neutron-producing device which no one knows how to make, and which may never even be possible to make, when at the same time [the recommendation is made for the] release of reactor technology information."[22] Second, a program had been instituted in early 1956 to grant to selected industrial firms access to classified reports in the field of thermonuclear research. It was anticipated that details of the work would leak out over the course of a year or two via this channel. This was a reason, for example, that Congressman Carl Hinshaw and Senator Clinton Anderson, of the Joint Committee, used in pressing for declassification.[23] Then in April had come Academician Kurchatov's speech at Harwell, which had constituted, in effect, a unilateral Soviet declassification of the Russian work on the pinch. Teller, Tuck, Spitzer, and Princeton physicist Eugene Wigner had been summoned to Washington to advise the commissioners on its importance. The four scientists "unanimously concluded that Kurchatov was apparently talking about genuine Russian accomplishments. . . . There was no question in their minds on the point that the statements made are impressive."[24]

Johnson and Bishop in the Division of Research had been consistently advocating declassification. They argued that the program needed the opportunity to recruit capable people and the free play for new ideas that only an unclassified program could guarantee.[25] Maneuvering within the changing political context, in April 1956 they had sent the commission a vigorous recommendation to institute an exchange of secret information with the British.[26] There was good reason to believe that the US program could benefit from British knowledge. The United Kingdom had stepped up its fusion energy program in 1954 and had, in particular, initiated construction of a very large toroidal pinch that the British called ZETA.[27]

In May, Johnson and Bishop had followed up their proposal for exchange with the British with the recommendation that Sherwood be declassified. To ensure that the West would learn from the Soviets, even as we told them what we knew, they had suggested that the declassification occur on the eve of an international fusion conference, to be held within 4 to 6 months. Sharing information with the British would itself add another goad toward declassification since this was favored by Sir John Cockcroft, director of Harwell Laboratory.[28] Commissioners Thomas Murray, Harold Vance, and Willard Libby had supported the May declassification proposal. Strauss had not, and he grumbled that the Kurchatov initiative might well have been intended precisely to trick America so that the Soviets could learn about and profit from US achievements.[29]

In September the Division of Research had come forth with a milder suggestion: All information was to be released except that of critical importance to the development of a working controlled thermonuclear reactor. Murray, Vance, and Libby had again favored Johnson's proposal. Strauss had again demurred. Unable to reach a consensus, the commission had taken the relatively unusual step of accepting the Division of Research's recommendation by a three-to-one vote.[30] By October 1956 a handful of Britons had visited Los Alamos, Livermore, and Princeton, and plans were under way for a trip by scientists and AEC staff members to Harwell and other British installations to inspect the British pinch work and to negotiate a joint declassification guide.[31] Thus in November 1956, when the commission had to face anew the selection of a Geneva exhibit, the commissioners were already committed to a disclosure of most of the program in conjunction with some sort of international conference.

Strauss was aware of the Sherwood community's hope that the stabilized pinch could be driven to thermonuclear temperatures. In addition, he had the still more optimistic opinions of his son, Lewis H. Strauss, whom he had recruited to travel to CTR sites and send him back direct and independent information on the program's progress. After listening to the papers on experiments with the stabilized pinch at Gatlinburg in June, the younger Strauss had reported to his father, "Now the pinch is maintained for about fifty microseconds,—already an interesting time . . . and I expect that milli-second pinches will soon be obtained. . . . I feel that [the pinch program] is sufficiently advanced for a major and immediate attempt at a machine to produce a few watts of fusion power. . . . It is quite clear to me that a successfully controlled thermonuclear reactor is not far distant."[32] At this point, therefore, Strauss suggested to his fellow commissioners an exhibit based upon Sherwood. But it was not Gardner's modest proposal for a showing of experimental fusion devices that he was advocating. Strauss wanted an exhibit featuring a reacting plasma with incontrovertibly thermonuclear neutrons.

Such an exhibit would be expensive, but it would not need funding from fiscal year 1957 funds, as the power reactor necessarily would. Strauss proposed requesting additional monies for it in the fiscal year 1958 budget. Thermonuclear plasmas, it was then believed, meant that reactors were a short step away. Strauss was envisioning pulling open the curtains, not merely on American progress, but upon a Sherwood Triumphant. The project leaders were not certain whether such an ambitious goal could be realized in the time available, and on their recommendation the decision on a Sherwood exhibit was delayed in order to see whether some breakthrough would occur.[33]

In the months that followed, a sense of rivalry with both the United Kingdom and the USSR swelled to a crescendo. In the case of the Soviet Union, the mood of competition grew in the form of a spreading fear among the Sherwood scientists and in the commission that the Soviets, whose current efforts were hidden from America, were outdistancing by many lengths the US workers.[34] In the case of Britain, competition was interlarded with cooperation. British work was now open to the US teams, and a welcome interaction was growing up between the nations' scientists. The team that visited Britain in mid-November 1956, composed of the Steering Committee augmented by pinch expert Colgate, showed clearly this mingled rivalry and collaboration. The members were impressed with ZETA, but simultaneously apprehensive that the British might be planning to exhibit it in advance of the Geneva talks in September 1958. And Deputy Director of Research Paul McDaniel felt constrained to canvas those present for a comparison between US and British progress that he could take back to the commission.[35]

Outside the commission, in the press and among politicians, attitudes of rivalry were exacerbated by secrecy. The Model C, for example, was just coming into the news again. In February 1957 the Division of Research had submitted to the commission a proposal to authorize the facility, in its modified form with helical windings and racetrack geometry, at a cost of $23 million.[36] The Model C was a very different kind of machine from the ZETA. ZETA was a variety of pinch—in fact, a stabilized pinch, for after their first contacts with the Americans, the British team had redesigned it, with both an additional, longitudinal magnetic field and conducting walls, to be the largest toroidal stabilized pinch in the Anglo-American world.[37] Model C was a stellarator. Furthermore, the Model C was at an entirely different stage than the ZETA. ZETA was slated to go into operation in the summer of 1957. Conceptual design on the Model C was only scheduled to begin around Labor Day. To the eyes of science reporters, however, viewing these two devices through the obscuring veils of secrecy, the Model C and the ZETA appeared as two competing versions of the same experiment.[38]

In August Britain's ZETA went into operation, and in September the Harwell laboratory team, led by Peter Clive Thonemann and Canadian William P. Thompson, reported their first results. A burst of neutrons of apparently thermonuclear origin had been observed each time the machine was activated. Ion temperatures were calculated from measurements on impurity spectra and buttressed by data on the total neutron yield. The Harwell group judged they had reached at least 1 million degrees centigrade and probably 5 million. The press announced that the British were one jump ahead of the scientific teams of other nations.[39]

The tone of British-American rivalry now grew strident. The British press played up the story. On September 7, "the *Financial Times* ran a 2-column feature. . . . Between then and the Press Conference [of January 23, 1958] there was a mean occurrence of at least 2 news features a week in the national press."[40] A joint British-American declassification guide had finally been worked out in June 1957. All information except that bearing on devices exhibiting a net power gain was to be opened (since no net-power-producing devices existed, this was tantamount to complete declassification), and the "international conference" at which the British and American results would be made public would be the Geneva conference, now only a year away. But although the British had ratified the new guide quickly, Strauss and his commissioners were dragging their feet.[41] The British press was indignant. British science was being muzzled, its legitimate achievements obscured by Strauss's sordid game of politics over international scientific prestige. The British press also took the opportunity to ridicule Strauss's expensive stellarator; the smaller ZETA was stealing victory from under its nose.[42]

The British claims exasperated Lewis Strauss. To his knowledge, the American effort was superior.[43] Linear pinches, constituted in accord with the Rosenbluth theory, had been further improved. Toroidal stabilized pinches—the category of machine to which the ZETA belonged—had given the Los Alamos group more trouble. Purely technical difficulties, so often the determining bottleneck in the fifties, had stood in the way of creating the very high currents needed for these pinches in the more complex geometry of a doughnut. By mid-1957, however, the Los Alamos group was getting, on a toroidal machine, bursts of neutrons that, if truly thermonuclear, implied a temperature of half a kilovolt (5 million degrees centigrade). They also had neutrons from Columbus II, a "superfast" linear, partly stabilized pinch. This machine, too, had been held up for almost a year by technological problems, in this case the lack of a spark gap switch capable of withstanding the enormous voltage that had to be put across it. By August 1957, however, the men at Los Alamos had invented a new and unusual switch and had recorded both the usual "instability" neutrons and another group, which they hoped might be products of thermonuclear fusions.[44] A quite new and promising route to thermonuclear neutrons, the Direct Current Experiment (DCX), had also opened out at Oak Ridge. Edward Gardner informed the commissioners in September 1957 that "there is a distinct possibility that either the machine at Oak Ridge or the one at Los Alamos will have confirmed by January 1958 the production of thermonuclear neutrons."[45]

Throughout 1957, even as the rivalry among nations sharpened, the leaders of the fusion research community were weighing the advantages and disadvantages of a Sherwood centerpiece. They were as eager as the commissioners to show off their achievements. They were also conscious that if Sherwood were to be chosen, their budget would rise from its total, for fiscal year 1957, of $11.6 million, and would be likely to remain high even after the conference was over. The show would also produce good publicity for the program. On the other hand, some of the scientists were apprehensive about the use of Sherwood for a propaganda purpose; a crash program to produce Geneva exhibits would distract the fusion teams from orderly progress toward their real goal.[46] By the fall of 1957 a still more weighty reason argued against the proposal. It was seeming more doubtful that thermonuclear neutrons could be produced in time for Geneva.

The commissioners, with the exception of Strauss, were also lukewarm. By this point, October 1, 1957, there were two new men on the commission. Princeton mathematician John von Neumann had died of cancer in the winter. Thomas Murray, a Truman appointee who had often been in disagreement with Strauss, had not been reappointed by Eisenhower for a second term. Instead, Eisenhower had taken Strauss's recommendation and chosen two lawyers with government experience to replace von Neumann and Murray. One was John F. Floberg, a feisty youngish Chicagoan, and the other was John S. Graham, a South Carolinian and former ambassador to Britain. Since Harold Vance, who had joined the commission in the fall of 1955, was an executive, the retired head of the Studebaker Corporation, Willard Libby was now the commission's only member with a science or engineering background.[47] The commissioners' opinions were solicited in early October in a memorandum by Gardner. He put before them the suggestion that the special exhibit feature thermonuclear neutrons and that "in the event that stable production of neutrons is not reached by January 1, 1958," a fall-back exhibit be prepared on the four laboratories and their hardware. Strauss concurred. Libby, however, who was by nature cautious, and who, moreover, as the scientist member of the commission was most directly responsible for Sherwood, wrote that "the state of the project hardly guarantees convincing performance." Floberg penciled a vehement "the key exhibit must be something we *know* is going to work."[48]

Then, on October 5, the commissioners and the public alike awoke to find the Soviet satellite Sputnik orbiting the earth. A shock wave went through the American government. American scientific prestige was plummeting, and many government officials feared that other nations would question the technological capacity underlying US military and economic power. Strauss and Libby arranged a special meeting between themselves and the leaders of fusion at the Division of Research and the laboratories. It was to be held on Saturday, October 19, at 4:00 P.M., in Strauss's office on the day following the regularly scheduled meeting of the Steering Committee at Princeton. On Friday, October 18, the Steering Committee discussion was held with a greatly expanded group of participants. A full complement of CTR people from the Division of Research was augmented by representatives from the Division of Classification and from the office handling the Geneva conference. Shipley from Oak Ridge attended, as well as the three founding fathers: Tuck, Spitzer, and York. L. D. P. King, the technical advisor to the conference office, "indicated that Saturday's meeting would probably make evident whether Sherwood was to be *the* U.S. exhibit or merely *an* exhibit." There was general agreement that "if a decision *has* to be made tomorrow, count us out."

The following day, the Division of Research went into Strauss's meeting with its position prepared. Normal exhibits should be prepared on Sherwood. In the eventuality that some breakthrough were to occur within the ensuing few months, the division would evaluate it and bring the commission its recommendations "as to whether such a machine could or should be exhibited."[49]

Strauss now faced conflicting pressures. With the exception of Shipley and Persa R. Bell from Oak Ridge, none of the laboratory scientists thought that a Geneva centerpiece of Sherwood neutrons could be brought off. His own commissioners, in addition, were unenthusiastic. On the other side was Sputnik, which had plunged US leaders into consternation and led them to look everywhere for projects that could offset the Soviet feat. Moreover, Strauss feared that the USSR was aiming for a second victory, this time in fusion energy, at the Geneva conference. As Paul McDaniel recollected, "we were certain that the British planned to break their information at the second Geneva Conference. Furthermore, the Russians were thought to be ready for a big show there." In addition, it was not so easy to find a suitable centerpiece. Atomic reactors were out, isotopes "were becoming old-hat. . . . What better than Sherwood?"[50] Capping all this was Strauss's pride in the US results, his assurance that the goal could be reached, and his own great ambition that thermonuclear power should be achieved under his leadership.

Strauss decided to go against his experts. "The Chairman, supported by Dr. Libby, stated that great efforts should be made to obtain 'thermoneutrons' and to demonstrate *at Geneva* a satisfactory thermonuclear device—one which would command world-wide attention."[51] This decision was a significant intervention in the procedure by which the fusion community and its Steering Committee governed themselves. The goal of quickly achieving thermonuclear neutrons, which had arisen internally among the pinch teams in mid-1956, had now been taken over by the highest AEC management and imposed upon all sections of the program. It was taken over, however, at a time when the scientists themselves had abandoned it as unrealistic.

Preparing for Geneva

At the end of October 1957 members of the Sherwood Steering Committee and the Division of Research met with representatives of the four fusion laboratories and AEC staff to orchestrate "an exceptional effort . . . to exhibit a device or devices producing thermonuclear plasma as a central showpiece at the 1958 Geneva Conference."[52] The crash program for a Geneva exhibit elevated Oak Ridge to one of the major fusion laboratories. Until the fall of 1957 Oak Ridge's fusion work had been a relatively minor effort. It had had a budget of $500,000 in fiscal 1956, about one fourth that of Princeton and Los

Alamos and one seventh that of Livermore. Oak Ridge's scientific staff was 12 in that year, while Princeton had 60, Los Alamos 47, and Livermore 70 scientists and technicians. In fiscal year 1957, the laboratory's CTR staff had risen, but only to 16. The Oak Ridge program had been an aggregation of small-scale, basic studies and lacked the backbone of an indigenous reactor concept.[53]

One of the Oak Ridge Sherwood projects, however, carried out for the Livermore group, had been the development of an apparatus that could create and accelerate ions and inject them into mirrors, producing a hotter plasma than the titanium washer stacks could provide. In pursuing this research, an inventive technician, John S. Luce, had hit on the idea of beginning with molecular instead of atomic ions. A singly ionized molecule of deuterium (D_2^+) consists of two nuclei, each of them a deuteron, together with a single electron. Luce pointed out that if the D_2^+ is injected from the accelerator into the mirror's static magnetic field, it will emerge again. But if it can somehow be dissociated into its constitutents while still within the region of magnetic field, in the reaction

$$D_2^+ = D_0 + D^+ \text{ or } D_2^+ = 2D^+ + \text{electron,}$$

then the D^+ will be trapped in the field. The reason is that the radius of the circle that a charged ion executes in a magnetic field depends on its mass, and the mass of the D^+ fragment is only half the mass of the original D_2^+. After this insight, in 1955 Luce had begun to experiment with various mechanisms for dissociation of the deuterium molecular ion, and by the summer of 1956 he had discovered the effectiveness of injecting the beam of ions into a carbon arc established along the length of the mirror tube. About 40% of the molecular ions dissociated on the arc. Luce proposed a facility, the Direct Current Experiment (DCX), to test the amount of trapping that could be achieved[54] (see figure 5.1).

Meanwhile, Oak Ridge theorist Albert Simon thought he saw a possible solution to the problem of charge exchange. One of the bugaboos of fusion work, charge exchange started with the release of sluggish neutral atoms or incompletely stripped ions into the body of the plasma. Such particles might sputter off the vessel walls, for example, and they would also be produced by the arc. A fast deuterium ion colliding with a slow neutral atom could transfer its charge and, becoming a fast neutral, fly into the walls. Simon had pointed out that for every charge-exchange collision that ended the life of a fast deuteron, there would be, on average, about 20 previous glancing collisions in which the ion would merely ionize a neutral by stripping off an electron, without losing its own charge. He reasoned that if

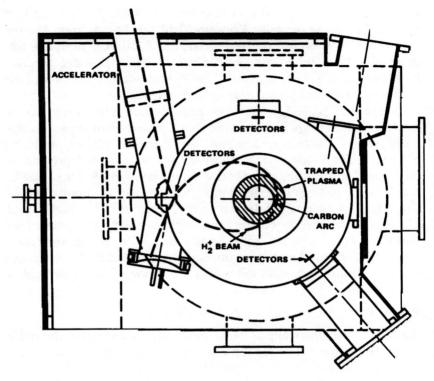

Figure 5.1
The DCX (later renamed the DCX-1).

enough ions could be fed in via the Oak Ridge beams, and the neutral particle background could be kept low enough, the neutrals could be "burned out," converted entirely into stripped ions and electrons. Simon's calculations had shown that, with the charge-exchange losses eliminated, plasmas energetic enough to be "thermonuclear" could be accumulated in the DCX.[55] Shipley and other Oak Ridge fusion leaders had begun to press for higher budgets and manpower to mount a major effort around the DCX.

In 1957 the laboratory had acquired its own man in Washington; Arthur E. Ruark had been appointed CTR branch chief to replace Amasa Bishop, who had left in 1956 to become AEC technical representative at the American embassy in Paris. As befitted the growing importance of fusion energy, Ruark was a more senior person than Bishop. At 57, Ruark was one of the grand old men of American quantum physics, holding a physics chair at the University of Alabama. He had first become associated with the Oak Ridge controlled thermonuclear program in 1952, when he had been invited to the laboratory as a consultant to the fission reactor program. Shipley, learning that Ruark had been sufficiently interested in the Richter episode in 1951 to make some fusion calculations of his own, had immediately drawn him instead into the CTR effort. From his peripheral position as a consultant, Ruark had had an opportunity to see at first hand Oak Ridge's strengths. The CTR group had at its disposal equipment from the World War II project for the electromagnetic separation of uranium isotopes. It included large, high-field magnets and superlative vacuum pumps. "Acres of magnetic field," Ruark recalled later, "high vacuum and ion [source equipment] on a scale that made the rest of the crowd look sick."[56] The laboratory also had exceptional access to electric power. Ruark believed that, given their equipment, it would make sense to increase Oak Ridge participation in the program.

The Geneva spectacular provided Oak Ridge with its opportunity. In preparation for the exhibit, the AEC staff decided to raise the Oak Ridge team to 56 members. The Tennessee group also received the largest of the supplemental budgets, $2.5 million, as against $1.35 million for Princeton, $1.15 million for California, and $0.8 million for Los Alamos. The Oak Ridge leaders now reoriented almost all their program around the DCX. They concentrated on "only one object, to make the DCX machine perform so as to produce 'burnout' of neutral particles."[57]

Meanwhile, the British were under steadily increasing pressure to publish their ZETA results. They could do nothing until the new classification guide was ratified. Once it was, they proposed to bring out their pinch data immediately, even if this meant publication would occur well in advance of the Geneva conference. It was difficult for the commission to delay ratifying the guide. Were the British to publish before Geneva, however, Strauss feared that they would be thought to be ahead in the thermonuclear field. "It occurs to me," he wrote in a memorandum to the general manager and commissioners on November 22, 1957, "that we have only one recourse which is to agree with the British that we will publish [on the pinch] simultaneously." The commissioners ratified the classification guide on November 27.[58]

Paul McDaniel was acting as division director, for Thomas Johnson had recently retired from the AEC for a job with Raytheon. McDaniel now dispatched a team of fusion leaders to Harwell for another look at Britain's ZETA. Stirling Colgate went to visit for a week. Spitzer, Tuck, Ruark, and the administrative head of the Livermore program, Chester Van Atta, spent two days. Their mission was "to attempt to determine whether the British have achieved 'thermonuclear neutrons,' " and to start arrangements for the simultaneous publication of British and American pinch results. The Americans found that there was a "major probability" that the ZETA neutrons were genuine. A fortnight later, the Perhapsatron S-3 and the Columbus S-4 also began to produce neutrons. December 1957 was the high point of the stabilized pinch, as far as the US scientists were concerned.[59]

When the American delegation returned from Harwell, a mood of increasing sobriety began to spread. It seemed to Spitzer that if the temperatures inferred by the British from the neutron yield were correct, then it was unlikely that the ZETA plasma was as stable as the British scientists believed it to be. The problem was that it was unlikely that the roughly 5 million degrees centigrade that was being inferred could be reached as quickly as the British measurements showed it to be, unless the heating was being brought about by an instability. In fact, already in late 1956, as a result of the earliest contacts with Harwell scientists, one of Princeton's scientists had experimented with similar plasma measurements on a stellarator and had deduced that the plasma temperature was due to "wriggle heating."[60]

Colgate was feeling his way toward a radically different objection. The high temperatures that the British claimed had begun to seem implausible; they were out of joint with other of the plasma's properties. Among his collaborators, Colgate now had Harold P. Furth. The Austrian-born Furth had nosed out the secret Sherwood project in 1955, while he was completing his doctoral thesis on high-field accelerator magnets at Harvard. Colgate and Furth were natural co-workers. They were both young (Furth in his mid-20s and Colgate in his early 30s), unconventional, and high-spirited. And they were both perceptive and productive scientists with talent equally for experiment and theory. In the end, the politically astute Furth was to outlast, and outclimb, the iconoclastic Colgate within the program's scientific leadership. Now, in early 1958, they joined with another colleague, John Ferguson, in a careful comparison of the results of all the neutron-producing pinch devices, including the small British Sceptre and their own Gamma. Colgate, Furth, and Ferguson took the electrical conductivity of the plasma, rather than the neutron yield, as a starting point, and made use of a well-known theory of the relation of conductivity to temperature. They concluded that the pinch temperature could not be much more than a tenth of what was being claimed. The neutrons were therefore, in their opinion, not coming from thermonuclear reactions.[61]

These growing doubts formed a background to the joint publication of data on Britain's ZETA and Sceptre, and LASL's Perhapsatron S-3 and linear pinches Columbus II and Columbus S-4, on January 25, 1958, in the British journal *Nature*. The press focused on the bitter international rivalry. Goaded by reporters hungry for some definitive statement after months of waiting, Harwell Director Sir John Cockcroft pronounced the British "90 percent certain" that ZETA had delivered thermonuclear neutrons. The ZETA team cautioned the press, as their American counterparts were doing, that practical fusion power lay at least 10 or 20 years in the future. But the team also laid out the simple scenario, typical of the 1950s, that a relatively straightforward extension of their machine could yield a commercial reactor. They outlined these steps. First, the scientists needed to verify the thermonuclear origins of their neutrons. They had measured the total neutron yield, and impurity spectra bearing on the temperature, but they had not yet judged that they had sufficient numbers of neutrons to make the decisive measurement of the neutron yields and energies as a function of direction. Second, they proposed to increase the temperature to 25 million degrees centigrade in the ZETA. Third, a larger version of the ZETA, to be in operation in 2 or so years, would bring them to break-even. A fourth stage would comprise a series of alterations to turn the device into a prototype power generator, and a further step would be commercialization.[62]

As the winter passed into spring, American attitudes were mixed, corresponding to the uncertain scientific picture. Chairman Strauss continued to press the staff into larger efforts. As he saw it, the Sherwood exhibit would be the making or breaking of the American presentation. He also wanted neutrons, be they spurious or true. The USSR might choose to exhibit a fusion device, and "Soviet scientists . . . might claim that their device was producing thermonuclear neutrons, although there would be no way to verify this claim during the Conference. . . . U.S. prestige would suffer, . . . if . . . the U.S. had no similar exhibit." In response to his direction, Gardner requested the UN to increase American exhibition space by 20,000 square feet, more than doubling the area originally bespoken. Accommodations for 100 extra scientific and technical personnel were also reserved.[63] Strauss's vision of an American exhibit that would dwarf the contribution of every other nation began to be translated into reality.

Between Los Alamos and the Livermore pinch teams, an eddy of controversy circled. Tuck and Columbus team leader Jim Phillips still estimated that possibly the Columbus II, and probably the Perhapsatron S-3 and the Columbus S-4, were sources of genuine neutrons. In

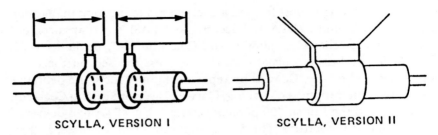

SCYLLA, VERSION I   SCYLLA, VERSION II

Figure 5.2
The early two-ring version of Scylla and the later single-ring version. [Adapted from W. C. Elmore et al., "Neutrons from Plasma Compressed by an Axial Magnetic Field (Scylla)," Geneva Conference (1958), vol. 32, pp. 337–342]

addition, the laboratory had a new entrant for the neutron show, the Scylla. Scylla (Tuck liked classical names) came out of a genre of device that used fast-rising magnetic fields to induce shock waves that produced compression and heating. An earlier Los Alamos device of this type, F. Robert Scott's Totempole, had given indifferent results, but the new Scylla let fly tens of thousands of neutrons in each pulse. Modified further by the shock-wave team, now led by Keith Boyer, Scylla's yield per pulse went up to tens of millions of neutrons[64] (see figure 5.2).

With its single-turn coil, Scylla was a kind of orthogonal variant of the pinch. In the ordinary pinch, longitudinal wires around the pinch tube excite longitudinal currents in the plasma, and these in turn create circumferential magnetic fields. In the Scylla, a circumferential, or, to use the technical term, azimuthal, current through the single-turn coil excited an azimuthal current in the plasma, which in turn created a longitudinal magnetic field (see figure 5.3). Scylla was therefore an azimuthal or "θ pinch"; the nomenclature reflects the fact that the Greek letter theta (θ) was often chosen to designate azimuthal position. At the same time, a new Sherwood team at the Naval Research Laboratory, headed by Alan C. Kolb, had independently proceeded from shock-heating experiments to a version of θ pinch and was also getting neutrons.[65] Scylla strengthened still further the hopes of Tuck and his men that they would have neutrons for Geneva.

Colgate at Livermore, on the contrary, was convinced that all the ordinary pinches (or, as they would now begin to be called, longitudinal or z pinches), including ZETA, Perhapsatron, and the stabilized Columbus, were giving spurious neutrons. He turned his team away from the planned construction of a new, larger pinch device and toward

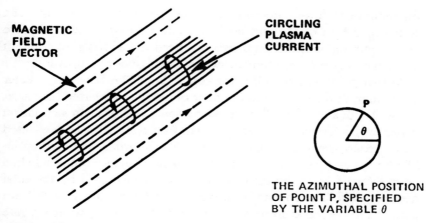

MAGNETIC FIELD VECTOR

CIRCLING PLASMA CURRENT

THE AZIMUTHAL POSITION OF POINT P, SPECIFIED BY THE VARIABLE θ

Figure 5.3
Current and field directions in a θ pinch.

an investigation of the pinch's difficulties, especially its low conductivity. Simultaneously, although in muffled tones, for the official US position was still one of hopefulness, Colgate challenged the strategic conception behind both the ZETA and American programs. That conception, that the achievement of thermonuclear temperatures would lead smoothly on to the stage of power production, was outlined by Ruark to members of the Congressional Joint Committee on Atomic Energy on February 10. Colgate, in contrast, told the committee, "The trouble is that in an actual fusion reactor one needs not only high temperatures but also long containment times of the hot plasma. [Long containment times can only be assured by high electron temperatures, as opposed to the high ion temperatures that give neutrons.] Many of us feel that progress toward a thermonuclear reactor of the stabilized pinch type depends far more vitally, at this point, on solving the electron temperature problem than on obtaining neutrons from real or spurious sources."[66]

In 1957 the British physicist John D. Lawson had published a hitherto classified calculation that gave a quantitative measure to the goal of long containment time. Lawson had shown from very general considerations that it was not enough for the plasma temperature to reach or exceed the ideal ignition temperature in order to get the breakeven condition in which the fusion power released is just sufficient to make up for the power supplied. It is also necessary to place a condition on the product $n\tau$, density times confinement time, a product which subsequently became known as the "quality of confinement." For a deuterium-deuterium fuel, Lawson derived a minimum quality of confinement of 10^{16}; for a deuterium-tritium fuel it dropped to 10^{14}. The shift urged by Colgate and his Livermore colleagues was therefore one of putting as much emphasis on meeting Lawson's $n\tau$ criterion as on meeting the requirement of the ideal ignition temperature.[67]

At that point, in April 1958, the competing ZETA bowed out. Peter Thonemann's facility shared the floor of a large, converted hangar at Harwell with a number of other experiments, one of them a cyclotron under the leadership of nuclear physicist Basil Rose. Rose had reflected on the controversy surrounding the ZETA neutrons. The maximum ZETA temperature of approximately 5 million degrees centigrade had been inferred from impurity spectra and corroborated by the total neutron yield. The spectra, however, could be coming as well from plasma turbulence as from high plasma temperatures, while Colgate and others had challenged the authenticity of the neutrons. The ZETA team had postponed measuring the neutrons' properties as a function of direction on the grounds that there were too few neutrons. But Rose thought that his team could carry out such a measurement. He and his colleagues trundled their high-pressure diffusion cloud chamber down the hangar floor and, in a careful set of experiments, determined that the majority of the ZETA neutrons were of spurious origin. The value of the ion temperature was still unclear, but it was now less interesting.[68] It was a humiliating episode for the ZETA group; the same instruments and methods had been available to them. Too much enthusiasm, too great a pressure from newsmen and politicians, and too low standards of care in the new thermonuclear specialty had combined to betray them. Shortly afterward, similar measurements on the Perhapsatron and Columbus showed that they also were producing false neutrons. DCX had still not reached burnout. Only the Scylla and the NRL θ pinch held hope, and there was not time left to do the measurements to prove them out.

The US exhibit at the second Geneva conference in September 1958 was a stunning success. It outdrew all other exhibits at the conference, registering 100,000 visitors.[69] The whole exhibit took up half the total floor space of the exhibition hall.[70] In the fusion section, guided by the scientists who had conceived and built them, visitors could see both demonstration models and fully operating experiments. Princeton was represented by the B-2 research stellarator, and models of the Model C, then under construction, and figure-8 and racetrack stellarators. There was an ingenious teaching model of Oak Ridge's DCX, in which the circuit traced out by the deuterium ions was made luminous by the addition of tiny carbon particles that the deuterons made incandescent. Livermore and Berkeley mounted models of a mirror machine and of linear pinches, as well as auxiliary equipment. Los Alamos had five experiments: a Perhapsatron made 10 million non-thermonuclear neutrons per shot; a Columbus produced over a billion; two other exhibits displayed a rotating plasma experiment and a gun for injecting plasma into a containment vessel; and the final Los Alamos device was the Scylla. Whether the Scylla's 20 million or so neutrons per shot were genuine thermonuclear products was uncertain.[71]

Scientists from nations that had barely begun to work on fusion research found the US performance dazzling.[72] The exhibit fulfilled brilliantly commission desires to enhance the prestige of US atomic science and technology. Yet the character and the significance of the Sherwood exhibit were entirely different from what Strauss had envisioned when he had first contemplated a fusion spectacular in late 1956. The very process of creating the exhibit, together with the effects of that international rivalry of which the decision to undertake the Sherwood show was one expression, helped to cause the change.

It was preparation for Geneva and competition with the British that led the Princeton group to their suspicions of the ZETA data and gave rise to the comparative study by Colgate, Furth, and Ferguson of all the extant stabilized pinches. This work led to the perception that the plasmas might be undergoing small-scale turbulence and that the theoretical understanding of plasma electrical conductivity might not be adequate. Both of these insights pointed to the need for more extensive scientific study of plasmas and their instabilities. It was political pressures from international competition together with personal ambition, an excess of enthusiasm, and the lack of normal scientific criticism that stems from secrecy that led to the ZETA episode. ZETA was sobering, even shocking, for American fusion scientists. The lesson drawn was, again, the need for careful professional experimentation. A rapid, slap-dash, by-hook-or-by-crook procedure of inventing one's way to reactors no longer seemed useful.

The exhibit Strauss had wanted would have had the character of a show of thermonuclear plasmas. In the event, however, the DCX had been unable to achieve burnout, and the neutrons of the Los Alamos Columbuses and Perhapsatrons had proved to be spurious. Only the θ pinch, Scylla, was producing neutrons that might be thermonuclear, but this could not be claimed in the exhibition hall because it had not yet been proved. Hence the exhibit over which Strauss presided was essentially a blown-up version of Gardner's original scheme for CTR experimental models.

Even had the United States presented an incontrovertible display of thermonuclear neutrons, however—and this is the important point—these would no longer have been widely received as a breakthrough. The belief that a reactor would follow thermonuclear neutrons by a short interval was fading. In its place, Livermore's opinion was beginning to represent the view of a growing number: "It should be recalled that the function of a magnetic bottle is just precisely the long-time containment of the plasma, not the achievement of fusion reactions. . . . Even if 100-million degree plasmas were eventually produced by means of pinch experiments, the failure to achieve long-time containment in a device of reasonable size would eliminate the pinch as a fusion reactor scheme."[3] Confinement, that is, stability, and not neutrons, was becoming the key issue.

At the end of June 1958, Lewis Strauss stepped down as chairman of the Atomic Energy Commission; his only remaining official action was to lead the American delegation at the Geneva conference. He had not succeeded in presiding over the invention of thermonuclear reactors. He had, however, raised the budget of the program from less than $2 million in fiscal 1954 to more than $29 million in fiscal 1958. He had forced the pace of research to a point where a new consensus on strategy was beginning to form. Within a few years, this new strategy was to dominate US research: Thermonuclear temperatures would remain a goal, but one that would be subservient to the central aim of a mastery of the physics of fusion plasmas.

Watershed

Among fusion scientists, 1958 is remembered as a year of discontinuities. The most abrupt change, and the one that had the greatest impact on the consciousness of the community, involved their perception of the state of fusion research. Where they once had thought that state to be well advanced, now, almost overnight, they saw it as primitive. Equally abrupt was the expansion of the Sherwood world. From a close, and closed off, group of Americans and Britons, the community distended, at a stroke, to include Europeans, Russians, and Japanese. Meanwhile the standing of thermonuclear research within the Atomic Energy Commission was changing: from a project promoted and supervised by the commissioners themselves, it was demoted to the status of any other research program. Finally, new institutions, notably universities and industrial corporations, were introduced into the fusion research establishment, and new organizations and journals were founded, while the research itself gradually began to show a substantial increase in quality and rigor.

The new leadership at the commission was one cause for the changes. John A. McCone was sworn in as chairman on July 14, 1958; unlike Strauss, who had been passionately engaged with CTR, McCone had no special interest in fusion. Another more important cause was the new policy on declassification. With Strauss gone, the last of the foot-dragging vanished. In a joint action with Britain, thermonuclear research was completely declassified on August 30, on the eve of the Geneva conference.[1] The unfettered presentation of research results at the conference allowed crystallization of the new consensus on the state of fusion research. In turn, the new consensus that the art was, as yet, rudimentary lessened the mood of competition and encouraged the internationalization of the field through the personal interactions that occurred at Geneva.

Declassification also made it possible for Sherwood to take up the practices that are normal to research and development activity. On the one hand, this change caused the infusion of new people and new organizations. On the other, it meant that Sherwood was now subject to the open scrutiny of the entire scientific community. In the mirror thus held up to their work, fusion scientists had yet another instrument to help them gauge more clearly the actual stage of knowledge that they had reached.

The Geneva conference

Even as the Sherwood delegates were arriving at Geneva in August 1958, they had some recent, sobering events to look back upon. Foremost among these, of course, was the revelation after the initial hoopla that the ZETA was not producing thermonuclear neutrons and the confirmation that the American z pinches were also failing to give this hoped-for result. In addition, a new and disquieting phenomenon was being observed at Princeton and earning the name of "pumpout."

Pumpout first became apparent in late 1956. The rebuilding of the Model B-1 stellarator as the Model B-1′ had solved the old difficulties with magnetic coil alignment, but the pumping systems of the B-1′ had proved inadequate. Impurities from the walls, gaskets, and other parts of the vacuum vessel had flooded the gas being used for the test plasma and had made it impossible to get reproducible results. In 1956, therefore, Princeton scientists undertook a major effort to improve the vacuum systems. They succeeded in reducing the base pressure (that is, the pressure before the gas, generally hydrogen or helium, has been introduced) by four orders of magnitude, from 2×10^{-6} to 3×10^{-10} torr (millimeters of mercury). So high a vacuum in so large a system was a considerable achievement.[2] When the redesigned B-1′ went into operation, however, and systematic observations for the first time became possible, it was discovered that plasma particles were diffusing to the walls, across the magnetic lines, at an inexplicably high rate. The loss rate expected from theory, called "classical diffusion," fell off as the square of the magnetic field. It was this favorable functional dependency that the Oak Ridge investigators had found in a mirror geometry in 1955.[3] In the strong fields contemplated for reactors, it would present no problem. Pumpout, however, was a loss that was very much greater than classical diffusion would have been. Its cause was completely unknown.[4] Experiments on the newest stellarator, the Model B-3, in early 1958 further verified the existence and seriousness of the particle loss.

As a consequence, Princeton's whole strategy for plasma heating had to be questioned. Spitzer and his colleagues had planned to raise the fusion fuel to an initial temperature of 1 million degrees centigrade by ohmic heating, and then to take it up to a full 100 million degrees centigrade by magnetic pumping. Magnetic pumping, however, requires time; it takes several milliseconds before the temperature of the plasma ions can be raised significantly. The ions in the B stellarator models were being hurled to the walls too quickly for the pumping to affect them. "I have been assuming that in the C Stellarator the pump-out rate would be materially less," Spitzer told Ruark on the eve of leaving for Geneva, "and that magnetic pumping would have a fair chance of heating deuterium to thermonuclear temperatures. The recent measures of pump-out on B-3 lead us to question this assumption. . . . [These measures] are so unexpected and so inexplicable theoretically that we feel very hesitant in predicting how pump-out will vary as the machine dimensions vary. While I still believe that the pump-out rate will be materially less in C than in our B models, I would not now exclude the possibility that it might even be more rapid!"[5]

No matter how somber the appraisals American scientists were making of their own programs in 1958, however, they had no grounds for being pessimistic about the general state of fusion. They simply knew too little about what was happening in the Soviet Union. Indeed, rumors and fears had been sweeping Sherwood in the months before the conference. It was rumored that the Russians had made some kind of extraordinary mirror machine. It was speculated that they had succeeded in breaking through to some marvelous piece of hardware that would encapsulate a solution to all the fusion reactor problems that were worrying the Americans. It was feared at every level of the program, from the laboratories up to the top management of the AEC, that the Russians would even arrive at the meetings with plans for a working reactor.[6]

Sherwood scientists came to Geneva in a mood of unusual excitement. The opportunity to stride out from behind the fence of security into a whole world of new colleagues was exhilarating. Work was going on in Germany, France, and Sweden, besides the Soviet Union, and there was also a scattering of effort in the other European countries and Japan.[7] The Americans had had a limited foretaste of this experience the previous year, when an international conference on ionized gases had convened at Venice, in June 1957; at that time, however, the bulk of controlled thermonuclear research was still classified.[8]

At Geneva, scientists from each of the Sherwood laboratories had arrived early to put together the apparatus piling up in shipping crates on the new cement floors of the temporary exhibition building. Taking up residence in the hotels and pensions in and around the city, they worked late into the night, 6 and 7 days a week. But they also found time to get acquainted with their foreign counterparts, as the other delegations moved into their hotels.

The American delegation waited with some agitation for the first, plenary session on thermonuclear research. The Soviet talk had been written by Moscow University Professor Lev A. Artsimovich, who was a winner of both Stalin and Lenin prizes and a collaborator of Kurchatov at the Moscow institute. It was read, in Artsimovich's absence, by the experimentalist E. I. Dobrokhotov. As he read, a feeling of relief came over his audience. The Soviets had done excellent work, but they had achieved no breakthrough. Their giant mirror, OGRA, existed, but it was an experimental device for injecting and trapping molecular ions, and not a reactor. They had got no further than the Americans. Dobrokhotov, for Artsimovich, pronounced from the podium a sentiment that judicious Americans under the Strauss leadership had preferred to express privately: Thermonuclear neutrons were not of central importance. "The question of whether a given neutron belongs to the noble race of descendants of thermonuclear reactions or whether it is the dubious offspring of a shady acceleration process is something that may worry the pressmen but at the present stage in the development of our problem it should not ruffle the composure of the specialists." He summed up, "The separate investigations . . . as yet have not brought us very much nearer to our ultimate goal. We do not wish to be pessimistic in appraising the future of our work, yet we must not underestimate the difficulties which will have to be overcome before we learn to master thermonuclear fusion."[9]

Also sobering was a theoretical study by B. A. Trubnikov and V. S. Kondryatsev, that exposed a hitherto unexplored mechanism by which a plasma could be drained of the energy it needed for fusion. Trubnikov and Kondryatsev calculated that the power lost in this way would escalate rapidly with temperature;[10] the effect would therefore be worst precisely at the temperatures needed for practical power production. The problem was especially serious for devices that employed plasmas at low densities. The Sherwood delegates were not sure that Trubnikov and Kondryatsev's calculation was believable, but if it were to prove true, it was important.

The general picture of fusion research that emerged as the delegates sat through one session after another at Geneva was summarized by

Artsimovich some years later: "At that time the main body of scientific information was, essentially, something that might be called a display of ideas. Most of these ideas were only thinly draped with rough and insufficiently verified experimental data, exploratory in character."[11]

The international competition that had flourished before August 1958 had been nourished on the hope that an early breakthrough in fusion development was a possibility. The international stocktaking that occurred at Geneva destroyed this hope and consequently dissolved much of the sense of rivalry among scientists. The delegates now felt the immensity of the task, and they also saw clearly the extent to which work had wastefully been duplicated. The keynote speakers were heartfelt in celebrating the new openness.[12] They also began to call, in general terms, for international cooperation of the sort that was being attempted in the sister field of fission reactors.[13] These first notes urging international collaboration were eventually to swell into an impressive, and even exemplary, melody.

The McCone commission

When the new commission chairman, John A. McCone, turned from the issues that interested him most—the civilian reactor program, including Hyman Rickover's nuclear-powered navy, and the possibility, which he opposed, of a nuclear test ban treaty—and looked at fusion, he laid great weight on the new scientific consensus. In a press conference he held with Willard Libby on October 30, 1958, he and Libby stressed the obstacles that had come to the surface, including stellarator pumpout and the Trubnikov-Kondryatsev energy loss. Fusion had been misjudged to be much farther along the road to development than was warranted by the actual state of the art. It was, in fact, still in the research stage.[14]

McCone and his attitudes were not the only novelties that the fusion leaders faced in 1958 in their agency environment. By the end of the year they also had both a new director of the Division of Research, and a new Steering Committee. Paul W. McDaniel had been serving as acting director since Thomas Johnson's retirement. In early 1958 McDaniel had allowed the Steering Committee meetings to lapse. He had come to feel that the personal relations among the committee members had become so acrimonious as to be an inconvenience. The new director, John Harry Williams, a Minnesota accelerator physicist and laboratory administrator, had taken over shortly afterward. In September Williams decided to reconstitute the committee as a group consisting of the directors of the four principal laboratories engaged in thermonuclear research, together with three distinguished outsiders.[15] "Now that this area of research has been declassified," Williams explained, "it will be possible to exercise a more considered judgment as to its importance with respect to other fields supported by the AEC." The new composition of the Steering Committee "would insure that general laboratory policy was represented and give consideration to the balance between parts of that laboratory's program." Williams's Steering Committee was made up of Alvin Weinberg, director of Oak Ridge National Laboratory; Edward Teller of Livermore; Norris Bradbury, director of Los Alamos; Henry Smyth, head of the Princeton Oversight Committee on Project Matterhorn; and scientists William A. Fowler, S. Chandrasekhar, and Robert R. Wilson.[16]

In July 1959 McCone called for a thorough review of Sherwood. He was subjecting each AEC program in turn to a searching scrutiny. The chairman was especially eager for a look at Sherwood funding. "I think we should recognize that we have reached our present level of expenditure as a result of something of a crash effort in connection with the 1958 Geneva Conference."[17] When the fusion budget had been tripled in November 1957, Sherwood leaders had stressed that the program should be maintained at that new level in the post-Geneva period. Once the American work had been revealed to the world at the conference, they had argued, sufficient funding should be provided so that the United States could retain international leadership.[18] But McCone did not accept this argument. He felt rather that funding should be determined by comparing the value of fusion research with that of the commission's other research programs. Additionally, he had adopted the principle for fission reactor development that the number of reactor types being explored should be pruned back to the handful that were most likely to succeed. He was inclined to transfer this approach to thermonuclear research. Was it not possible, he asked, to focus on "the most promising concept"?[19]

The matter was referred to the new steering committee, which unanimously, and effectively, opposed a constriction in the number of concepts. On the other hand, the Steering Committee assented to McCone's principle of funding: "The proper level of support . . . must . . . be arrived at in much the same way as is the proper level of support for any other area of basic research in which the Commission has a serious interest." A 10% cut was being proposed across the board for research activities. With the exception of Teller, who entered a vigorous dissent, the Steering Committee was willing to acquiesce to the same cut for CTR.[20]

The actions of the Steering Committee, Williams, and McCone all tended to the same result: to remove fusion from the anomalous and favorable position it had enjoyed under Strauss's patronage, and to restore it to the position of a normal research project. Indeed, it was a research activity that was decidedly peripheral to the commission's main concerns. New men were defining commission priorities now, and they were basing their decisions on a new, more pessimistic evaluation of the program. As the Steering Committee told McCone, "The character of the thermonuclear project has changed since pre-Geneva days. What was once believed to be a clearly defined path to controlled fusion has turned out to require much investigation and basic research in plasma physics."[21] Under McCone, fusion settled down to normalcy under a commission leadership whose attitudes toward it ranged from mildly supportive to indifferent.

At this time, President Eisenhower's desire for international cooperation in the peaceful uses of atomic energy provided ever increasing opportunities for contact between American and Soviet fusion scientists. Having unveiled CTR at Geneva, the administration continued to regard it as one of the prime areas for collaboration. In September 1959, during the visit of Premier Nikita S. Khrushchev to the United States, McCone and his Soviet counterpart, Vasily S. Yemelyanov, negotiated an agreement that included an interchange of visits by scientists.[22] The first exchange involving fusion took place in 1960; a team of Russians traveled through American laboratories in May, and an American group lead by Ruark returned the visit in July. Specially planned lectures and informal outings and inspection of apparatus allowed the teams to begin to know each other's programs with an intimacy not possible through reading alone.[23] New publications, however, also contributed to the exchange of information. The Soviets had presented to the Geneva delegates a four-volume compendium of the research they had conducted during the years when fusion had been classified. In the years from 1959 to 1961, these volumes were made available to Sherwood personnel through translations. In 1960 the International Atomic Energy Agency established the journal *Nuclear Fusion*, to be published in English, Russian, French, and Spanish.

Controlled thermonuclear research was attracting some of the best among the Soviet scientists. One reason for its popularity in the USSR may have been precisely the unusual degree of international cooperation that was emerging in the field. Fusion research offered the Soviets exceptional access to their Western colleagues. Soviet interest, in turn, was doubly welcome to the Americans. At that time, when support within their own agency was lukewarm, the Russians' energetic prosecution of fusion became one of Sherwood's chief arguments to Congress and the administration for the importance of the American program. In addition, a broad streak of idealism ran through the Americans' attitudes. In collaborating with the Russians on energy, Sherwood scientists saw themselves as setting an example for East-West cooperation on other global problems.[24]

Sherwood enters the scientific mainstream

In arguing for declassification in the years before Geneva, Johnson had emphasized the negative effect that secrecy had on the recruitment of capable personnel. In actuality, declassification's most important impact was that Sherwood became subject to the normal standards and procedures of the scientific community. Declassification altered the nature of conferences at which Sherwood scientists presented their findings, allowed open publication, and inspired graduate programs in universities. Collectively, these changes engendered a higher quality of research.

Before Geneva, most CTR reports were given at closed AEC meetings, so that Sherwood scientists in the main exposed their work only to the scrutiny of other insiders. They depended upon the same agency and, indeed, competed vigorously with each other for their own reactor schemes. "You thought that you were stepping into a family quarrel," remarked one outside scientist who had attended the meetings. "You'd hear one man after another speak like a salesman," and, he added, even the good work suffered because "it didn't have the ferment of the scientific community as a whole around it."[25]

After Geneva, other forums gradually supplanted the Sherwood meetings. As early as the last classified Sherwood conference in February 1958, Princeton's Melvin Gottlieb had called a special meeting to initiate moves for the establishment of a Division of Plasma Physics under the auspices of the American Physical Society. Gottlieb believed it was extremely important to open the research to the judgment of the scientific world. Response from the Sherwood community was immediate and enthusiastic. About 20 people attended the special meeting, and the division was able to schedule the first of a continuing series of annual meetings for December 1959 in Monterey, California. In addition, the executive committee began to plan a series of joint meetings with other APS divisions for the regular society meetings; the first of these was held with the Division of Fluid Dynamics in January 1960 in New York City. The annual division meetings were largely taken up with talks on Sherwood subjects by Sherwood personnel, though the audience was much wider.[26] The joint APS sessions were more catholic, however, and provided an opportunity to hear, and exchange ideas with, astrophysicists and fluid dynamicists.[27] Meanwhile, the International Atomic Energy Agency was organizing an important series of international meetings on fusion research, as a follow-up to Geneva, with the first scheduled for 1961 in Salzburg, Austria.

Before Geneva, little information about Sherwood projects had found its way into the journals, although Sherwood workers were as anxious as any other scientists to publish their work. After declassification, they began to send in a steady stream of articles. *Physical Review* and *Physical Review Letters*—journals of the American Physical Society—were one choice. The still more important American vehicle for fusion articles was the newly founded *Physics of Fluids*. All these journals employed the referee system; papers submitted to them were screened by the authors' scientific peers, who could recommend revision or rejection for substandard work. François N. Frenkiel, editor of the *Physics of Fluids*, was the more anxious to enforce high standards because his journal was new and because fluid dynamics and plasma physics were not yet universally accepted as proper fields of physics.[28]

Before declassification, there was no regular university training in high-temperature plasma physics and engineering. Those who were cleared to receive Sherwood information faced the barrier that it was a violation of secrecy regulations to teach many aspects of the subject; those who were not cleared were unaware of scientific developments that might otherwise have interested them. After Geneva, several schools quickly moved to set up graduate programs. The freedom provided by the end of classification was reinforced by the new consciousness that fusion was going to be a long-range program. There would be plenty of time to train new people, and there would be more than enough for them to do when they had completed their education.

The most extensive programs were developed at Princeton and at the Massachusetts Institute of Technology.[29] Plasma physics had become the largest research program on the Princeton University campuses by 1958. Almost immediately after Geneva, the Matterhorn group advertised postdoctoral appointments. The response indicated widespread interest: 36 applications were received within a few months. In the spring of 1959 Spitzer and the university moved to establish graduate training under an interdepartmental committee representing astronomy, physics, aeronautical engineering, electrical engineering, and mechanical engineering. Instruction was given in the several departments, but all the lecturers were drawn from Matterhorn. The years just after Sputnik were a propitious time to seek private money for financing education in science and technology. Both the industrial firm of General Dynamics and the Ford Foundation indicated their interest, and the Ford Foundation became the funding agency in October 1959.[30]

An influx of graduate students necessarily affected the character of the work done at a facility like Matterhorn that had housed a long-standing AEC program. The students' work had to satisfy the standards of the larger physics community for an acceptable doctoral thesis. Experiments had to be "clean"; that is, the critical phenomena had to be isolated and admit detailed comparison with theory. Sherwood work before Geneva did not always have this quality. The need to find such projects for graduate theses influenced the laboratories to concentrate more of their efforts on basic research.[31]

The very assimilation of the fusion programs into the administrative machinery and the university community enforced a move toward more basic science. In September 1960 the Matterhorn Review Committee "suggested two possible courses Matterhorn might conceivably take. The first . . . would be to consider Matterhorn as being engaged primarily in a 'project' for the AEC to determine if the C Stellarator could be made into a thermonuclear reactor. The second alternative was that Matterhorn should be engaged primarily in basic plasma physics research. . . . The Committee strongly recommended that Matterhorn, as a part of Princeton University, follow the latter alternative and that at least half the effort be devoted to basic research in this field."[32] Shortly afterward, the Project Matterhorn Executive Committee dropped both the word "project," with its connotation of mission orientation, and the code name Matterhorn, and renamed their installation the Princeton University Plasma Physics Laboratory.[33]

Massachusetts Institute of Technology, unlike Princeton, did not possess a Sherwood facility. The MIT Research Laboratory for Electronics (RLE), which had been founded in 1946 with the facilities of the wartime Radiation Laboratory, did have a plasma physics group. Its leaders, William P. Allis and Sanborn C. Brown, had maintained contact with the CTR program as consultants. AEC proposals that MIT actually join Sherwood had been rebuffed, however, on two grounds. First, MIT scientists were not impressed with any of the basic CTR approaches of the fifties and did not wish to build a program around any of them. Second, because they judged that fusion was a field needing more basic research of an open nature, the MIT scientists did not **want to bring the program onto their campus while it was** secret.[34]

In 1958, when Sherwood was declassified, Allis, Brown, George Bekefi in the Physics Department, and Louis D. Smullin in Electrical Engineering, who had all been working through the RLE, started fusion research in their home departments. Meanwhile, Manson Benedict had been organizing a new Department of Nuclear Engineering for the institute, and he decided to recruit a professor of nuclear fusion. The man he invited was an MIT graduate, David J. Rose of the Bell Telephone Laboratories. Rose had done a thesis on gas discharge phenomena under Allis and Brown for his doctorate in physics. His wartime experience with the Canadian radar effort and his years with the Bell System had accustomed him to think of developing technologies as whole systems, with both engineering and physics features. As he took up his new position within a department of engineers, this approach was, not surprisingly, strengthened. He began to stake out for himself and his students the special area of fusion reactor studies. Within the American CTR program, Rose was to assume a position like that of Harold Grad. Because of the vigor of his intellect, and perhaps also because of his independent position as a professor, Rose was, like Grad, immune to the "party line." He was, instead, to develop his own line, which he would argue for over a decade: A major criterion for deciding which approaches to pursue should always be that of engineering practicality.[35]

The groups that formed within the three MIT Departments of Physics, Nuclear Engineering, and Electrical Engineering established informal cooperative relations. They were funded individually; the physics group first through the Joint Services, then through the National Science Foundation and, to some extent, the AEC; the electrical-engineering group through the Joint Services and later the NSF; and Rose's group through the NSF. They worked on systems design and small experiments. Like Princeton, the MIT program provided a mechanism for cutting across the boundaries of more traditional disciplines and creating an education in a new specialty of "plasmology."[36]

Increasing numbers of plasma specialists began to emerge from Princeton, MIT, and other, smaller graduate programs. There were 14 PhDs produced in 1959 and 1960, 31 in 1961, 56 in 1962, and 99 in 1963.[37] As they entered the fusion program, they ameliorated some of the difficulties in communication and problems of inadequate knowledge that stemmed from the diverse backgrounds of early fusion workers. Education of plasma scientists over the whole nation was judged to be somewhat inferior to general physics training in these early years; nevertheless, from the vantage point of Sherwood, it contributed to a distinct rise in the level of work.[38]

In summary, Sherwood took one step after another to constitute itself as a normal scientific discipline after declassification. Each step— public meetings, professional organizations, refereed publications, graduate education—reacted back upon the program to enforce more rigorous standards. As a cumulative result, the quality of work began to rise toward prevailing scientific levels.

At the same time, each step toward normalcy supported the re-orientation from trial-and-error invention to systematic scientific investigation already taking place. Graduate education created the need to provide problems in basic research for students. Open meetings and refereed publications gave Sherwood scientists an audience that was more interested in scientific and engineering advances than in fusion reactors. The very fact that research was spreading into the universities made a difference. In a criticism of program secrecy in February 1958, accelerator physicist M. Stanley Livingston had pointed out that different institutions are likely to reward different aspects of fusion. The "large laboratories" had a "natural tendency" to put emphasis on "scale models, apparatus and practical pieces of hardware. Thereby, allocations for laboratory financing and budgets can be more easily justified through administrative controls."[39] In contrast, research in a university setting was bound to be skewed away from equipment and toward more fundamental investigation.[40]

Industrial programs

There was irony for industry in the events of 1958. Before Geneva, it was inconvenient to arrange access to information about Sherwood, whereas afterward, material about CTR quickly became available in the open literature. Yet before Geneva, practical fusion reactors seemed only a decade or two in the future, so that it made sense to set up privately financed industrial programs, whereas afterward, fusion was revealed as, at best, a long-range and expensive investigation, more appropriate to government than private laboratories. It is not surprising, then, that the two major industrial CTR programs, at General Electric and at the new firm of General Atomic, were both inaugurated before 1958.

The watershed for industrial participation in Sherwood must be placed in 1956. Until 1956 industrial involvement in controlled thermonuclear research was limited to the loan of company scientists and engineers to AEC projects. In early 1956, however, shortly after it had been persuaded to make public the fact of, and the sites for, controlled thermonuclear research, the commission also took action to include CTR "among the categories of information which may be made available to holders of Secret Access Permits." Industries were eligible to apply for such permits. It was now possible for them to gain access to Sherwood information without the necessity of making personnel available to the AEC program.[41]

It is important to keep in mind that in 1956 the era of nuclear generation of electricity was still in the future. As they looked ahead to it, some American businessmen saw fission as problematic. First, the utilities had no immediate need for fission since fossil fuel supplies were still adequate. Second, it was not clear that the first generation of fission reactors, due to come into operation in the sixties, would be economically attractive, while the waste disposal problems that they posed would be enormous. At the same time, fusion scientists were optimistic that controlled thermonuclear generation could be demonstrated within 4 or 5 years. Extrapolating from the fact that 15 years were elapsing from the first self-sustaining fission reaction in December 1942 to the Shippingport power reactor of 1957, the scientists calculated that a demonstration fusion plant could be available in 1975. There thus seemed to some a possibility that utilities could leap over the fission reactor stage altogether and go directly to the fusion engine.[42] The electrical manufacturers, those firms that build electrical generating equipment for the utilities, could not afford to ignore fusion, if controlled thermonuclear reactions held the potential of becoming a major source of base load central station electricity.

In fact, General Electric had already done some in-house work, even before access to Sherwood results was opened. In 1955 Kenneth H. Kingdon, head of GE's Knolls Atomic Power Laboratory, assembled a small group of researchers, including Willem Westendorp, who had recently returned from the Model D study group, and steered them toward the evaluation of an approach based upon work he and a collaborator had done in the thirties. Kingdon's idea was to set up a number of simultaneous gas discharges around the periphery of a sphere. The plasma thus formed would flow to the center of the sphere, converging to a density sufficient for fusion energy production. But this approach encountered the problem that no sound means could be devised for keeping the plasma confined in the center, and by late May of 1955 Kingdon had concluded that they had "run out of non-magnetic ideas for containing the plasma."[43]

In 1956, when access to Sherwood information was opened to industry, General Electric decided to start a substantial research program."[44] There were several reasons. General Electric had been involved in atomic energy research since World War II, and the study of fusion represented a natural evolution for the firm. Second, plasma confinement by magnetic fields, and plasma heating by electrical and magnetic means, both involved intricate electrical engineering, and General Electric believed it had the special engineering expertise to make an important contribution. Finally, fusion might transform the electrical equipment industry. "It would hardly be fitting for members of that industry to remain on the sidelines while its future is being determined solely through the initiative and diligence of other organizations."[45] In these pre-Geneva days, General Electric shared the general optimism that fusion would work, without, however, giving voice to the idea that it would be possible to leap over the stage of fission energy.

General Electric chose one of its leading scientists to head the new program. Henry Hurwitz, Jr., had been at Los Alamos for the last year of the Manhattan District Project. He had joined the staff of GE's Knolls Atomic Power Laboratory in 1946, and by 1954 he had been named by *Fortune* magazine as one of the ten most prominent scientists in industry. Hurwitz brought to his new job a background in gas discharge physics, with its connection to plasma physics, and fission reactor science, with its connection to the engineering side of fusion energy. He brought, too, a concern, atypical in the fusion world of 1956, for the economic dimensions of the problem. "In the final analysis the important question is not whether problems can be solved but rather how inexpensively they can be solved. . . . The advantage which fusion plants may have in regard to radioactivity [for example] . . . must ultimately be weighed on the basis of the resultant savings in mils per kilowatt hour and balanced against any added costs which may accrue from other less favorable features. . . . Similarly . . . the expected negligible fuel cost is not an absolute one. . . . It too must be evaluated on the basis of mils."[46]

The two other major electrical manufacturers, Westinghouse and Allis-Chalmers, did not establish research programs, but each, in its own way, did keep a hand in. In 1956 Westinghouse already had two men on loan to Project Matterhorn. One was Don J. Grove, whom Princeton had originally borrowed for the Model D study and who had stayed on for the conceptual design that Princeton carried out for the Model C installation. Another was John L. Johnson, whom Westinghouse had groomed for in-house fusion research and who had instead been sent to augment the Princeton theoretical group in the summer of 1955. Now, in 1956, Westinghouse managers reasoned that there were several machines contending in the race for a successful reactor, that there was no way to predict which one would finish first, and that to set up their own program around a single contestant would be extremely expensive. They therefore elected to keep their men on permanent loan to Princeton. Grove and Johnson were made, in effect, into expert consultants to the Westinghouse laboratories.[47]

Allis-Chalmers became involved in CTR in 1957, through the Model C. The firm was one of a group that had responded in spring 1955 to an invitation from Spitzer to submit proposals for detailed engineering work on the installation. Allis-Chalmers had sent in a joint proposal with the Radio Corporation of America. By the time the Model C had been postponed in the summer, the AEC selection committee had arrived at a short list of three: Allis-Chalmers and RCA, Westinghouse, and General Electric in collaboration with the firm of Jackson and Moreland. The selection process was resumed in March 1957. Spitzer had initially favored Westinghouse. The AEC, however, wanted to spread knowledge of new technologies over as many firms as possible, and probably for this reason the project went to Allis-Chalmers and RCA.[48] Choosing the name of C-Stellarator Associates, they set up a group at Matterhorn in September 1957, which they soon built to a strength of 60 members.[49]

The one firm that followed General Electric in establishing a privately financed fusion program was a new one, General Atomic. Its parent company, General Dynamics, was created after World War II by financier John Jay Hopkins to specialize in weapons technology. In late 1955 Hopkins decided to enter the atoms-for-peace field; he persuaded his board of directors to found General Atomic to explore nuclear applications to civilian purposes. Hopkins enlisted Frederic de Hoffmann, physicist and former aide to Edward Teller at Los Alamos, to serve as president of General Atomic, and Edward Creutz of Carnegie Institute of Technology, at that time just finishing a survey of Sherwood for the AEC Division of Research, was chosen as director of research. Creutz, in turn, hired a number of physicists from the national laboratories, among them plasma theorist Marshall Rosenbluth.[50]

In the summer of 1956 the staff of General Atomic gathered to decide which projects to undertake. They were visited by a deputation from the Texas Atomic Energy Research Foundation (TAERF), a new consortium of 11 Texas utilities that had banded together to support fission research. The Texans did not feel, given their location in an oil-rich state, that they needed fission plants, but they were worried about the consequences for private utilities if fission should come to be dominated by public power. De Hoffmann and his scientists per-

suaded them to interest themselves instead in the longer-range technology of fusion reactors, and in 1957 a 4-year, jointly financed fusion
research program got under way, with TAERF and General Atomic
each putting up $5 million.[51]

The man that General Atomic hired to lead their fusion project was
Donald W. Kerst. Kerst had won an international reputation with his
invention of the electron accelerator, known as the betatron, in the
forties. He had also taken part in the wartime discussions on fusion
energy at Los Alamos and had watched the program in the intervening
years while continuing his own work as an accelerator physicist. Kerst
perceived his own contribution to accelerator physics to have been
the introduction of theoretical analysis and careful engineering design
to replace the rough empirical methods of the thirties. He hoped to
bring the same techniques to Sherwood research.[52] At General Atomic
he found that the CTR group had picked the stabilized, toroidal pinch
as their first major experimental device; the choice was a natural
outgrowth of Rosenbluth's interests.

De Hoffmann, Creutz, Kerst, and theoretical group leader Rosenbluth
all had most of their experience in universities or national laboratories.
This fact alone favored an attitude that was more academic than
industrial. In addition, De Hoffmann deliberately tried to foster the
atmosphere of a university. "To us," he explained to Congress, "research is not just an adjunct to our business, but really is the foundation
stone of the business. I think we have demonstrated that we have a
research team that can indeed carry out even the most basic kind of
research in the traditions of academic freedom and with the vigor that
an industry can supply.... Research is a business to us, just as fuel
elements."[53] The difference between General Electric, where the ultimate touchstone for research was its contribution to profitable commercial production, and General Atomic, which maintained an
academic atmosphere and desired to sell research as well as reactors,
was to make itself felt in subsequent years as the goal of fusion energy
receded into the more distant future. GE was to abandon fusion as
not economically competitive with fission, while the General Atomic
project was to become absorbed into the AEC program.

The commission's expansion of CTR in 1953 and after had cemented
a fusion community. Fusion scientists and engineers acquired an institutional base, regular meetings, publication via the distribution of
classified conference reports, and their own name, Project Sherwood.
The transition from secrecy to openness in 1958 permitted a further
step, the emergence of a new discipline for the scientific, as opposed

to the still rudimentary engineering, half of fusion energy research. This new discipline, the study of high-temperature, fully ionized plasmas, had a scientific society, the Division of Plasma Physics of the American Physical Society; refereed journals like *Physics of Fluids* and *Nuclear Fusion*; and, in the new graduate programs, a system of professional training. The structure of rewards was now somewhat altered; good work could be recognized by scholarships, university appointments, publications, and public honors, as well as by promotion through the ranks of government or the acclaim of insiders.

New institutional viewpoints were also brought in by declassification. The views of universities were shaped by such concerns as the research that would be appropriate for training graduate students; the views of industrial laboratories were shaped by such market considerations as the possible cost advantages of fusion reactors. These new concerns, however, were destined to remain peripheral. As the fact that fusion was a long-range program percolated through the AEC management and the Congress, the budget, which under Strauss had gone from $1.8 million in fiscal year 1954 to $29.2 million in fiscal year 1958, stopped increasing.[54] At the very time when declassification was making it possible for the Division of Research to place contracts with universities, or with industries that, unlike General Electric and General Atomic, were unwilling to commit their own funds, budgetary constraints were making it difficult. In the 1960s as in the 1950s, therefore, the history of the American fusion program was to be written largely in the government laboratories.

The Mirror Confronts Instability

The first "stable" plasmas

From 1955 to 1960, CTR theorists produced ever more rigorous demonstrations from ideal magnetohydrodynamic theory that mirror confinement is unstable.[1] At the same time, the mirror group at Livermore produced ever clearer cases of apparently stable plasmas. In January 1960, Richard Post, Marshall Rosenbluth, and two coworkers submitted a report "providing evidence for the existence of a stably confined plasma, observed under circumstances where the simplest hydromagnetic theory predicts instability."[2]

The mirror machine that Post and his colleagues used—a variant of the Table Top—consisted of a single cylindrical chamber. Plasma was injected at one end, and the magnetic field was then increased several hundredfold. By this means, the plasma was compressed and the temperature of the electrons was raised by a factor of 1,000.[3] Ideal magnetohydrodynamic theory predicted that under such circumstances a flute instability would form and grow at such a rate that the amplitude of the flutes would nearly triple in 0.14 microsecond. In contrast, the Livermore team was typically keeping its plasmas intact for 2,000 microseconds, and in some cases was reaching confinement times of 30,000 microseconds. Thus, the plasmas were as much as 100,000 times more stable than predicted.

Post and Rosenbluth thought that the most probable explanation of their results was "line-tying." A flute instability grows because the fluted disturbance at the plasma surface, once established, causes a separation of charge, which, in turn, causes electric fields. The electric forces then act to enlarge further the crests and troughs of the flute pattern.[4] Mirrors, however, have an open magnetic geometry. The lines of magnetic force do not close upon themselves, but stream out of the containment region to end upon the walls of the encasing vacuum chamber. If these walls are metal, as they were for the Livermore Table Top, and if the hot plasma is surrounded by a colder, more tenuous plasma that extends to the walls, then an electrically conducting path is established along the magnetic field lines, consisting of the plasma and the metal wall. The magnetic lines are "tied" to each other electrically. The electric fields that cause the flute instability to grow could conceivably be shorted out via this path. Post and Rosenbluth could not prove that this was happening in Table Top, but it would explain the experiments. "Although we cannot as yet provide a documented theoretical explanation," they concluded, "we believe we have adequately demonstrated, experimentally, a case of stably confined hot plasma. . . . Hopefully, the experiments provide an opening wedge into solving the problem of long-time magnetic confinement."[5]

While Post was directing the Table Top experiments, another Livermore scientist, Frederic H. Coensgen, was taking a different tack in a sequence of apparatuses called Toy Top. Coensgen had built the first Toy Top in 1954, shortly after he had joined the mirror group as a young experimentalist with a doctorate in high-energy physics. The machine had been designed to test a new method for the construction of high-field, fast-pulsed magnets. Coensgen and his team had gone on to use Toy Top in experiments on heating by magnetic compression.[6] In the technique of magnetic compression, the greater the final value of the magnetic field, the stronger the compression and the higher the final plasma temperature. The problem was in filling a standard mirror container with a strong magnetic field. The total energy contained in a field depends upon both the field's strength and the container's volume. This energy must be generated and stored in some system—a flywheel, perhaps, or a bank of electrical capacitors—and switched into the mirror magnet coils at the appropriate moment. The storage and switching of such large amounts of energy posed severe technological problems.

Coensgen wanted higher temperatures without an increase in stored energy. His solution was to replace the single mirror with a set of three connected chambers: "The single compression is replaced by a number of compression stages, such that only a weak magnetic field is established over the largest volume and each successive compression establishes a larger field over a smaller volume."[7] Coensgen's multistage experiment, Toy Top III, created a plasma in which the ions were hotter than the electrons. In the experiment that Post and his collaborators had reported in January 1960, the electrons had the higher temperature.[8] It was important to see whether the hot-ion plasma could also be stable. Coensgen and his colleagues went on a double shift. At the end of October 1960 they had the first two stages working and were able to report that they, too, had produced a "stable" plasma, outlasting the lifetime allotted to it by hydromagnetic stability theory by a factor of 1,000.[9]

The results that Table Top and Toy Top achieved in 1960 were extremely encouraging to the fusion community. If the mirror machine could confine plasma stably, there was real hope that a reactor could be developed out of it. One could take advantage of the economy entailed by the simplicity of its magnets. Moreover—and this was of great importance in Post's estimation—the fields of the mirror magnet had symmetry about the central axis of the containment vessel. The strength of the magnetic force lines in the simple mirror varies with the distance that one moves radially outward from the axis, and this spatial variation of the field causes the ions and electrons to drift away from the magnetic line about which they gyrate. The drift could be disastrous in an asymmetrical field, but in a symmetrical magnetic field, it is innocuous.[10]

A briefing of the commissioners on the entire Sherwood program was scheduled for January 10, 1961. Ruark called on the project leaders to submit scenarios for reaching net power production in their respective machines. In response, Post brought to the session a highly articulated and buoyantly confident 5-year plan.[11] Post had been heartened, not only by the stability found in the Toy Top and Table Top plasmas but also by the results of a study he had made in 1960 of the specifications for a mirror reactor. He had begun the study with assumptions about both the plasma behavior and the technology. As far as the plasma went, Post assumed that the stability they had been finding would persist as they went to reactor temperatures and higher pressures; he contemplated keeping the maximum plasma pressure low to improve the chance of maintaining stability. He also made the assumption—highly optimistic in view of the little understood about plasmas in 1961—that no new mechanisms for the loss of particles or energy beyond what was then known would come to light. On the technological side, Post supposed that it would be possible to develop very low temperature (cryogenic) magnets so as to cut down on the power that the magnets would consume, and that the methods of injecting particles into mirror geometries could be significantly improved. Based upon these postulates, Post estimated a power balance ratio—that is, Q, the ratio of nuclear power produced to power lost or consumed by the reactor subsystems—of 5:1 or better for deuterium-tritium fuel mixtures.[12]

A power balance of 5:1 appeared in 1961 as a reasonable foundation for a reactor. Post's 5-year plan envisioned completing two stages of the march toward that goal. The first stage was to increase the temperature and the plasma pressure, as reflected in the parameter β. Temperature was to be increased from the 4 keV (thousands of electron volts) already achieved to 100 keV, and β was to go from 0.5% to 15%. As the increases were made, the Livermore team would test the assumptions of continued stability and the absence of new loss mechanisms. This first part of the work would make use of machines already available. These were the one-stage Table Top, the multistage Toy Top, and a machine called ALICE (adiabatic low-energy injection and confinement experiment), then in the process of construction and designed to test the injection and trapping of beams of neutral atoms. In the second stage, the group would proceed to experiments whose emphasis would be upon testing cryogenic coils and injection methods. This part of the plan would make use of new devices. These were an ALICE II, a new Toy Top-like multistage compression apparatus, and, sometime in the first part of 1965, as a culmination of the experiments, a reactor prototype. The prototype would reach reactor values of temperature, pressure, and confinement time. It would produce power, and it would be, in part, dedicated to engineering studies such as the testing of structural materials and heat transfer systems.

Post told the commission, "We think it likely that the end result will confirm our expectations that a positive power balance can be achieved."[13] Thus, Post envisioned that scientific feasibility, for the mirror at least, could be established by the mid-1960s.

In April, when Coensgen presented his results at a meeting of the American Physical Society, Post told reporters that they had proved "nature doesn't abhor a stable plasma." The third stage of Coensgen's multistage compression device was being readied at this time, and he and Post expected confinement times of 5 milliseconds and a record temperature of 80 million degrees centigrade to be attained.[14] Mirrors were widely regarded in both the United States and abroad as the leading reactor concept in early 1961, and the administrators in Washington were also excited by the successes of Toy Top and Table Top. In late 1960 and the first part of 1961, the Livermore mirrors, with their apparently stable plasmas, were the AEC's controlled fusion showpieces.[15]

The Salzburg conference

In September 1961 the International Atomic Energy Agency sponsored the first international conference on controlled thermonuclear research since the Geneva meetings of 1958 in the city of Salzburg on Austria's Bavarian border. Just before the conference, Coensgen and his co-workers had succeeded in putting the third stage of their magnetic compression experiment into operation. The data had been unexpected and disappointing. Instead of a confinement time extended beyond the 1 millisecond that Livermore had reported at the April meeting of the American Physical Society, the life of the plasma had shrunk by a factor of 10. The plasma was moving as a whole to the wall of the second segment of the apparatus in slightly under 0.1 millisecond. Whether this pathology was due to asymmetries in the fields created by the magnets, whether the Teller instability had at long last made an appearance at Livermore, or whether some entirely different, unknown mechanism was responsible, Coensgen did not know.[16]

A team of Soviet experimentalists at the Kurchatov Institute in Moscow, lead by M. S. Ioffe, had also been working with mirror machines. They had found, contrary to the Americans, that flute instabilities do develop in mirrors, but roughly 100 times more slowly than predicted by ideal hydromagnetic theory. This result was in contradiction with the abstract for the Toy Top III paper, which had been distributed before the meeting and did not include Livermore's last-minute results. Thus, even before the meeting, the Soviets had had some reason to believe that the optimism over the results of the multistage compression experiment, as expressed in the abstract, was wrong. This was confirmed for them at the conference. The figures Coensgen gave orally agreed with those of Ioffe's group. Coensgen and Lev A. Artsimovich, the senior member of the Soviet delegation, were able to agree on the source of error. It had been a regrettably "elementary mistake."[17] A number of neutrons had been found emerging from the Livermore plasmas 1 millisecond after the start of the experimental runs. Coensgen and his team had interpreted them as "prompt" neutrons, that is, neutrons emerging from concurrent deuteron-deuteron reactions in the plasma. In fact, they were neutrons from much earlier reactions that had been slowed down by the very scintillation counters that were being used to measure them, and hence were being recorded late.

Artsimovich was caustic. "I want to say," he remarked, referring to Post's paper on the outlook for mirror reactors, "that Ioffe's results are in sharp contradiction with the attractive picture of a thermonuclear

Eldorado...drawn by Dr. Post. After the initial assertions of Dr. Coensgen that the plasma containment time was about 1 msec, have proven erroneous, we now do not have a single experimental fact indicating long and stable confinement of plasma with hot ions within a simple magnetic mirror geometry."[18] Artsimovich's comments could serve a second purpose, in addition to clarification of the truth. To denigrate the Americans was also to set in relief the excellence of another Soviet experiment that had been done at the Kurchatov Institute. Ioffe's group at Moscow had done more than study instabilities; they had corrected them. They had added a set of six conducting bars to the mirror. When current was sent in the proper directions through both the bars and the ordinary mirror coils, the magnetic field created by the two types of conductors together took the form of a "magnetic well." The magnetic field at the center was finite but a minimum, and the strength of the field increased in every direction out from that minimum. Ioffe and his collaborators could increase the confinement time by a factor of 35 simply by switching on the "Ioffe bars."[19]

Magnetic wells were no novelty to the American delegation. The same ideal MHD theories that had predicted the inevitability of the flute instability had also shown that the magnetic well configuration would cure it. There had been several reasons, however, why—in contrast to Ioffe's group—the California mirror scientists had not elected to install coils that would produce wells. For one thing, they had not observed macroinstabilities until the mysterious sideways drift had shown itself in Coensgen's multistage apparatus shortly before the conference. Moreover, they had had some reasonable explanations for the absence of the flute instability. One was line-tying. A second was a theory just developed by Marshall Rosenbluth, Nicholas Krall, and Norman Rostoker at General Atomic to show how finite gyration orbits might be stabilizing the plasmas using the fact that the electric field that causes flutes to grow is not uniform, but varies from point to point in the plasma vessel. Since the ion's radius of gyration around a magnetic field line is larger than the electron's, the ion samples a larger portion of this varying field. The average fields that the two kinds of particles "see" therefore differ, and their differing response, the General Atomic group theorized, is what keeps the flute from growing.[20] A final reason for not installing wells was that the Livermore team hoped to extrapolate its machines into reactors and hence wanted to keep its system of magnets simple, and thus less expensive.

The Salzburg conference changed Livermore's outlook. The Ioffe team's experiments showed that magnetic wells could correct macroinstabilities, not only in theory but also in actual apparatuses. Experimental proof was vastly more persuasive than theoretical prediction in controlled thermonuclear research, where an unexpected experimental outcome was more the rule than the exception. The story of their rather simple mistake had unfolded in an international arena, drawing to it both publicity and the considered reflection of the scientists of other nations. The Americans had come out looking badly in comparison with their Soviet competitors at a moment when rivalry with the Soviet Union for technological preeminence was at a peak. Indeed, only a few months before the Salzburg meeting, President Kennedy had authorized the Apollo program to land men on the moon before the Russians. The episode moved the entire question of stability from the sidelines to center stage at Livermore. Post was determined that his group should not be surprised by unexplored instabilities again.[21]

A change in direction

Returning from Salzburg to his laboratory, Coensgen turned immediately to the task of determining the origin of the drift of the plasma to the wall in the second stage of the multicompression experiment. Was the Toy Top showing the flute instability or was it merely a field asymmetry, introduced when the magnets used to transfer the plasma to the final stage were activated, that was dumping the plasma to the walls? The previous Toy Top goal, incorporated in Post's 5-year plan, of reaching energies of 100 keV, was set aside. "When a system is found which exhibits gross stability the emphasis will again shift towards achieving high ion temperatures."[22] The first task was to eliminate magnetic field asymmetries. Coensgen's team reduced the asymmetry below 0.5%. Nevertheless, in those experiments where line-tying did not intervene, the plasma was unstable. Where line-tying did occur, the plasma was confined more than four times longer. Coensgen reasoned, however, that with line-tying, not only electricity but also heat was being conducted along the magnetic lines to the conducting chamber walls. This was an effect that he did not want.[23]

Post, after the Austrian meetings, turned to a searching investigation into the hot-electron plasmas of Table Top. Why was the hot-electron plasma in the Table Top stable, in apparent contradiction to the experiments of Ioffe and Coensgen? Together with a collaborator, Walton A. Perkins, he slowly uncovered a complex situation. Plasmas of a wide variety of motions and configurations could be formed; some were stable, some were quasi-stable, some were unstable. Neither the ideal MHD theory nor finite-orbit theory nor line-tying, taken alone, was adequate to explain the multiplicity of effects. One thing was quite clear, however. The flute instability, hitherto undetected in the Table Top, could definitely be triggered under the proper circumstances.[24]

Post also joined in the work on ALICE, in collaboration with Charles C. Damm, its team leader. The ALICE scientists began experimentation in 1962, following a major development effort to achieve the extremely high vacuums and intense neutral beams that they wanted.[25] By early March 1963 the team had verified the existence of a collective effect: The plasma was emitting electromagnetic oscillations of characteristic frequency. Oscillations alone do not always mean plasma loss. But a few months later the ALICE scientists perceived that these oscillations were indeed causing so much loss that the plasma density could not rise above the very low value of 2×10^7 particles per cubic centimeter.[26] It would clearly be impossible to use the machine as it stood to implement Post's 5-year plan. In that plan, ALICE was projected to retain a high-density plasma for several seconds, and ALICE II was projected to show the same long confinement time for a plasma that was 10 times more dense still.

By summer 1963 Post faced squarely the question whether to make a major change in hardware to incorporate magnetic wells. This is how he viewed the situation: In favor of magnetic wells was the fact that they actually worked to overcome the flute instability, as evidenced both by some preliminary Livermore runs of the Table Top with quadrupole Ioffe bars and by Soviet and British experiments. In favor also was the fact that the elimination of macroinstabilities would clear the way for experimentation on the more slowly acting microinstabilities. Microinstabilities are so called because they play themselves out on a scale comparable in dimension to the particle radii of gyration, and hence are outside anything the magnetohydrodynamic theory of macroscopic, continuous fields can predict or explain. Mirrors were expected to have greater problems with microinstabilities than other types of devices because mirror plasmas are more anisotropic than other types of plasmas. Consequently, mirror plasmas have an extra degree of departure from normalcy and might be expected to exert extra effort, by means of instabilities, to move toward a more comfortable state. Various kinds of mirror microinstabilities had been discussed theoretically, but no incontrovertible evidence of their presence had yet been found. Post had been optimistic about microinstabilities when he had first arrived at the Salzburg conference. He had been willing to take the position that none of the ones that had been explored on paper would introduce a serious limitation on the mirror machine.

The conference events, however, caused him to reconsider. He now began to see the study of microinstabilities as a pressing priority. For Post was now far more worried about these less understood effects than he was about macroinstabilities.[27]

Against the use of the magnetic well was the fact that the Ioffe bar conductors reduced the access into the plasma chamber for diagnostic equipment and the neutral beams being used on ALICE. Also, the more costly magnetic well magnets did not look promising for use on economically competitive fusion reactors. Finally, and most telling for Post, the Ioffe bars destroyed the symmetry of the magnetic fields. This feature complicated the process of measurement, made it difficult to develop theories to compare with measurement, and threatened to introduce new sources of instability into the plasma.[28]

Post's problems were compounded by the situation in Washington. In 1962 Paul McDaniel, now director of the Research Division, had requested the Commission's General Advisory Committee to conduct a review of the whole Sherwood program.[29] The subcommittee formed by the GAC to carry out this task, headed by Carnegie Institution scientist Philip H. Abelson, submitted its report in August 1962. It contained some strong conclusions. Although it praised the Livermore mirror program, the subcommittee advised that no new large machines be funded until extensive preliminary work had been done on simpler devices. Now Livermore was in the process of constructing a larger machine, the 2X. Coensgen, when he had returned from Salzburg, had made some quick and qualitative trials with Ioffe coils to stabilize Toy Top. The attempt had foundered on the poor vacuum conditions. The trials had convinced Coensgen that a magnetic well sequel to the Toy Top experiments would have to be twice as large as Toy Top, to allow for faster pumps and to increase the distance of the confined plasma from the chamber walls. The Abelson committee, however, opposed the 2X: "As long as such cheap, interesting and straightforward experiments are available for elucidating Toy Top behavior, the Committee cannot endorse proceeding with the expensive Toy Top 2-X program." Moreover, the subcommittee, building on some of the conclusions being reached in David Rose's MIT course on fusion reactors, took the view that the mirror, in contrast to closed machines like the stellarator, could never be made into a reactor, and must be viewed as only a tool for plasma study, a tool that was, however, "indispensable . . . at this time."[30] The General Advisory Committee had endorsed its subcommittee's report and was insisting to the commissioners that the report's recommendations be implemented.[31]

The Division of Research was able to convince the commissioners that the report had been hastily done and had overlooked or mistaken some cardinal features of the program. Some of the recommendations, among them that of halting construction of the 2X, were therefore rejected. Nevertheless, as a direct consequence, the commissioners decreed that a more thorough programwide review be scheduled for 1965, and the implication was that this review might be used to make a stop-or-go decision on the entire controlled thermonuclear effort. Sherwood as a whole was under the gun, and so was Livermore's 2X.[32]

At this point, Harold Furth added a new element to the balance sheet of the pros and cons of the magnetic well. Furth was still working with the Livermore pinch group in 1963. By this time, that group's search for the cause of instability in the "stabilized" pinch had led them to experiments with "hard-core" pinches. A linear pinch had been constructed first; it had a metal conducting rod down its center, from which the hard-core pinch drew its name.[33] While the linear pinch was being tested, a toroidal version of the same idea was constructed. The inner, conducting tube in the toroidal hard-core pinch is, of course, a closed ring. The Livermore group had decided on an arrangement that, at the time, seemed somewhat outlandish. Instead of suspending the ring within the conducting vessel by material supports, they programmed the fields so that the magnetic forces, interacting with the current in the ring itself, kept the ring floating during the 20 or so milliseconds of the experiment. Even more strikingly than the linear hard-core pinch, this "Levitron" was clearly not a reactor idea but a research apparatus for testing stability theories. In looking about for a machine that *was* capable of being escalated into a reactor, Furth thought that the mirror, with a magnetic well added to undermine MHD instability, offered the brightest possibility. Many theorists had returned from Salzburg and begun to explore magnetic well configurations. Furth had been doing computations aimed at finding configurations of magnetic coils that could produce wells but would avoid some of the disadvantages of the cumbersome Ioffe bars. In the spring of 1963 Furth succeeded in showing that a coil could be designed that at one and the same time provided a magnetic well, albeit a shallow one, *and* axial symmetry.[34]

Furth's discovery tipped the balance for Post. In late summer the mirror group leader abandoned the plan of basing his program largely on the ordinary mirror magnets and called for a major revision of the ALICE machine. An entirely different coil system and a new vacuum chamber would make it possible to test the axisymmetrical magnetic well. Finite-orbit stabilization might well be more economical in the long range, but "we . . . consider the 'minimum B' [magnetic well] approach to be the one most likely to succeed in the short run."[35] Post anticipated that the ALICE alterations would total $1 million over several years. Associated preliminary magnetic well investigations would add $350,000. Post estimated that if an additional $650,000 and a staff increase of 31 persons could be provided during fiscal year 1964, and an intense effort were mounted, the modified ALICE could be ready for experiments at the start of 1965. As for future plans, the ALICE II would follow ALICE at some unspecified date. The mirror reactor prototype was no longer mentioned.

There is a parallel between the stories of the mirror and the stellarator. Like Spitzer and his team, Post and his coworkers had an abrupt and unexpected confrontation with hydromagnetic instability. Spitzer's confrontation came in October 1954, with Teller's speech of warning at the Princeton gun club. Post and Coensgen's came in mid-1961, with the discovery of the lateral drift to the wall of the Toy Top plasma. Like Princeton, the Livermore group had started out with plans leading in a straightforward way to a reactor prototype. Like Princeton, the mirror scientists had to abandon those plans in order to explore a new, more complex, but, it was hoped, more stable field configuration. In the case of Princeton, helical windings were added to the original coils in early 1956. In the case of Livermore, Post called for a revised ALICE with added coils that would create a magnetic well configuration.

In both cases, the effect was to shift the strategy of investigation further toward an emphasis on basic research. The thinking at Livermore was well expressed in a blueprint for the fusion program that was offered by Post and Furth at the end of 1964. "After 10 years of paving the way . . . we are presented with the opportunity of an important new orientation in controlled-fusion research. . . . Today . . . the critical scientific [question] . . . is whether a reacting plasma can be confined *sufficiently stably* by a magnetic field. . . . Recall . . . the most straightforward and economical way of answering questions in physics is to attack them with apparatus designed for the purpose. . . . Whether any of the tools of research that will be used could be pictured as reactor prototypes, or how they might compare with others in respect to any features other than the clarity and relevance of their contributions to the stability question, is of secondary importance for present needs."[36]

In one of its aspects, the whole history of the decade from Teller's lecture in 1954 to these statements of 1964 may be grasped as a shift from the tactic of technological innovation to the tactic of fundamental studies of plasma physics. The new orientation permeated ever more widely and deeply into the physics community. By 1964 the second phase in the development of fusion research, marked by an emphasis on science, was at its zenith.

Skepticism in Congress

In 1962, the House Appropriations Committee had made a first attempt to trim the fusion budget in its action on the fiscal year 1963 appropriations. The toll for the program was minor, however. Only $300,000 was cut off, and the total of operating funds for the year was $24.2 million. In the fall of 1963 the ax struck closer to the bone. Acting on the funds for fiscal year 1964, then partly over, the House committee expressed its displeasure with the CTR program by reducing its operating funds by $3.75 million, leaving only $20 million in this category. By promising a "breakthrough"—a promise based on the rosy picture of the prospects of axisymmetrical magnetic wells painted by Livermore—the commission leadership managed to persuade the Senate Appropriations Committee to add back $2 million. The final bill emerged from the conference committee with $21 million in operating money for the year in progress.[37]

Congressional opposition to controlled fusion spending had its roots in part in a general attitude; the post-Sputnik science spending spree was increasingly being questioned in the early sixties.[38] Within the frame of this general concern, however, members of Congress had some very particular questions to raise about the controlled thermonuclear program. First of all, a dozen years had passed since the program had been inaugurated. Not only had no reactor been developed, but no knowledgeable scientist was able to assert that such a reactor was possible. Was fusion a wild-goose chase? "I am wondering in my own mind," said Senator John Pastore of the Joint Committee at the hearings in the spring of 1964, "how long do you have to beat a dead horse over the head to know that he is dead?"[39]

Second, the money that Congress had allocated for technology had in effect been redirected to research. As the people involved in Sherwood had become increasingly aware that the trial-and-error empirical methods of the early days were not working, they had come to apply their efforts, and the funds they had been given, to a purely scientific enterprise: the physics of high-temperature plasmas. At the same time, the AEC began to place emphasis in its congressional testimony on the contributions of the program to basic knowledge. Congress, however, was more interested in technological development than research.[40] From 1960 to 1964 Sherwood's spending had run between $24 and $32 million a year.[41] "Is this not indeed a very expensive way of getting this basic knowledge?" Pastore asked at the hearing. "We can build these machines until the cows come home. Somewhere along the line somebody has to think that this is a lot of money and maybe we ought to be putting it into some other place where it may be more productive."[42]

Congress also questioned why such a superabundance of confinement schemes were being tested. There were five main categories of machines in the budget in 1963: the stellarator, the mirror, the DCX of Oak Ridge, a machine called the Astron at Livermore, and various forms of pinches. The House Appropriations Committee wanted to know why so many avenues were being pursued. Surely, after so many years, some sorting could be made between concepts that were promising prospects for reactors and those that were not.

In trimming the budget in 1962, the House Appropriations Committee had explained that it wanted some approaches dropped. In making the more devastating reduction of 1963, the committee had stated, "The Committee indicated last year that it expected that the number of concepts in this program would be reduced. Nothing was done about it. It is expected that this reduction in funds will achieve the objective." At the same time, the committee admonished that the selection be guided by "greater emphasis on the overall usefulness of the potential results."[43]

Congress's astonishment that in more than a dozen years the scientists had failed to determine the feasibility of fusion energy was a product of the predictions of the 1950s. It was from the lips of the CTR scientists themselves, 5 or 6 years earlier, that members of Congress had heard the promise that the program would be at the stage of prototype reactors by the early sixties. As the program leaders discreetly explained to the Joint Committee in 1964, they previously had completely misjudged the extent of scientific and technological knowledge that was needed. Congress's bewilderment over the large number of alternative approaches that the fusion scientists were pursuing was also rooted in the testimony they had heard at an earlier time. When various classes of machines had been selected for study in the fifties, they had in fact been conceived as differing approaches to a reactor. This had changed, however, as the fusion scientists had shifted from a development to a research psychology. Machines proposed as reactor prototypes had come to be regarded by their scientists either solely or additionally under the aspect of experimental apparatus for basic physical research. What Congress saw, for historical reasons, as multiple approaches was really a mixture of reactor approaches and experimental devices. The AEC budget categories had failed to reflect the evolution that had occurred in the uses to which the machines were put.

The early optimism was now returning to haunt the fusion scientists. The Joint Committee on Atomic Energy was adding its voice to the demand for a thoroughgoing review of the Sherwood endeavor for its perusal by early 1966.[44] The fusion leaders could now see vividly revealed the political dangers of premature promises. The impression that the budget reviews of the early sixties made on them was to last more than a decade.

Astron

When McDaniel received the final CTR budget for fiscal year 1964, in December 1963, he judged that he had a clear mandate from Congress to reduce the number of fusion concepts. He proposed to the commissioners to eliminate the Astron, which in fiscal year 1963 had absorbed $1.4 million, close to one fourth of the total budget for fusion research at Berkeley and Livermore.[45] Cutting out Astron had been one of the recommendations of the Abelson committee, the year before. This project's maverick leader, Nicholas Constantine Christofilos, had always been something of an outsider within the fusion community, both scientifically and by his background. An American of Greek parentage, Christofilos had been brought back to Greece as a child in 1923 and had lived there until 1952. He had studied electrical and mechanical engineering in Athens at the National Technical University; in physics he was self-taught.[46]

Shortly before returning to the United States in late 1952 for a job at Brookhaven National Laboratory, Christofilos, who at that time had no contact at all with the secret CTR programs in America and elsewhere, elaborated a scheme for a thermonuclear reactor that he called Astron. Early in 1953 he went down to Washington to talk over his idea with AEC officials. Classification Director James Beckerley afterwards wrote "[Christofilos] is truly a remarkable individual, a regular 'idea factory.' . . . The device he has dreamed up is quite different from that being studied by Spitzer, Tuck or Post. . . . It has a kind of a Rube Goldberg (or should it be Rubinos Goldbergopolis?) air about it. . . . If a successful thermonuclear reactor is ever developed, it may well be a composite of several basic ideas, including this one of the remarkable Nicholas."[47] The Research Division was impressed enough to ask Christofilos to continue his calculations while machinery to obtain a security clearance was set in motion.

The heart of Christofilos's Astron was a rotating cylindrical shell of electrons at the extraordinarily high relativistic energy of 20 MeV (millions of electron volts), about 1,000 times higher than the 20-keV energies that ions would have in a reactor plasma. This layer of electrons (Christofilos called it the "E layer") would be formed within a magnetic mirror machine. The layer would act like a solenoidal coil, creating a magnetic field opposite in direction to the field of the external coils. When a sufficient density of electrons built up in the E layer, "field reversal" would intervene (see figure 7.1). That is, the net magnetic field at the axis would reverse to oppose the field that would have been created by the mirror coils alone. At this point most of the lines of force would become closed curves, like the magnetic lines in toroidal devices.

Once the E layer had been formed, a neutral gas of deuterium and tritium would be injected into the chamber. The E layer would ionize the gas into a plasma and then transfer still further energy to the plasma ions, heating them to thermonuclear temperatures. All the while, the plasma would be confined by closed magnetic force lines. Some diffusion would take place across the field lines, as in any toroidal device. The special and undesirable loss *along* the field lines that is characteristic of mirrors would not take place, however, except in the outermost portion of the vacuum chamber. In that part of the chamber, mirror loss was actually expected to be an asset, allowing the Astron to preserve the mirror property of being self-cleansing. Since the electrons of the E layer would be continually degraded in energy as they first ionized and then heated the plasma, new electrons would have to be continually fed in. For this purpose, an electron accelerator, providing a high-energy particle beam of unusually great intensity, would have to be a permanent part of the installation.[48]

The Astron idea was complex both scientifically and technologically. Scientifically, whereas other devices had only the rich phenomena exhibited by a single confined plasma to investigate, the Astron had both the plasma and the E layer. Thus, it had the disadvantage, as CTR Branch member Hilliard Roderick was to express it, that "our unknowns have been increased. Added to (a) our ignorance of plasma are now (b) ignorance of dense relativistic electron rings and ignorance of the interaction of (a) plus (b)." In addition, since the same E layer was responsible for both containment and heating, it would be difficult to disentangle experimentally the problems due to each.[49] Technologically, the Astron was complicated because before the point of

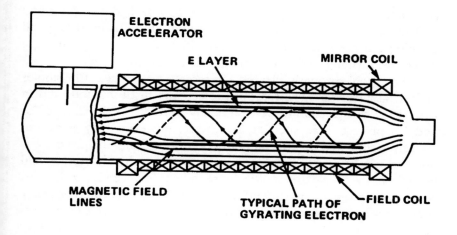

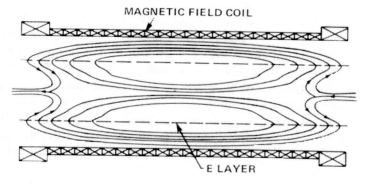

Figure 7.1
The Astron system. (Top) Gyration of electrons in E layer. (Bottom) Formation of pattern of closed magnetic field lines. [Adapted from S. Glasstone and R. H. Lovberg, *Controlled Thermonuclear Reactions* (New York: Van Nostrand Reinhold, 1960), pp. 405, 408]

confining and studying a plasma could be approached, two prior and difficult technical tasks had to be completed. One was inventing and building the novel electron accelerator that was necessary, and the other was forming and maintaining a dense electron layer.

In late 1956, properly cleared for secret work at last, Christofilos arrived at Livermore, where the project was to be housed, and in early 1957 construction for Astron finally started.[50] While Christofilos and his project were sited at Livermore, however, they were not entirely assimilated into the laboratory. For one thing, the project was not exclusively devoted to thermonuclear research. Christofilos had brought in the Department of Defense to help finance his accelerator. A contract with the department's Advanced Research Projects Agency supplied a large part of the funding. In return, Christofilos was preparing to use the accelerator, when it became ready, for military research—probably the exploration of particle-beam weapon principles.[51] For another thing, Christofilos had a personality that did not make it entirely easy for him to fit in. He was not adept at team research, and he tended to be a loner. He was sometimes abrasive, and "his own ideas were urged with a single-minded gusto that at times made him impatient with the system."[52] Furthermore, he had the credentials of an engineer, in a world of scientists. "He still has no degree in physics, and his Greek accent, Greek volubility and love of passionate argument keep him an outsider," *Time* reported. "One top U. of C. scientist remarked frostily, 'Well, my contacts have been with other members of the scientific fraternity, and Christofilos really isn't a member.' "[53] Most important, however, Christofilos was an inventor in the mold of the heroic age of fusion of the fifties, who was suffering the misfortune that his device was only first coming into being during fusion's age of science. He did not really do experimental physics. He did not, for example, seek to break the difficult problem of establishing the E layer into discrete physics questions that could be attacked in a series of small and simple experiments. The whole focus of his effort was on the Astron as a complete system—on testing it, revising it, pushing it politically, and making it work.[54]

It was against this background that, in January 1963, McDaniel recommended to the AEC the termination of Astron as one step in adjusting to the actions of the House Appropriations Committee. The commission that had taken shape after the inauguration of John F. Kennedy in 1961 was distinctly more favorable to research than the McCone commission had been. As the new chairman, Kennedy had selected Glenn T. Seaborg, Nobel laureate in chemistry, chancellor of the University of California at Berkeley, and a former associate director of the Berkeley Radiation Laboratory. The lead commissioner for the CTR program was Gerald F. Tape, a physicist whose route to the AEC had led through the University of Illinois and the administrative staff of Brookhaven National Laboratory. Robert E. Wilson, who had joined the commission in April 1960 and in late 1963 was on his way out again, had been a research engineer turned industrial manager. He was heavily involved in energy industries, having been, for example, chief executive officer of Standard Oil of Indiana. Of the lawyers on the commission, one, James T. Ramey, appointed by Kennedy in the summer of 1962, had come from the Joint Committee on Atomic Energy, where he had served as the executive director since 1956. Ramey had experienced the euphoric optimism about fusion of the midfifties and, later, had participated in the growing congressional skepticism about the program's soundness. The other lawyer member, John G. Palfrey, who had been a Harvard classmate of Kennedy's, was an authority on the legal problems of nuclear energy.[55]

Seaborg, who was close to the California Radiation Laboratory, shared the point of view of John S. Foster, director of the Livermore branch of the laboratory, that decisions on the technical contents of programs should be reserved for scientific leaders in the field. Already in February of that year Foster had made clear his opposition to a termination of Astron dictated from Washington: "The establishment of research in broadly defined areas, and the funding for such areas is the responsibility of the AEC. The technical decisions within an area of research or development rests with the Laboratory. Forced compliance with technical direction from outside the Laboratory would destroy our technical integrity, and is therefore unacceptable to the Laboratory."[56] Seaborg therefore requested that the commissioners first consult with Foster; on December 20 they decided that Astron should be retained in the program.[57] Livermore assured the AEC and the Joint Committee that fiscal year 1965 was to be Astron's year of decision. If field reversal were attained by then, the experiments would be continued; if not, the project would be eliminated.[58]

Had Astron been terminated, $630,000 would have been available at Lawrence Radiation Laboratory to distribute among remaining projects. As it was, the mirror program had to take its licks with the other programs. Washington imposed a roughly uniform cut on the four major laboratories and the Washington office. California, which had just requested an additional $650,000 in operating funds for the prosecution of an attack on the flute instability by means of minimum-B magnets, was instead cut back from $6.8 to $6.2 million.[59]

The actions of Congress and the subsequent budget adjustments at the Washington office left the mirror leaders in a dilemma. The attitude shown on Capitol Hill made it more important than ever to concentrate on installing magnetic wells. For it was even more politically expedient than before to come up with results that would have the appearance of positive, forward steps on the road to a fusion reactor. Magnetic well stabilization almost surely could give the longer confinement times that would look good to Congress. Furthermore, but for slightly different reasons, magnetic well stabilization would advance the program in actuality. It would provide experimental tests for hydromagnetic theory. It would also open the way for an evaluation of the severity of the threat posed by microinstabilities; this in turn would be a step toward meeting Congress's demand for a decisive answer to the question whether fusion energy was feasible. But Congress had also plucked away some of the funds on which the Sherwood leaders had been relying, and that made the testing of the new magnetic field geometries, as well as every other program advance, more difficult.

Post was already prepared with his position: "The basic policy in readjusting the program to a reduced budget is that (1) All possible effort will be applied to prepare for crucial minimum-B experiments." The Ioffe bar windings for ALICE and Table Top for preliminary minimum-B tests were to be installed "full speed." The quadrupole Ioffe bars for the 2X were also to be put in "as soon as possible." As for the windings for the axisymmetric fields that Furth had invented, however, because of the budget cuts these were to be designed only, with fabrication delayed.[60] In the event these magnets, so important for the decision to proceed at full throttle with magnetic well geometries, were never to be made at all.

Microinstabilities

In March 1964 the first results of Livermore's experiments with magnetic wells appeared. Walton Perkins, working with William L. Barr, reported that by adding a set of 6 Ioffe bars to the Table Top, they had reduced the macroscopic flute instability a thousandfold.[61] Next came the report from the ALICE team. Damm and his colleagues had added 12 Ioffe bars to their apparatus, and in the summer of 1964 they announced that the flute instability was suppressed and that the density, previously limited by the instability, was increased by a factor of 10.[62] It did not increase more; as Post and his colleagues had feared, a new loss mechanism, hitherto masked by the more virulent flute instability, was preventing any further density rise. Investigations showed that the density limitation was correlated with a hallmark of microinstability: radiowave emission of high frequency. Simultaneously, Perkins and Barr were also finding that the suppression of the flute instability in Table Top was allowing previously hidden microinstabilities to become manifest.[63]

The ALICE's 12 Ioffe bars had been planned as a preliminary experiment, to precede installation of the radically new coils designed according to Furth's scheme. The very act of working with Ioffe bars, however, made the scientists more comfortable with asymmetrical field configurations. They had always recognized, moreover, that Furth's magnetic fields had the drawback that the wells they incorporated were extremely shallow. At this time, news of a different magnetic coil design reached Livermore from Culham Laboratory in Britain. The Culham coil was shaped like the seam on a tennis ball, and it gave, by means of one continuous conductor, both the field of a mirror coil and the field of quadrupole Ioffe bars. The tennis-seam field was asymmetrical, but it had several important advantages. Unlike Furth's design, it gave a deep well, and, unlike the Ioffe bars, it provided good access for neutral beams, vacuum-pumping equipment, and diagnostic devices into the containment chamber. A similar design had been discussed by Perkins and Furth in the summer of 1963, but had been submerged by the interest in the axisymmetrical field geometries. In late 1964 the mirror group reviewed the issue of the ALICE magnet and decided to adopt the British design rather than the axisymmetrical one. Livermore americanized the British scheme by naming their revised ALICE apparatus the Baseball.[64]

The other major US mirror program, the DCX at Oak Ridge, was also coming face to face with microinstabilities. The path that had led Oak Ridge to this point was very different from the one Livermore had followed. The underlying idea of the DCX had been to create a plasma by injecting a stream of very energetic molecular ions into a mirror containment geometry and dissociating the ions on a carbon arc. The DCX team had succeeded in this way in trapping protons of unusually high energy (hundreds of kilovolts) for intervals of around 10 milliseconds. As work resumed after the Geneva conference, however, it became evident that the plasma density was stuck at 10^9 particles per cubic centimeter. Clarence F. Barnett, who was responsible for DCX operation, enlisted two new PhDs in experimental physics, Herman Postma and Julian L. Dunlap, and in a careful series of experiments the three men discovered the problem. The carbon arc was responsible, not only for the creation of the trapped protons but for their rapid destruction. After first being split out of the ionized hydrogen molecules by collisions in the arc, the free protons were going on to capture electrons from the incompletely ionized carbon atoms and, thus neutralized, to fly out of the containment vessel.

One way out was to eliminate the arc, allowing the incoming beam to dissociate simply by collision with the neutral background gas. When the DCX team attempted this strategy, they found that dissociation did occur but that protons were being lost by the same mechanism of charge exchange that had vitiated the carbon arc experiment. In this case the protons were exchanging charge with the neutral background. Redesigning the vacuum system, the scientists pushed the background gas density ever lower. The mechanism for dissociation was now simply the electric and magnetic forces within the containment vessel, which ripped apart a number of the incoming particles.[65] By early 1964 the DCX physicists had reduced charge-exchange loss to a tenth of its previous magnitude. Still the density did not go up. Since charge exchange could not be responsible, it was clear that they faced some other, new mechanism of plasma loss.[66]

Dunlap, Postma, and the DCX team now turned to correlation of plasma particle losses with the bursts of radiowave emission that were also occurring. On this basis, they were able to rule out the presence of a macroscopic instability; thus, a magnetic well would not be effective for the DCX. Instead, they showed that the operative instability was connected with a charge asymmetry. Rather than being uniformly distributed around the storage ring, the plasma protons were agglutinating and rotating as a lump. It was a nice piece of investigative physics. The conclusion, however, was less nice; they had discovered a microinstability for which no cure was then known, and which led to severe particle losses.[67]

By this time, late 1964, DCX had been renamed DCX-1 to make room in the laboratory nomenclature for a different attack on the problem of injected-ion dissociation via a new machine, the DCX-2. The DCX-2 was based upon an obvious idea. A second way to deal with the deleterious effects discovered in the carbon arc experiments was to use a very much weaker arc, so as to substantially lessen the amount of charge exchange on the arc. To compensate for the proportionately weaker amount of dissociation (from 50% the dissociation would decrease to about 1%), the machine would have to be constructed so as to force the molecular ion beam to intercept the arc many times. The design chosen, out of a field of two, was by Persa R. Bell[68] (see figure 7.2).

DCX-2 was significantly more taxing technologically than DCX-1 had been. The long uniform central magnetic field that the design

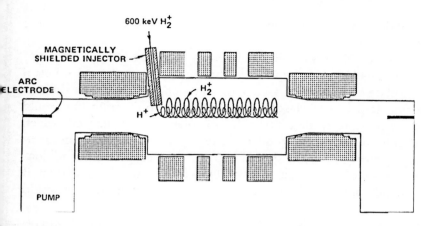

Figure 7.2
The principle of the DCX-2. [From P. R. Bell et al., "Proposal for a Thermonuclear Experiment...," March 6, 1960, ORNL central files]

required called for great care in the construction and placement of
field coils.[69] Once the equipment was built and debugged, the task of
determining the properties of the plasma proved vexatiously time-
consuming. Here, Bell was joined by George G. Kelley and Norman
Lazar, who had followed him in transferring from the Physics Division
electronics group. It was only toward the end of 1964 that the group
began to form a clear picture of a surprisingly complex plasma. One
large component actually consisted of ions whose energies had been
pushed—no one knew how—way above the energies of the injected
particles. An analysis of the radiowave emission showed that microinstabilities were present in DCX-2 as in DCX-1. Their relation to
plasma loss and to the high-energy ion component had not been
elucidated by late 1965; nevertheless, it was becoming obvious that
some phenomenon was intervening to keep the plasma density below
10^{10} particles per cubic centimeter.[70]

The response of Sherwood scientists to the new results of mirror
experimentation was a mixture of satisfaction and disappointment.
The verification that macroscopic instability could be controlled by
magnetic wells in one machine after another was certainly a triumph.
Lyman Spitzer, speaking at the second International Atomic Energy
Agency meeting on fusion research in Culham, England, in September
1965, called the verification "the most significant milestone of the
present meeting." It represented a major new area of agreement
between plasma theory and experiment.[71] The uncovering of micro-

instabilities on the other hand, though it was not surprising, was disappointing.

Microinstabilities were less dangerous than the MHD instabilities because they developed more slowly. Richard Post therefore began to articulate an approach based upon a taxonomy of microinstabilities. The first step was to classify them according to severity; this would make it possible to determine "which ones can be ignored, which need only to be limited to tolerable levels, and which must be completely surpressed."[2] The second step was to investigate the conditions for the onset of the subset of instabilities that required elimination; if these conditions could be prevented, the associated instabilities would be avoided.

A major step in this direction was made by Post himself, in collaboration with Marshall Rosenbluth of General Atomic, in 1965. They treated three theoretical mirror instabilities that looked as if they might be particularly ominous, and that had not been predicted at the time of the Salzburg conference. They found that onset could be prevented by reshaping the plasma into a rough sphere.[3] Unfortunately, the radius of the sphere would be so large that the conditions for finite-orbit stabilization could no longer be satisfied. Finite-orbit stabilization of flute instability had been held in reserve well into 1964 on the assumption that it might prove superior to the method of magnetic wells. It now became unusable. Rosenbluth and Post also implicated, as causes of instability, the severely monodirectional and monoenergetic character of the beams of particles that the mirror groups were injecting into the DCXs and ALICE. If these beams could somehow be tailored to give particles with a spread of directions and energies, stability should also be improved. Thus there began to emerge a program for adjusting the fine features of the plasma to control, one after another, the various modes of instability. This approach was a far cry from that of the first year or two of the CTR program, when it was assumed that any device that did not give a completely quiescent plasma would be abandoned.

By 1965 both Oak Ridge and Livermore had made important technological strides, for example, in eliminating particle loss by charge exchange. Both had made impressive progress in uncovering and clarifying new plasma phenomena. Yet these advances showed precisely that the mirror plasma was plagued with microinstabilities. It seemed that the more that was learned, the less favorable was the outlook. The prospect for a mirror *reactor* at the end of 1965 looked substantially worse than it had 5 years earlier. "I think myself," Tuck wrote to Washington, "though my views may and doubtless should be discounted to some extent by my admitted bias in favor of pulsed high β, that [mirrors] are reaching the end of the road."[4]

8

The Doldrums

Hard times at the Model C

In late May of 1961, for the first time after 41/2 years of design and construction, the Model C stellarator was sufficiently complete to enable the engineers to release three evenings a week of machine time to the scientists for experimental studies.[1] Don J. Grove led the experimental team for the Model C; in the echelon above him, Melvin B. Gottlieb was simultaneously taking over the administration as the new director of the Princeton Plasma Laboratory, leaving Spitzer with only the scientific leadership. There was a lot to do on the new machine — debugging instruments, checking the magnetic field configuration, and measuring the levels of the impurities that had been a major problem on the B machines and were also flooding the C. By August, despite the press of other work, Grove and his team were able to institute preliminary experiments to measure pumpout.[2]

By this time, the group examining pumpout on the B-3 stellarator had succeeded with the difficult task of varying one machine parameter at a time, and had obtained quantitative, functional relations between plasma loss and B-3 parameters. Over a range of magnetic fields from 5 to 40 kilogauss, diffusion had varied directly as the electron temperature and inversely as the field $(D \propto T_e/B)$. Thus, the equivalent of Bohm diffusion was occurring in the stellarator. The Princeton group had also run the B-3 both with and without its helical stabilizing windings and had found that pumpout persisted whether or not the windings were activated.[3] The same functional dependence of particle loss upon T_e and B was found by Grove and the Model C team in 1961. It is true that the Model C confinement was 10 times higher than the B-3 confinement for any given values of T_e and B, but this was only because the Model C plasma radius was 4 times larger; the Model C plasma was also being lost by Bohm diffusion.[4]

This was an inauspicious time for pumpout to make its appearance in the C machine. The Abelson committee, in its mid-1962 review, treated the Model C stellarator even more harshly than the Livermore 2X. The committee claimed that the quality of the Model C engineering work was below the standard set by other laboratories, and that the work was being starved of sufficient numbers of able personnel. In addition, the committee insisted that the AEC "require that the Princeton Stellarator Project produce a plasma in the kev temperature range with a ratio of material pressure to magnetic pressure, β , of at least 1 percent within three years, or, if unable to do so, the Model C portion of the Princeton program should be closed out."[5] It was a recommendation, Princeton pointed out, that was physically impossible, and in itself constituted evidence that the committee had not examined Sherwood sufficiently closely. Model C was designed to reach a maximum β of only 0.001%.[6] Princeton had, however, set the goal of an ion temperature of several kilovolts within 3 to 4 years, and pumpout, by limiting the confinement time, could make it very difficult for any heating mechanism to raise the plasma temperature. The problem was therefore as unfortunate politically as it was technically.

One of the few bright spots on the Model C's horizon at this time came from research into the phenomenon of superconductivity. By 1960 it had become obvious to Ruark and the fusion scientists that fusion power would be out of the question if fusion reactor magnet coils had to be manufactured from a standard conducting material such as copper.[7] The amount of power consumed by the magnet would be so large as to preclude an overall favorable power balance. The situation was particularly serious for the stellarator because, of all the concepts in the field, it was being projected to run with the largest ratio of magnetic power expended to thermonuclear power produced.[8] Fusion scientists had therefore begun to look at the possibility of using refrigerated (or cryogenic) coils; at temperatures of roughly 250 degrees centigrade below zero, copper coils consume less than a tenth of the power that they need at room temperature. At the same time, the scientists looked wistfully at superconducting coils. Superconducting metals present no resistance at all to the passage of electricity, and hence the only power they consume is the negligible amount required to cool them to the very low temperatures at which they take on the superconducting property. The obstacle to using superconducting coils in fusion reactors, however, was that none of the superconductors known could withstand the magnetic field strengths necessary for reactors; as field strengths rose, these "soft" superconductors went "normal."[9]

This situation changed in 1960. Scientists at Bell Laboratories and at MIT began to drive the permissible magnetic fields in superconductors up toward 20 kilogauss for new types of materials. Then, at the beginning of 1961, Bell Laboratories' researchers announced that they had discovered a "hard" superconductor. In a properly prepared compound of tin and niobium, they had maintained superconductivity at fields up to 88 kilogauss. The reasons for their success were murky; existing theory did not embrace the data they had discovered.[10] Moreover, a major development effort would be required to go from the tiny Bell Laboratories magnets to the large magnets required by fusion reactors and to build into reactor magnets reliable safety features to prevent them from suddenly going normal in the midst of operation. Nevertheless, the discovery of hard superconductors hugely brightened the long-term outlook for stellarators.

The short-term outlook, however, depended critically upon understanding, and eliminating, Bohm diffusion. A first task was to decide whether the phenomenon was a micro- or macroinstability. The initial hypothesis that Spitzer and his associates proposed in 1960 implicated microinstabilities. The Princeton scientists pointed out that, in the experiments on the B-3 to that date, pumpout had taken place during the entire ohmic heating pulse, and they argued that the heating current was probably causing localized electric field fluctuations. They supported their case with the fact that pumpout on the B-3 went on whether or not the helical windings were turned on. The combined effect of the external windings, helical and ordinary, on the stellarator was to create helical magnetic force lines. The pitch of the helices was not constant but increased with increasing distance from the axis. Ideal MHD theory predicted that this change of pitch, called "magnetic shear," would suppress macroinstabilities. Therefore, "since the observed rate of particle loss is unaffected by whether or not the helical windings are used, one may infer that the observed effect is not due to a purely hydromagnetic instability."[11]

Any comfort engendered in the Princeton group by the sense that they understood pumpout soon dissipated, however, as the hypothesis that ohmic heating was at fault became increasingly untenable. In 1961 and 1962 experiments on the B-1 and B-3 suggested that pumpout could occur even when the conditions placed upon the ohmic heating current by the theories of Spitzer and his Princeton theoretical group were not met.[12] Then, in 1963, the cogency of the whole Princeton line of reasoning was called into question by theories put forward on the West Coast.

The Livermore Levitron team had been evaluating the possible causes for the disappointing performance of their hard-core pinch, and one candidate was ideal magnetohydrodynamic theory itself. This theory assumes that the plasma has a negligible resistance to the passage of electric current. Suspicious of this assumption for the actual cases being encountered in experiment, Furth and John Killeen of Livermore began working with Rosenbluth of San Diego on a theory that incorporated finite resistivity. They found that instabilities ruled out by the ideal theory were permitted by their more realistic treatment. Against some of these new "resistive" instabilities, their theory predicted, the shear provided by helical windings would be impotent.[13] Princeton's experimental finding that helical windings did not prevent pumpout could therefore no longer be taken to show that pumpout was not a hydromagnetic problem.

Hypotheses on the cause of pumpout continued to multiply. In one view, for example, the problem was that the amount of magnetic shear created by the Model C's coils was inadequate, for the simple fact was that it was extremely difficult to build coils that could provide really high shear, and none had yet been built. Sentiment was also swinging away from the opinion that Bohm diffusion had a single cause to the view that it was due to a synergistic combination of causes. Meanwhile, Don Grove and his Model C stellarator team—now joined by Munich physicist Wolfgang Stodiek as an additional principal investigator—continued their experimental investigations. They were measuring particle loss for three distinct methods of heating, of which ohmic heating was one, and also for the case where there was no heating at all. Pumpout loss, conforming to the Bohm relation, was found under all four operating conditions. Perhaps it was "an inherent property of the specific confinement scheme [the stellarator] employed?"[14]

By this time, September 1965, the International Atomic Energy Agency's second international meeting on controlled fusion research had convened; the site was Culham Laboratory in Great Britain, to which the British program had been moved from Harwell. Lyman Spitzer, summarizing the experimental results, had occasion to survey the whole field of toroidal devices.[15] There were four main classes. The first was the stellarators, but with the exception of Munich's "Wendelstein," these all showed Bohm diffusion. The second was the hard-core Levitron, which was also losing its particles anomalously through pumpout. The third was the British ZETA, which by and large was beset with hydromagnetic instabilities. Recently, however, because the British had succeeded in running the machine at higher pressures,

ZETA was exhibiting an interesting "quiescent" period during each pulse, in which instability fluctuations were very much muted.[16] The fourth class was a Soviet variant of the toroidal z pinch, called tokamak, which had evolved from work begun in the fifties at the Institute of Atomic Energy in Moscow.

The z pinch had been virtually abandoned in the United States by the time Spitzer delivered this summary evaluation. Tuck's interest in the Perhapsatron and Columbus had dropped off sharply in the wake of the Geneva conference. One of the beauties of the earliest pinches had been that they were high-β machines. The switch to the stabilized pinch had spoiled this, for theory showed that that machine bought stability at the price of the amount of plasma compression that was possible, and hence of the plasma density. The energy loss to which Trubnikov and Kondryatsev had called attention at Geneva was most serious for low-β machines. During the year following Geneva, the Soviet scientists' calculations had been probed and debated, and by late 1959 it had become clear that the Russians were right. Tuck had therefore begun to deemphasize the diffuse pinch and to look about for other confinement ideas.[17] At General Atomic also, the z pinch had lost ground. First, the theoreticians had turned their attention away from pinches to mirrors, and subsequently the experimentalists had found an exciting new concept in the multipole, which they preferred to pursue.[18] At Livermore, as we have seen, the pinch team had constructed the hard-core toroidal pinch, or Levitron, in its effort to understand the instability of the "stabilized" pinch. In exploring the question whether the current through the plasma was creating instabilities, the team had begun to run the Levitron with very little plasma current. Now a strong plasma current is the essence of a pinch device. This current, the magnetic field that it creates, and the force produced by the interplay of current and field are indispensable to the pinch concept. The evolution of the Livermore pinch program had therefore carried it to experiments that transcended the z pinch.

In contrast to the United States, the Soviets had continued to put vigorous effort into diffuse pinches. They had run their pinches, however, in a way different from the West. In the American and British experiments, the stabilizing longitudinal magnetic fields were weak, on the order of several hundred gauss, and almost completely captured within the constricting plasma, forming a "backbone." The Soviets' longitudinal field was strong (10–15 kilogauss), was present outside as well as inside the plasma, and had nearly the same value in both regions. The field external to the plasma cord was therefore a superposition of the longitudinal field and the circular self-field of the pinch current, and hence had a helical shape. A prominent theoretician, V. D. Shafranov, had predicted that this configuration, with its magnetic "corset" surrounding the plasma, would have greater MHD stability.

Lev A. Artsimovich, then director of fusion at the Moscow laboratory and national coordinator of the Soviet program, had become interested in the tokamak's exceptional stability. After the death of the tokamak project leader, Artsimovich had placed the machine under his direct supervision. By the time of the Culham conference, the Soviet scientists were claiming that their machine was more resistant to micro- as well as to macroinstabilities. That is, for their best runs the Soviets reported that their confinement time was 10 times larger than the Bohm formula permitted.[19]

Spitzer, in his review of toroidal experiments, looked upon the Soviet claim with skepticism. The values of temperature and confinement time upon which it was based had not been measured directly but had been inferred from experimental measurements of other quantities. Spitzer thought that there were inconsistencies between the value of the temperature and some of the experimental measurements; a fairly acrimonious debate occurred between Spitzer and Artsimovich in the course of the meetings, but neither succeeded in convincing the other.[20] Under these circumstances, Spitzer did not accord the assertion of 10-fold improvement over Bohm time much significance. "I would like to emphasize not the differences between [stellarators and tokamaks] . . . but rather the general agreement that anomalous particle loss is present and that it is roughly within an order of magnitude of that predicted by the Bohm formula." Reviewing all the data on toroidal devices, therefore, Spitzer found ground for deep pessimism: "Early hopes that pumpout would disappear together with ohmic heating seem to have been dispelled. . . . The instabilities seem to be an instrinsic property of a heated plasma in a toroidal chamber."[21]

Spitzer's gloom over toroidal devices was in resonance with the growing sobriety over the prospects of mirrors. In 1965 the θ pinches were not thought to be reactor candidates; mirrors and toroidal machines were the two main hopes, and a crowd of instabilities, many newly discovered and/or poorly understood, were threatening the viability of both.

The 1965 program review

In the midsixties there was almost no advocacy for the fusion program outside the CTR community itself. One might suppose that a natural constituency for the fusion effort might have existed among American physicists since almost all of the Sherwood project leaders at that time had been trained in physics or closely related fields. But the fact was that plasma physics itself was not well regarded. "I am considerably worried, as perhaps you also are," Rosenbluth wrote to Ruark in May 1964, "that plasma physics has not yet attained respectability in the eyes of those [physicists] working outside the field."[22] The subspecialty of fusion physics was even less esteemed, in part because of the ad hoc, string-and-sealing-wax image it had acquired in the fifties.[23]

The general public certainly could not be counted upon by the CTR Branch. The very real progress that had been made in fusion could scarcely be translated for the layman. Higher temperatures, longer confinement times, greater particle densities, all these were comprehensible to the public. But how was the nonscientist to understand, for example, the remarkable strides that had been made in the ability simply to determine what the plasma temperature or density was? Ingenious experiments and innovative equipment had been designed for diagnostics, but they did not yet show up in the march toward higher parameters.

Finally, within the AEC and the upper layers of its staff, support was tepid. To be sure, the commissioners felt that thermonuclear energy research was a good thing, as long as it did not impinge upon other projects. But their hearts were elsewhere. Milton Shaw was taking over the post of director of the Division of Reactor Development and was beginning to focus on the Liquid Metal Fast Breeder Reactor (LMFBR). The LMFBR would steadily become the major energy technology issue before the commissioners.[24] Even Paul McDaniel was perplexed over the direction that the CTR Branch within his Division of Research should take.

An additional trouble came in the form of a General Electric Research Laboratory report on fusion, completed in July 1965. General Electric, which was running one of the two major industry-financed fusion research programs, had begun to reconsider its commitment. The machines that would be required to pursue future CTR problems would be substantially more expensive than the present generation. The breakthroughs for which the GE research management had hoped when it embarked on its project in 1956 had not materialized. The success of fusion was uncertain and, at best, far off. Henry Hurwitz, Jr., running the GE team, saw three possibilities: General Electric could raise its fusion budget sharply to finance the new generation of machines; it could enter into financial partnership with the AEC; or it could terminate its program.[25]

Since 1957 the laboratory's Research Analysis Group, led by Leslie G. Cook, had regularly reevaluated major programs. In the first half of 1965 the fusion project was selected for such a study. The timing was unfortunate from the fusion community's viewpoint because GE was just then uniting its Research Laboratory and its Advanced Technology Laboratories into the GE Research & Development Center, for the express purpose of "shorten[ing] the time between laboratory discoveries and the availability of their advantages to customers in new or improved products." Not only the fusion program itself, but the whole question of the desirability of large-scale, long-range programs was up for review.[26]

The GE report was written in the atmosphere of exuberant optimism about fission energy that marked the midsixties. General Electric had made a contract for a fission plant with the Jersey Central Power and Light Company at the end of 1963.[27] The company expected to be able to construct the plant for $110 per kilowatt, and Jersey Central was counting on an average cost of 3.7 mills (0.37¢) per kilowatt-hour of electricity over the 30-year projected life of the installation. The members of Cook's committee assumed that breeder reactors would also cost about 3 mills per kilowatt-hour when they came into operation.[28] The costs of strengthening the safety features of fission reactors, which would become important as the environmental movement took on strength, were as yet hidden. The Cook committee was also bullish on coal-fired plants. It drew attention to the fact that the cost of coal plants had been decreasing over the past 20 years. It estimated typical prices for 1965 at $100 capital cost per kilowatt of installed capacity, and less than 4 mills per kilowatt-hour of energy. In sum, the committee concluded, a fusion plant would have to produce electricity in the range of 3–4 mills per kilowatt-hour (in 1965 dollars) and, for a privately financed plant at least, cost not more than $150 per kilowatt to construct in order to be competitive.

Looking at one feature of the reactor after another, the Cook committee decided that it was unlikely that fusion-based reactors could meet these criteria. Any estimate was highly speculative since so many facets of plasma performance and reactor design were unknown. The members recognized that "this lack of definitive knowledge can be regarded optimistically or pessimistically as one may be inclined." Over the next 25 years, at least, the committee inclined to pessimism: "The likelihood of an economically successful fusion electricity station being developed in the foreseeable future is small."[29] As a result of the Cook committee report, General Electric decided against the first of Hurwitz's alternatives, that is, raising their fusion budget. In late

1966 Hurwitz was to try the second alternative, asking the CTR Branch to finance his team in a joint effort. The AEC turned down the request in mid-1967, and after that General Electric proceeded to phase out its program.[30]

The General Electric report came during the preparation for the 1965 program review. The commissioners had ordered the review in 1962, after they had received sharp criticism of the fusion program from the Abelson panel of the General Advisory Committee. In 1964 fusion leaders had pledged to the Joint Committee on Atomic Energy that the review would be thorough and unsparing, and that its results would be ready by 1966, in time for Congress's consideration of the fiscal year 1967 appropriations bill.

In sharp contrast to the Abelson committee review of 1962, the Washington CTR Branch had a central role in organizing the 1965 review. Arthur Ruark threw himself into the work with energy. He wanted it to be "the best review of a broad scientific field that has ever been turned out for the Federal Government."[31] The panel, constituted in May 1965, was certainly prestigious. Samuel K. Allison, director of the University of Chicago's Enrico Fermi Institute, was in the chair. Raymond G. Herb, a University of Wisconsin experimental physicist and accelerator designer, was the member who was to take over as chairman when Allison suddenly sickened and died while in England for the Culham conference of September 1965.[32] Their mandate was to recommend whether the program should be decelerated, accelerated, or killed and to subject each individual project to the same kind of evaluation.[33]

The committee toured the research sites, heard presentations, read papers, and posed questions. The minutes record the intensity and range of disagreement of the discussions. They also record the consensus the members finally hammered out: the CTR program was worth preserving. In its final report, the committee, first of all, cited the advances that the program had made. It then offered an assessment of the political and economic situation. The committee argued that, should fusion power prove possible, the United States would not be in a position to develop the new technology rapidly unless it had active research teams doing CTR work. It held that spin-offs from fusion research would raise the general economic level and that the nation should not let the trained technical personnel working in fusion drift off into other fields, thus forfeiting their expertise. It also maintained that US prestige would suffer if it were not the first to demonstrate the feasibility (or infeasibility) of fusion power. This would be a defeat fully comparable to losing the race for mastery over space. In contrast with General Electric's Cook committee, and after some vehement discussion, the members of the Allison committee agreed among themselves that to try to estimate the costs of an immature technology was irresponsible.[34]

If fusion research was worth supporting, the Allison-Herb committee argued, it was worth supporting adequately. The committee recommended that the budget be increased by 15% per year; this was the standard number that scientists everywhere in the United States were urging upon the federal government in the midsixties as a reasonable rate of increase for research and development. In addition, the members urged the construction of new machines to replace a generation of obsolete devices.[35]

With the panel's review completed, it was now the task of the Division of Research to prepare a policy paper for action by the commissioners in January 1966. Ruark turned the normal activities of the office over to his staff and retired to write it. The report with which he finally emerged, however, was unsatisfactory.[36]

Paul McDaniel, faced with the need to submit a satisfactory report by the deadline imposed, turned to Amasa Stone Bishop, who had been working at Princeton since in 1961 as an experimentalist on the Model C. In two weeks of intensive work, Bishop prepared the report with the aid of staff member Stephen Dean and Richard Post. McDaniel had wanted for some time to replace Ruark, now in his midsixties. He was grateful to Ruark for having undertaken a difficult job at a time when most scientists had shunned it, but he believed him to be an ineffectual manager, insufficiently dynamic, and unable to coordinate a field of fusion teams, each having its own direction.[37] Ruark, moreover, was innocent of the political acumen that was necessary for keeping a research and development program nourished with funding in the Washington environment. The very impartiality of his goal for the program and the honesty of his technical evaluations were political disadvantages to the program. The goal, in Ruark's view, was to "determine the possibility or impossibility of fusion machines producing net power. . . . Statements concerning the probability of attaining net power production, [however], or concerning the production of economical power, lie beyond the limits of our present knowledge."[38] What he did not see was that program managers had also to be program promoters. Ruark's failure to deliver a satisfactory report gave McDaniel his chance. Shortly after the policy paper was ready, McDaniel persuaded Bishop to replace Ruark as assistant director for controlled thermonuclear research.[39]

When Bishop took over, the program was in the doldrums. Morale in the field had plummeted as declining budgets led to scrimping on equipment and the firing of personnel. Many Sherwood scientists were also discouraged by the state of research; scientific progress had been good, but progress toward their technological goal had not, and there was a widespread sense of stagnation.[40] Outside Sherwood, there was little support in industry or among the public, and no great enthusiasm in the AEC or the Congress. The favorable review of the Allison-Herb committee, however, provided one platform upon which to build. And the very fact of Bishop's appointment stirred enthusiasm among the members of the General Advisory Committee and within the commission and thereby strengthened his position as he came into office.[41]

Fighting uphill

There was a fundamental difference in approach to the fusion program between the wry, idiosyncratic Arthur Ruark and the urbane, hard-driving Amasa Bishop. Ruark was noncommittal on the question whether fusion power could be attained. Bishop, who had a clearer perception of the Washington system, expressed himself as "convinced of [the controlled thermonuclear program's] eventual success."[42] He saw his role as aggressively advocating the program within the government. "[We have, in the policy paper,] a reasonably satisfactory document considering the two-week time limitation for its preparation," he wrote to review panel chairman Ray Herb. "The task of defending it, of gaining . . . official approval . . . and then of implementing it . . . will be an uphill fight."[43] Bishop moved into the fight with vigor.

The first hurdle was the AEC. The authors of the policy paper — Bishop, Dean, and Post — had proposed that the percentage contribution by the United States to the world CTR effort should double within the coming 5 years. This, they estimated, "would entail an increase in the number of scientists and engineers . . . by a factor of three over the same period."[44] On February 8, Bishop, Herb, and the project leaders met with a supportive Chairman Glenn Seaborg and Commissioner Gerald Tape and a skeptical James Ramey. Ramey, by now a partisan of the breeder reactor, doubted that either the rate of recent progress or the urgency of fusion power was sufficient to merit such a rapid acceleration as the policy paper proposed. All the commissioners, as well as General Manager Robert E. Hollingsworth, agreed that some increase in budget was essential. After considerable discussion, they fell back on the smaller, standard 15% per year R&D growth rate.[45]

Strong opposition to even this rate of increase now came from the President's office. The Johnson government, trying to juggle the financing of the Great Society and the war in Vietnam, was facing growing inflationary pressures.[46] Donald F. Hornig, Johnson's science advisor, saw no possibility at all of expanding the $22.6 million CTR operating appropriation contained in the budget that the President had just sent to Congress for fiscal year 1967. Moreover, the administration did not find the panel's arguments for fusion compelling. "The [Policy and Action] report appears to rest its conclusions largely on the statement that the U.S. share of the world effort is declining," wrote Bureau of the Budget Director Charles L. Schultze. "I do not quarrel with this factual finding . . . but I do not agree with the inference drawn from it."[47] Hornig and Schultze asked to have the policy paper, already revised by the commission, further reviewed by the President's Science Advisory Committee and the AEC's General Advisory Committee.

A commission policy on CTR had been promised to Congress in time for the Joint Committee on Atomic Energy's consideration of the fiscal year 1967 budget. Instead, the commissioners sent a text without affixing its own judgment. It was an unpleasant state of affairs for the fusion branch. Congress was suspicious: "As I understand it now," expostulated Representative Craig Hosmer, "the Herb Report, having been massaged in the Commission, is now going to be . . . remassaged by the General Advisory Committee. . . . When that is completed who is going to review the reviews? . . . Is it actually a process of refining . . . or is it a process of delaying . . . ?" Moreover, the Joint Committee wanted to know whether the Herb-Allison panel had really sought to determine the value of continuing the program or simply had engaged in an exercise of a posteriori justification.[48]

While congressional debate was going on, Bishop went to do battle with the President's Science Advisory Committee (PSAC) and the AEC General Advisory Committee. He found that PSAC was "definitely friendly, but not as enthusiastic as I had hoped. . . . They were generally sympathetic to the desirability of moving ahead rapidly, but seemed to feel that it could (and should) be financed—at least in part—from funds saved by tightening up the present effort."[49] As for the General Advisory Committee, its members did not see the need for a new-equipment allowance on top of the general 15% yearly increase: "We do not believe that the urgency of this work is so great as to justify a total increase in the budget greater than the 15 percent per year for five years recommended by the AEC staff paper."[50] Given the AEC drive for the breeder, the committee members saw fusion mainly as a backup program.

Finally, in June 1966, the questions came round again to the commissioners. The split remained; Gerald Tape continued to support the 15% increase in operating funds and an extra $3–4 million a year for new equipment, and James Ramey continued to hold that the program did not merit such a large increase.[51] The result was a compromise. The commissioners decided that the operating funds were to rise 15% in the first year and then level off, rising only something over 6% between the fourth and fifth years. Equipment was to be budgeted at an average $3–4 million per year over the half-decade.[52]

Though Bishop's initial attempts to promote fusion's financial fortunes bore little fruit, he was more successful at implementing his mandate to improve the program's management and to harmonize the separate projects. The Allison-Herb panel, for example, had placed greater coordination high on its list of objectives and had specifically recommended reconstituting a Sherwood Steering Committee.[53]

Bishop responded by creating a two-tiered system. At the top was a standing committee. Like the old Steering Committee, the Standing Committee included the administrative heads of the fusion divisions in the major laboratories. To guard against mere horse trading, Bishop complemented the laboratory leaders with four "civilian" members chosen from outside the program. Since the project heads participated from the outset in the process of arriving at consensus, the laboratory leaders felt a strong obligation to honor committee decisions. Bishop viewed the committee as the program's central policy-setting organ, where problems could be tackled with maximum participation from the laboratory fusion scientists.[54]

At a lower level, Bishop envisaged a sequence of ad hoc panels to provide a detailed review for proposals for major new experiments as they came up. The Standing Committee would then determine the relative priorities of the experiments that the ad hoc panels endorsed. Bishop intended to staff these panels with a mix of leading scientists expert in the different kinds of machines, both to secure broad commitment for each new device from the whole community and to foster communication among fusion's intellectual leaders.

There had been, up to that time, serious fragmentation in the program for purely scientific reasons. The plasmas created by the different machines were so different in character that the order in which their properties were attacked had necessarily been different. For low-density plasmas like those of the stellarator, the first question to be attacked was the shape of the orbits of single particles. For high-density plasmas like those in the pinches, the first question was the equilibrium of the plasma, considered as a fluid, and the calculation of orbits could be put aside for the moment. As a consequence, in the first years of the program, there was often little common ground. "The intuition that goes with the stellarator, that goes with mirrors, that goes with pinches [were different. For experimentalists in particular] there was almost no way that they could communicate with each other."[55] Bishop was prepared to constitute the ad hoc panels in a way that would facilitate scientific exchange. In setting up the Standing Committee and the ad hoc panels, Bishop was creating an organizational framework that would allow a shift of decision making from the leaders of the individual projects to a programwide fusion leadership.

A number of proposals for new experiments were coming out in the spring of 1966. Among them were proposals for Livermore's ALICE II, a new accelerator for the Astron, and a variety of multipoles. Bishop decided to cut the teeth of the ad hoc panel system on a Los Alamos request for a machine named Scyllac. Scyllac, for "Scylla closed," was to be a toroidal machine replicating the plasma parameters of the linear Scyllas.

Until the midsixties, neither James Tuck nor his scientists at the laboratory had conceived of the θ pinch as a candidate for a reactor. At this time, however, Fred L. Ribe, leader of the Scylla group, began to suggest that his machine could make a practical energy generator. For one thing, Ribe was personally interested in the design of fusion reactors. As his administrative duties had increasingly taken him away from "hands-on" laboratory work, he had begun to take up theoretical reactor studies as one creative outlet. For another, Ribe understood the political importance of offering a plausible reactor scheme as a way of "protecting your concept." He had a deep interest in, and commitment to, the θ-pinch concept, and wanted to win the fullest possible exploration of its potentialities. He understood that the θ pinch had to compete for operating funds against the mirror and the stellarator and that these machines were being promoted as usable both for testing scientific feasibility and for reactor cores.[56] The most recent Scylla, Scylla IV, was in fact giving outstanding results. It had reached ion temperatures of 80 million degrees centigrade at densities of 2×10^{16} particles per cubic centimeter. The one drawback was the short confinement time, about 2 microseconds, which the team believed to be due to the loss of particles out the ends of the machine.[57]

In 1965 Ribe had collaborated with Warren E. Quinn and Thomas A. Oliphant, Jr., to show that a linear θ pinch could form the basis for a reactor. The fundamental innovation incorporated in their design was to put a pulsed compression coil within the blanket instead of outside it. This placement lowered overall expense because it greatly decreased the volume of magnetic energy required.[58] The linear device, however, would have to be extremely long to reduce the effect of end-loss. Ribe and his collaborators estimated that a length of half a kilometer would be required to achieve a 3-millisecond confinement time. At the stage they had then reached, the construction of a capacitor bank large enough to energize a half-kilometer coil was out of the question.[59] It was therefore a natural suggestion that the next step on the path to a θ-pinch reactor be a toroidal machine.

Although Scyllac was a natural outgrowth of previous work, it was by no means secure from all objection. For one thing, any torus had to face the old problem of toroidal drift, to counteract which Spitzer had originally introduced the pretzel shape of the stellarator. In the high-density Scyllac plasma, the drift would show up as an expansion of the plasma ring into the outer wall of the toroidal chamber. The Scyllac group at first proposed to solve this problem by adding Ioffe bars. This would not give an absolute magnetic well as in the mirror. No closed system of magnetic lines could have an absolute well because the force lines in a well all curve away from the plasma, whereas in a closed system they return upon themselves. It is intuitively clear that force lines cannot both close and simultaneously everywhere bend outward. On the other hand, it was possible to have toroidal machines with fields that were magnetic wells on average (also called minimum average-B fields). Los Alamos theorist Werner B. Riesenfield had already explored the behavior of a dense plasma in such a field, however, and had convinced himself that the plasma would be unstable. While he was pointing this out to his colleagues, the choice of field configuration shifted to the so-called Meyer-Schmidt corrugated field.[60] The Meyer-Schmidt configuration was recognized to be unstable. Ribe and his group proposed to correct the instability by the method of "dynamic stabilization." It was not clear, however, whether the time-varying fields used in dynamic stabilization could successfully be made to penetrate into the plasma and, if they could, whether they would not themselves excite new instabilities.[61]

It would seem that Scyllac might also have been opposed on strategic grounds. Since its inception, the CTR program at Los Alamos had been a collection of relatively small experiments. The largest projects, including the Perhapsatron and the Scylla IV, had not commanded more than three or four scientists at a time. There was also a fluidity of organization, with scientists moving from one experiment to another as their interests or the technical problems dictated. Tuck had been convinced that this was the way fusion had to be approached: "I have maintained ever since the beginning of Sherwood one cardinal point in policy. This is that I would try to continue in what seemed the best possible scientific direction no matter how it twists and turns; until one sees a clear line ahead, a policy of being scientifically light on one's feet is the best in my opinion."[62] Scyllac, however, would require a rigid organization, a large research team, sustained planning, and a range of engineering effort unprecedented in the laboratory's fusion history.

Tuck, however, had been jolted into questioning his own policy of small, cheap experiments during the congressional budget cut of fall 1963. At that time, McDaniel, irked by a display of what he saw as Tuck's excessively competitive attitude vis-à-vis other laboratories, had revealed to the Los Alamos leaders that he had been entertaining the idea of closing down their Sherwood program completely. To do so would be one among a number of possible ways to satisfy the congressional demand to eliminate a "concept." And Los Alamos was the one laboratory with no large investment in expensive machines that had to be protected.[63] Tuck had come away from that episode with a healthy appreciation of large experiments. "We resisted the temptation to build huge machines or hire large staffs," he told Congress in 1964. "This sounds very virtuous, but I have now come to realize that it was suicidal."[64] By 1966, Los Alamos Director Norris E. Bradbury had replaced Tuck as administrative leader by Richard F. Taschek, head of the experimental physics section, and it was Taschek who took the Los Alamos slot on the new standing committee. But Tuck, Taschek, and Bradbury were by now united in the view that Los Alamos needed a large central experiment in order to secure its position within the national program.[65]

The panel that Bishop assembled to evaluate the Scyllac proposal was chaired by Hans R. Griem, a University of Maryland physicist who had collaborated in the θ-pinch research at the Naval Research Laboratory. Its other members were drawn from the GE θ-pinch team and from stellarator, mirror, and Levitron work. Harold Grad represented the Courant Institute group of mathematicians.

The panel did a thorough technical review that both served the Standing Committee as an adequate basis for a decision and brought new and valuable advice to the members of the Los Alamos team. But it also did more. Although the panel members approved of the Scyllac proposal, they included caveats and suggestions that amounted to a reorientation of the program that Ribe was suggesting. The members did not agree with the Scylla IV team that its data conclusively pinpointed end-loss as the main culprit for particle loss. They did concur that many unsolved problems inhered in the scheme to over-come instabilities with dynamic stabilization. They therefore suggested, as an essential first step, an intermediate experiment: a new linear pinch 15 times as long as the Scylla IV. The 15-meter device would serve as a testing ground to elucidate the loss mechanism of the Scylla IV and would, as well, make possible preliminary studies of the Meyer-Schmidt equilibrium field and of the dynamic stabilization scheme. The panel recommended Scyllac only with the understanding that the 15-meter device would precede it. At the same time, the panel spelled out their view that Scyllac should be approached, not as a reactor candidate, but as a physics experiment.[66] As one correspondent ex-pressed it to Griem, this changed orientation "does not tie the initial Scyllac expenditure to a presumption of merit of a particular (and rather debatable) scheme or 'machine concept,' but ties it rather to the merit of high-β research as such, and proven Los Alamos capability in this area."[67]

The Scyllac panel thus fulfilled the potential inhering in the new organizational forms that Bishop had created. In the name of the fusion community as a whole, it recommended decisions that affected the experimental strategy of Los Alamos and aimed the Scyllac program more firmly in a research direction. Bishop's reforms had set in motion a movement that was shifting the locus of strategic decision making.[68]

Multipoles

To the invigoration brought to the program by Bishop's administrative reforms was added, in 1966, the exhilaration of the success of a device called a multipole.

The origin of the multipole went back to 1960, when Donald Kerst, at General Atomic, began systematically probing behind the "apparent differences and novel characteristics" of all the various fusion devices in an attempt to resolve them into their basic types and to illuminate their virtues and shortcomings.[69] Working with him was the Japanese physicist Tihiro Ohkawa, who shared Kerst's background in accelerator design, and whom Kerst prized as a truly inventive physicist. Ohkawa had collaborated with Kerst as a visitor to the University of Illinois when Kerst was exploring the fixed-field, alternating-gradient accel-erator, and in 1960 he had accepted Kerst's invitation to join the GA fusion group.[70]

As part of their survey, Kerst and his colleagues had studied the "very important suggestion" of James Tuck for caulking the leaky cusp geometry of the 1954 picket-fence concept. Tuck had proposed introducing a metallic ring inside the vacuum chamber. The resultant field from currents through both the ring and the external, picket-fence coils had the disadvantage of being an average minimum-B rather than an absolute minimum-B field, but the advantage that the cusps vanished, leaving all the field lines closed upon themselves.[71] They had also studied the hard-core pinches, including the Levitron, which similarly incorporated an internal conducting ring. Kerst and Ohkawa had then designed an optimized internal ring device. It was simplified in that it eliminated the external conductor, which was a feature of both the caulked picket fence and the hard-core pinch, and it was generalized in that the two authors worked out their scheme for an arbitrary number of pairs of internal rings. It gave a field with the property of average minimum-B, and it might therefore be expected to contain a plasma stably against MHD disturbances. Since the shapes of the magnetic fields formed in their scheme were identical to the shapes of the fields that would be formed by suitably placed pairs of north and south magnetic poles, the Kerst-Ohkawa concept came to be called the multipole[72] (see figure 8.1).

The multipole was not conceived as a protoreactor. A multipole could be used as a reactor only if the rings could be made superconducting. This would mean, however, keeping rings at temperatures close to absolute zero in proximity to a plasma 10 times hotter than the interior of the sun. Kerst did not intend to deal with problems like that. On the contrary, he believed that the state of fusion research was such that it did not make sense to be concerned seriously with prototype reactors. There was too much to be learned about plasma properties. The multipole, with its MHD stability, would make an excellent apparatus for studying the new microinstabilities then coming to the fore. For Kerst, "useful experience and interesting plasma results" were the object.[73]

Kerst's enthusiasm for the multipole was not shared by the theoreticians at General Atomic, who were at that time most interested in mirror machines. Kerst therefore moved to the University of Wisconsin in the fall of 1962, and with $50,000 in seed money from the university and an AEC contract set about building a machine. At General Atomic, the experimentalists split into two teams; one pursued mirror experiments and the other, led by Ohkawa, inaugurated its own multipole program.[74]

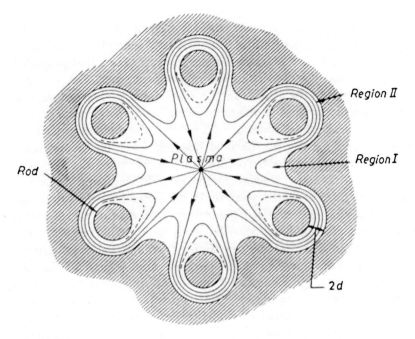

Figure 8.1
Cross section of the multipole configuration. [From T. Ohkawa and D. W. Kerst, "Multiple Magnetic Field Configurations for Stable Plasma Confinement," *Il Nuovo Cimento 22* (1961), 786]

Kerst and Ohkawa both selected toroidal octopoles, devices containing four internal conducting rings, as their principal experimental apparatus; these octopoles gave the magnetic configurations with the deepest wells and also with the shortest connections between regions where conditions for stability were unfavorable and the rest of the plasma. By 1965 both teams had obtained good results. Not only were the predictions of MHD stability confirmed, but in addition, once the injection process was completed, the plasmas were quiescent, showing almost none of the fluctuations in density and electric potential that usually accompanied Bohm diffusion. It was impossible to make a direct check whether the diffusion did in fact deviate from the Bohm formula because the plasma was being lost too quickly. Ohkawa, Kerst, and their coworkers judged, however, that the cause of the loss was relatively trivial; most probably it was due to the fact that the plasma was intercepted by the set of supports that held the rings in place. The level of fluctuations was one that would be appropriate to a diffusion rate hundreds of times smaller than Bohm diffusion.[75] Then, in November 1966, Kerst and his students presented new results at the Boston meeting of the AIP Division of Plasma Physics indicating that in those parts of the plasma where the conditions for MHD stability were met, and where low levels of fluctuations obtained, "the maximum turbulent diffusion coefficient was three to four orders of magnitude below the Bohm value."[76]

The multipole results of 1965 and 1966 gave the Washington CTR office new progress of the type that nonscientists could easily understand. To be sure, the plasma densities and temperatures were not high. In the octopole of Kerst and his group, for example, the density was only 10^9 particles per cubic centimeter and the temperature only 0.1 keV (about 1 million degrees centigrade).[77] But the confinement time was up dramatically, and the data showed that Bohm loss was not an unavoidable property of toroidal confinement.

The good scientific news was accompanied by disappointing institutional news. The Texas Atomic Energy Research Foundation was withdrawing its funds from the General Atomic's fusion program, with the financing to end completely in April 1967. Physicist William Drummond of the GA group had taken a professorship at the University of Texas in Austin, and plans were under way to build up fusion research at the university. TAERF preferred to bring its money back into the state, to help create indigenous research capability. General Atomic therefore asked the AEC to pick up its entire fusion group, at an annual cost of more than $1 million.[78] This group, which spread over to the nearby La Jolla campus of the University of California, where some of the GA theoreticians (Rosenbluth, Norman Rostoker, and Bruno Coppi) had appointments, was one of the strongest, liveliest, and best-integrated fusion teams in the country.

The request put Bishop and the Standing Committee in a quandary. Funding General Atomic would, in effect, be equivalent to setting up a new AEC fusion laboratory. This course was not provided for under the policy paper that the fusion leadership was trying valiantly to have implemented. There were serious doubts that such a new laboratory should, in any event, be housed in industry. Bishop even wondered whether keeping the group intact or allowing it to disperse to fertilize other programs would be more advantageous to the US effort as a whole. Most important, the CTR budget was extremely limited, so that it would be difficult to find the money.[79] On the other hand, the Standing Committee did wish to support both GA's theoretical group and Ohkawa's multipole program. In the end, the committee decided to provide them with $500,000 a year for a limited term.

At the heart of the problem was the fact that neither the good multipole results nor the improved program management was being translated into financial advantage. In late 1967 Bishop was still constrained to put to the fusion leaders the problem of "what might be done to insure that, as a result of the present budget squeeze, the intensified effort called for in the CTR policy paper does not wither away but indeed becomes a reality."[80]

Tokamak

When the third international conference on plasma physics and controlled thermonuclear fusion convened in August 1968 in the Siberian city of Novosibirsk, a major focus of attention was Lev A. Artsimovich's report on the most recent tokamak results. The values of temperature, density, and confinement time that Artsimovich claimed for the Soviet T-3 and TM-3 tokamaks were dramatically better than the parameters so far recorded on any other toroidal device. He reported electron energies of up to 1 kilovolt, corresponding to temperatures over 10 million degrees centigrade, and ion energies of up to 0.5 kilovolts. At the higher temperatures, the energy confinement times exceeded the Bohm time by factors of up to 50. The quality of confinement, $n\tau$, was reaching toward a new high of 10^{12} cm^{-3} sec. Previously, the parameters for closed devices had typically been 1 million degrees centigrade and 10^{11} for the quality of confinement.[1]

When the Soviets had claimed tokamak confinement times 10 times better than Bohm times at the Culham conference in 1965, they had been greeted by incredulity. By the Novosibirsk meeting, the situation had changed. Bohm diffusion was no longer feared to be a universal property of toroidal systems. The multipole machines had already achieved confinement times greatly exceeding Bohm times, albeit at low densities and temperatures, and even the Model C stellarator seemed to be beating Bohm diffusion by a factor of 4 or 5. Moreover, both the confinement and the temperatures of the tokamak plasma had been greatly increased over 1965. The Soviets' 1965 tokamak results had been open to the interpretation that they were essentially identical with stellarator results, if one allowed for experimental error and the intrinsic differences of the machines; they had been so interpreted by Spitzer at the Culham meeting. The new data had to be treated more gravely. To be sure, the temperature measurements upon which the whole edifice of Soviet conclusions was based were still indirect. Nevertheless, the tokamak paper was received by many of the assembled fusion scientists as one of the most important events of the conference.[2]

Sitting among the auditors, Amasa Bishop found the tokamak paper thought provoking. He had already been concerned about the direction being taken by the American toroidal program. He was worried about the stellarator project, whose large and expensive Model C had drawn much outside criticism. He was worried about duplication in the Livermore and Princeton multipole projects. Livermore's multipole program was a continuation of its Levitron program. Princeton's was guided by Shoichi Yoshikawa, a physicist and nuclear engineer who had felt Princeton must explore at first hand why multipoles were not hampered by Bohm diffusion while stellarators were. At the moment, Livermore and Princeton were taking a disturbingly similar direction in their attempts to investigate the limitations on multipole confinement times. Both laboratories were concluding that the trouble with multipoles was the mechanical supports holding up the internal conducting rings and the electric field necessary to maintain the currents in these rings. They were both planning to build single-ring devices with a levitated, superconducting ring. Two such experiments seemed unnecessary. Moreover, Bishop was by no means sure that the ring supports, which levitation was designed to overcome, and the driving electrical force in the internal conductor, which superconductivity was designed to overcome, were really the factors that were limiting confinement times and temperatures in closed devices.[3] Bishop had already seen the Moscow tokamaks and talked with the Kurchatov team during a visit to Russia the previous year, and he had been impressed with the device.[4] The question that naturally arose, therefore, was whether it would be wise to reorient the American toroidal program and institute a tokamak effort.

As Bishop probed this possibility with his American colleagues, several issues emerged. The first was scientific. At this time, Princeton was the undisputed leader in American toroidal fusion research. Gottlieb had recruited Harold Furth away from Livermore to serve as coleader with Thomas Stix of the laboratory's experimental division, and he had persuaded Marshall Rosenbluth to move to the nearby Institute for Advanced Study. And Princeton scientists, especially Harold Furth and Wolfgang Stodiek, did not believe the Soviet temperature figures. In principle, it was possible to make direct measurements of the temperature of the plasma electrons by shining light on the plasma and observing the way it was scattered. The scattered light is so small a proportion of the incident light, however, and the plasma itself is so highly luminous, that an exceptionally intense light source must be used. This method therefore depends upon the use of lasers. In 1968, lasers had just recently joined the battery of diagnostic techniques commanded by the American and European fusion scientists, and they were not yet in use in Russia. Instead, the Soviets inferred the plasma pressure from measurements of the magnetic fields and then used this pressure and a measured value of the plasma density to derive a value for the electron temperature.[5] A crucial assumption was that the plasma electrons had an equilibrium, or "Maxwellian," distribution of velocities.

The Princeton scientists, first of all, believed that the magnetic effects were not a sufficiently trustworthy basis for inferring the pressure. The second sticking point was the Soviet assumption of a Maxwellian distribution. Furth and Stodiek thought that the distribution was probably not Maxwellian, but rather that a small group of very high velocity electrons—so-called runaway electrons—was causing the measured values of the pressure. This explanation seemed the more probable in that there was clearly something anomalous about the behavior of the pressure in the TM-3 machine.[6] The hypothesis could also help answer the question, still unresolved, of how to reconcile the plasma resistance with its temperatures. Runaway electrons could be expected to give strange values for electrical resistance.

A second issue inherent in undertaking tokamak projects was financial. Tokamaks were expected to cost roughly $500,000 per machine. The CTR research budget, however, was barely keeping pace with inflation. Although Congress had authorized $26.9 million for fiscal year 1968, the Bureau of the Budget had pared the figure down to $24.7 million, as the Johnson administration struggled to reduce nonmilitary expenditures so as to finance the Vietnam War. Bishop expected $26.775 million for fiscal year 1969, but this figure was the more uncertain in the fall of 1968 as the Washington bureaucracy awaited the inauguration of a new president, Richard M. Nixon.[7] It was certainly possible to ask the AEC management for supplemental funding. If such money were not granted, however, it had to be asked whether it was worthwhile to divert dollars from the projects to which the laboratories were already committed.

Finally, the limited finances of the program threw into sharp relief the issue of governance. The Standing Committee and ad hoc panels had been eroding the old system whereby each laboratory set its own research directions, untrammeled by any coordinating body. But the issues surrounding the tokamak showed that a struggle was still very much in progress. In December 1968, at a meeting devoted to analyzing the effect of the Novosibirsk conference on the American program, Bishop raised the question "whether our present path [in the toroidal program] makes best use of the limited resources available to us."[8] In particular, he wanted to explore the duplication between the Livermore and Princeton multipoles. In response, Van Atta and Fowler from Livermore defended the traditional mode of decision making. "Charles [Hartman] wants to build [the superconducting Levitron]; he is a member of our staff and thinks it best. . . . If a good man wants it, that's enough." Gottlieb asked in exasperation, "Do you think that this committee can out-think the scientists?"[9] Was the California superconducting Levitron, or the Princeton multipoles, which had been born out of the ingenuity and experience of the American groups, to be put aside because the Washington management had its eye on an imported, Russian, idea?

Ormak

Oak Ridge, among the four major fusion laboratories, had a particular reason to be interested in the tokamak. The core of the Oak Ridge program was the DCX machines, in which energetic molecular ions were injected into simple magnetic mirrors. And for several years preceding the Novosibirsk meeting, these crucial DCX experiments had been going badly. The situation had already reached the proportions of a crisis in 1965. The experiments with DCX-1 were coming to an end as the instability that had kept the particle density to an unacceptably low 10^9 particles per cubic centimeter was finally elucidated. At the same time, the second machine in the series, the DCX-2, was also revealing a fundamental limitation in its particle density.[10]

The DCX-2 was the seat of unusually intricate and interesting plasma phenomena. The crucial strategic question was whether to devote program resources to explaining these phenomena. "There are essentially two courses of action open now. One is to study the present plasma more deeply in the hope of establishing the reason for the density limitation. . . . The second . . . would be to drop this plasma and look for a more worthy recipient of large-scale thermonuclear attention."[11] Oak Ridge Director Alvin Weinberg noted, "In trying to produce a stable plasma, one is working also for the *dullest* of plasmas. . . . Many of the instabilities as we see them are fascinating . . . [but]

as we are under contract to try to produce the dull kind of plasmas, we must not let ourselves wander too widely . . . into other attractive pastures."[12] The DCX-2 was clearly not going to give the kind of plasma that would be needed for fusion reactors.

As plans for machines to succeed the DCX-2 evolved at the end of 1965, it became clear that the course being charted by Oak Ridge pointed straight to a collision with Livermore. Both the experience with the DCX-1 and recent work by Oak Ridge theorist T. Kenneth Fowler had suggested that a new machine, tentatively called DCX-3, should inject neutral atoms, rather than molecular ions, into its magnetic mirror. Fowler also recommended that a magnetic well be built into the mirror, to secure MHD stability. Neutral-atom injection and magnetic wells, however, were fundamental features of the Livermore design called Baseball I, which the California group was about to put into operation. The specter of redundancy was enhanced by the fact that the British and Soviets each had similar machines.[13]

In view of the widespread competition to machines of the DCX type and Bishop's concern with duplication, Weinberg and Arthur Snell, director of Oak Ridge's Thermonuclear Division, began to speed up work on the design of the DCX-3. In August 1966 Snell announced to the Standing Committee Oak Ridge's plan to start fabrication in 1967; he also apprised the committee of the laboratory's recent decision to make the magnet coils out of superconducting materials. This did nothing to alleviate Bishop's worries, for the Baseball II at Livermore was also planned to incorporate superconducting magnets. Moreover, Livermore had submitted a proposal for a still larger machine, the ALICE II, while Oak Ridge had been floating the idea of a DCX-4, and these two designs looked as if they might be even more similar than DCX-3 and Baseball II.[14]

Snell was actively casting about for other projects that might serve Oak Ridge if the DCX-3 and DCX-4 had to be abandoned. One line of research that eminently satisfied the criterion of indigenousness was a hybridization of the DCX sequence with electron-cyclotron heating. The Oak Ridge program in electron-cyclotron heating had been the creation of Ray A. Dandl, who had originally come to the division on loan from Instrumentation and Controls to aid in the development of diagnostics. Dandl had reasoned in the late fifties that it must surely be a step in the right direction to get at least the electrons of a plasma heated up. Tough-minded as well as inventive, Dandl had pursued his investigations in the face of considerable skepticism from colleagues. He had fed microwave radiation into mirror plasmas at frequencies matching the electron-cyclotron frequency (the frequency with which the electron completes a circle about the magnetic force line) and had succeeded in producing plasmas consisting of very hot electrons and cold ions. In 1965 Dandl had constructed a new piece of apparatus, which he called Elmo. The name referred, in the first instance, to the electric discharge phenomenon called St. Elmo's fire, but the contrast between this homely name and such names as Astron and Phoenix also appealed to Dandl's cast of mind. Dandl's previous devices, PTF and EPA, had used comparatively low-frequency microwaves. Because the cyclotron frequency is tied to the magnetic field strength, PTF and EPA therefore employed fields too weak to contain hot ions effectively. The Elmo, in contrast, was to use microwaves of sufficiently high frequency that the corresponding magnetic fields could provide hot-ion containment. This was where the DCX experience could be brought to bear. The Oak Ridge group reasoned that it could use Dandl's hot-electron plasma as a target for high-energy injected neutral atoms. The atoms of the injected beam should charge exchange with the cold plasma ions, resulting in hot trapped ions and cold neutrals. An experiment of this type, INTEREM, was started in 1965.[15]

In September 1966 a special meeting was convened in Washington to consider the convergence of the Oak Ridge and Livermore programs. Snell defended Oak Ridge's priority vehemently, but he was acutely conscious that the DCX sequence did not meet the condition of uniqueness. In February 1967, after a long and searching discussion, the Standing Committee advised Snell that Oak Ridge should give serious consideration to terminating its DCX-3 effort. The situation was all the more ominous in that the Bureau of the Budget was suggesting to the AEC in the spring of 1967 that, rather than increase the CTR budget by 15% as the Herb-Allison panel had recommended, they would prefer to cut one of the laboratories out of the program and distribute the savings to augment the other project budgets. It was clear to everyone that the bureau had Oak Ridge in mind. Snell now brought forward the target plasma program. Dandl and theorist Gareth E. Guest presented calculations predicting a buildup of the trapped-ion population and stability of the resulting plasma. Oak Ridge was reprieved, and in 1967 a second target plasma experiment, IMP (injection into microwave plasmas), using the magnetic well configuration and superconducting coils originally developed for the DCX-3, was begun.[16]

By this time, the Oak Ridge Thermonuclear Division was undergoing a change of leadership. Snell's reign had been from the first a kind of temporary receivership. Snell was not himself a plasma physicist; moreover, he held the additional job of assistant director, which entailed many nonthermonuclear demands on his time. Snell had followed a consistent and deliberate policy of recruiting vigorous new people over the years, and by the midsixties some of them had achieved scientific and managerial maturity. Snell and Weinberg chose Herman Postma from this group, promoting him first to assistant director, then to associate director of the division in 1966, and finally to director in 1967. Postma's experience in plasma physics was complemented by a lively interest in managerial technique; in addition, he had an acute political sensibility.[17]

Postma, taking over in the middle of December, was presented with a triumph that was simultaneously a difficulty. The reason for the basic limitation on the DCX-2 density was finally being illuminated by George Kelley, in collaboration with John F. Clarke, one of the young and enthusiastic newcomers Snell had been at pains to recruit. The trouble was an instability similar to the one that had been found in the DCX-1. It was, furthermore, a type of instability that seemed peculiar to devices like the DCX-1 and DCX-2 and hence not general enough to merit prolonged investigation. A strong case could therefore be made for terminating DCX-2. DCX-1 was finished, and DCX-3 had been ruled out as redundant.[18] The line of experiments that had provided the backbone of the Oak Ridge program for over a decade was coming to an end. Neither Postma nor some of his senior scientists judged the target plasma program as sufficiently promising to replace the DCX series.

Some of the members of the division staff were so discouraged over the situation at Oak Ridge and, more generally, over the national CTR program, that they were considering abandoning the field of fusion altogether. George Kelley was one; John Clarke another. It seemed to them that a dullness, a lack of creativity, permeated the entire field. No present apparatus was getting results that would ultimately lead to energy, and no new ideas were coming into view.[19]

Then, in August 1968, six of the Oak Ridge group traveled to Novosibirsk. They heard Artsimovich's talk without attending to it too closely. Oak Ridge was a mirror machine laboratory, and almost none of the group had the background in toroidal devices that would have enabled them to evaluate the degree of progress represented by the tokamak claims. The exception was Michael Roberts, who had come to Tennessee in 1966 after studying plasma physics at Cornell. Snell, who was continually seeking possible new directions for the Oak Ridge program, had put Roberts to work building a small toroidal multipole under the guidance of Igor Alexeff, coleader of the plasma physics group. Roberts, however, was still too green in the fusion physics field to be captured by the tokamak results. It was Arthur Snell, who had not attended the meeting at all, who took notice. As the laboratory administrator directly above Thermonuclear Division head Postma, Snell was still searching for alternative fusion experiments. He asked John Clarke to make an intensive study of the possibilities inhering in tokamaks.[20]

As Oak Ridge leaders began, in the fall of 1968, to consider the tokamak as a possible new direction for their laboratory, a split began to develop among the division staff. Guest, Dandl, and some of the other Oak Ridge scientists in the INTEREM-IMP and electron-cyclotron heating programs did not favor the tokamak. They reasoned that Oak Ridge had years of expertise in mirrors and in the production of hot-electron plasmas, experience it would be foolish to jettison. Conversely, a jump into the field of toroidal machines, which was virtually unknown to them, would be full of dangers.

Herman Postma and George Kelley, on the contrary, were inclining toward tokamaks. Kelley was impressed with their achievements and saw in them the new idea he had been longing for. Postma observed that there were almost no tokamaks then in existence outside the USSR. Ohkawa at General Atomic had been steering his multipole program in the direction of tokamaks, and there was a small tokamak device in Australia. The one US facility with obvious expertise in toroidal geometry, the Princeton Plasma Physics Laboratory, had turned up its nose at the concept. There was no reason to believe, given the Princeton scientists' conclusions on the role of runaway electrons in producing the Soviet data, that the New Jersey scientists would want to install tokamak experiments. Postma saw a chance for Oak Ridge to become the leading American tokamak laboratory and in so doing to provide the unique and unifying backbone to its program that it so badly needed.[21] In early 1969, therefore, Postma set up a small group to learn more about tokamaks. He assigned Clarke to the group, and also Roberts because of his toroidal multipole experience. To complement the youth and enthusiasm of Clarke and Roberts, he added the seasoned and judicious George Kelley as group leader.

Several principles guided the thinking of Kelley, Clarke, and Roberts as they read, talked, and consulted with tokamak experts.[22] First of all, it was politically wise to tie the features of any machine designed for the laboratory to the particular strengths of the Oak Ridge facility. Second, the major task for an American tokamak in the first part of 1969 would necessarily be to test the validity of the Soviet claims, since these were still being debated. But third, somewhat contradictorily, the Oak Ridge physicists also wanted to incorporate novel features into their tokamak. They perceived that novelty was at a premium in this period when innovative ideas were rare, and was likely to help them in getting the proposal accepted. They wanted, in addition, a machine on which they could do good, searching physics.[23]

The Oak Ridge trio saw a need for a tokamak with enhanced symmetry. Empirical studies were showing that the more symmetric the tokamak, the longer the measured confinement times.[24] The Soviets had for some time been emphasizing as one of the virtues of the tokamak the perfect symmetry of its torus-shaped chamber about a vertical axis through the center of the doughnut hole. In contrast, the stellarator, with its racetrack shape of two curved end sections joined by two straight side sections, lacked this feature. But the geometrical symmetry of the tokamak plasma vessel needed to be complemented by magnetic coils wound sufficiently carefully and placed sufficiently accurately so that the fields they produced would have a comparable axisymmetry. The Russians had been sloppy in winding and placing their coils. The precise placement of coils so as to give exactly the required magnetic fields, however, had been brought to a superlative art by Kerst and Ohkawa in their work on the multipole. It was just this art that Roberts had mastered for the Oak Ridge multipole program; he had visited Wisconsin as part of his duties and had initiated himself into Kerst's technique.[25] The Oak Ridge team decided to make use of his expertise in this area.

To achieve symmetry, Kelley, Clarke, and Roberts devised a novel way of producing the toroidal magnetic field. Instead of coils, they decided to use a solid copper shell. The shell was to be fed from a transformer located just below it, and they would need to ensure that the joints that carried current from the transformer secondary to the copper shell were indeed transmitting the current in a completely uniform manner (see figure 9.1). The wire leads were to be exquisitely engineered in the style of the multipoles. The team decided upon an unusually high magnetic field, 50 kilogauss as opposed to a maximum in the largest of the Soviet machines, the T-3, of 35 kilogauss.

To meet their goal that their device both test the Soviet results and go beyond them, Kelley, Clarke, and Roberts decided to run the experiment with two copper shells of different sizes in succession. The second shell would have an "aspect ratio," the ratio of the radius of the torus to the radius of the plasma cross section, of roughly 2, in contrast to the aspect ratio of nearly 7 in the Soviet T-3. Naming their Oak Ridge tokamak the Ormak, the team called its two planned manifestations Ormak I and Ormak II. The Russians had made the prediction—viewed, however, as uncertain by many American sci-

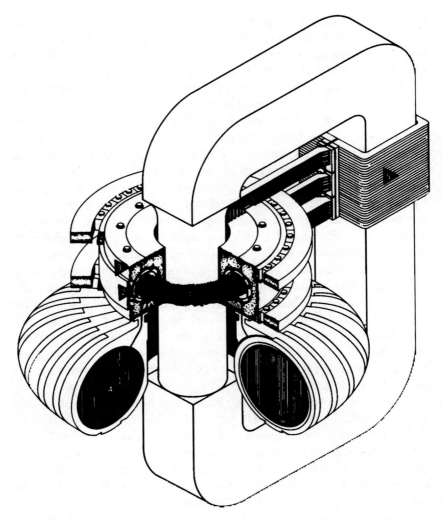

Figure 9.1
The Oak Ridge Ormak I: (1) plasma; (2) vacuum chamber; (3) solid copper shell, producing toroidal magnetic field; (4) transformer; (5) plasma current-driving winding.

entists—that a low aspect ratio would improve tokamak performance. Ormak II would allow the team to attack the unsolved issue of the effect of aspect ratio on machine performance.[26]

Oak Ridge had another area of expertise that Postma and Kelley would have liked to turn to Ormak's advantage: the technology of particle beams. There was only one practical way to heat a tokamak plasma in 1969, and that was to use ohmic heating. Artsimovich had warned at Novosibirsk that "one should not forget that at high temperatures, this method of [ohmic] heating is very inefficient and . . . it is hardly likely that the plasma temperature will be raised above a few kilovolts. . . . One of the main tasks for the future will therefore be to devise new methods of plasma heating in . . . Tokamak and similar systems."[27] It was a commonly held hunch that one plausible method would be to inject beams of very energetic neutral atoms, which would become ionized within the plasma and then share their energy with the other charged plasma particles. Oak Ridge had been working on neutral beams, and the closely associated technology of charged-ion beams, for decades. But the high-energy beams that the Oak Ridge scientists were accustomed to send into their various mirror machines had currents of the order of tens or hundreds of milliamperes. To double the temperature of a tokamak plasma from the level to which it could be brought by ohmic heating would require a current of 3–12 amperes.[28] To bring the temperature to reactor levels of, say, 100 million degrees centigrade would require about 100 amperes, a 1,000-fold increase over the beam intensities then available. Postma and Kelley did some preliminary calculations, but they did not see a way to extrapolate the techniques to meet the tokamak's requirements.[29]

The United States adopts the tokamak

In spring 1969 Lev Artsimovich came to the Massachusetts Institute of Technology. His trip had nothing to do with thermonuclear research. Through the Pugwash conferences on disarmament, the Soviet scientist had come to know MIT Provost Jerome Wiesner and MIT Physics Professor Bernard Feld. Wiesner and Feld had issued the invitation. Artsimovich had intended to give a few lectures and to devote some quiet weeks to an elementary physics text on which he was collaborating. A serene visit, however, was not in the cards; by this time, excitement over the tokamak was too widespread in the United States.[30]

In April Artsimovich was prevailed upon to give several lectures on the Kurchatov machines.[31] Artsimovich was not an impartial reporter, but an enthusiast for his machine, as were most project leaders both in America and abroad. Nevertheless, he was an excellent scientist, and his talks were lucid and convincing.[32] Moreover, he had brought important new data with him. The most impressive was a new measurement of the temperature of the plasma ions. Previous Russian results had been based upon the determination of the energy spectrum of charge-exchange neutrals.[33] The interpretation of the spectrum, however, was not free from objections. The Soviets had to explain away the fact that too many neutrals with very low energies were being measured; they did so by claiming that these particles had been formed from the less energetic ions at the plasma's edge, rather than the ions in the bulk of the plasma. The new data were obtained by filling the tokamak with deuterium and measuring the yield of fusion neutrons. The temperature measured in this way was even higher, by 20%, than the 5 million degrees centigrade found by charge exchange. Artsimovich also reported that the energy confinement time had been pushed up still further, to 100 times the Bohm value.[34]

In addition, Artsimovich gave audiences. It was impossible to glean the whole process of making a new confinement device from papers or lectures. Essential technological information, the "go, no-go" tricks, did not all find a way into the literature. Thus, teams began to make a path to Cambridge, Massachusetts, to sit at a long table opposite the tokamak guru and question him at first hand. Clarke and Roberts made their pilgrimage with Roberts's chief, Igor Alexeff.[35] A team from Texas led by William Drummond and British pinch expert Alan A. Ware also went.

Drummond had for some years been interested in raising plasma temperatures by "turbulent heating." In this method, a strong electric field is sent through the plasma in a deliberate attempt to excite collective motions; the turbulent oscillations evoked then dissipate their energy as random motion among the particles. Turbulent heating had been accomplished successfully in linear devices, but no test had been made to determine whether the turbulent plasma could also be contained for useful time intervals. Drummond wanted to build a tokamak the size of T-3 to decide this issue. If successful, his Texas Turbulent Tokamak would solve the problem of heating tokamaks to reactor temperatures.[36]

One crucial effect of Artsimovich's lectures was to catalyze an interaction between the plasma physicists and the magnet engineers at MIT itself. The Francis Bitter National Magnet Laboratory at the institute was the world center for the fabrication of magnets of exceptionally high field, using a novel design invented by Bitter. Benjamin

Lax, director of the laboratory, was himself interested in plasma physics. Nevertheless, the Bitter laboratory had never built magnets for CTR devices. Indeed, from the midfifties on, when MIT had first declined to house a major Sherwood machine, the MIT plasma community had not thought it worthwhile to build any of the current large fusion devices. MIT did, however, have an active community of plasma physicists. A few months before Artsimovich's visit, the institute had also acquired a theorist with wide experience in toroidal devices. This physicist, Bruno Coppi, had received his doctorate in Italy, at Milan, but had spent much of the sixties at the Plasma Physics Laboratory at Princeton.[37]

Coppi had not planned to work on tokamaks; he had rather taken the job at the institute with the intent of shifting his emphasis from fusion to astrophysics. He had, however, come with an idea in his intellectual baggage, one that he had been playing with since 1967. The highest density of current that is permissible in a tokamak is the so-called Kruskal-Shafranov current density, proportional to the ratio B_t/R (B_t is the toroidal field and R the major radius). The power delivered to the plasma by ohmic heating depends upon the resistivity times the square of the current density ($P = \rho j^2$). Thus, if B_t/R is made large, the ohmic heating per unit volume increases, and temperatures rise. Coppi's idea was to create high-temperature plasmas by building a small, fat tokamak torus, with small R. He also reasoned that if the current density could be made high enough to excite microinstabilities, but not so high as to exceed the Kruskal-Shafranov limit and thus excite macroinstabilities, then the plasma resistivity could be expected to increase beyond its classical value. This anomalous resistivity would jack up the ohmic heating still more, as the formula indicates.

Before he arrived at MIT, Coppi had thought only of decreasing the major radius, R, in order to get the high value of B_t/R that would be necessary. Once at MIT, however, he became acquainted with the Bitter laboratory and its facilities. At the same time, the MIT plasma physicists were being inspired and persuaded by the Artsimovich lectures; the tokamak might finally be a machine of sufficient promise to be worthy of the institute. Under this stimulation, Lax introduced Coppi to D. Bruce Montgomery, the Bitter laboratory's leading magnet designer: Could Montgomery design a high-field Bitter magnet that would have a toroidal shape, rather than the customary linear one? Montgomery thought that he could. The machine that Coppi and Montgomery now began to design would have an exceptionally high B_t as well as a small R. Coppi hoped that this tokamak—by the summer it would be christened Alcator, from the Latin for "high-field torus"—would attain such high densities and temperatures from its combination of large field, small major radius, and anomalous resistivity that it could reach reactor-level conditions from ohmic heating alone.[38] The strength of Alcator, in his view, was precisely as a better instrument for studying high-density, high-temperature plasmas; practical reactor problems "belong to a later phase of the thermonuclear research program."[39]

In Washington, there was as much worry about possible sequelae to the T-3 tokamak as excitement over its good results. The Soviets were planning a sequence of machines culminating in a very large T-10 tokamak, and they expected the T-10 to be capable of reaching full reactor plasma conditions. They would thereby snatch from the United States the glory of being the first nation to demonstrate scientific feasibility. Bishop, testifying to the Joint Committee at this time, found the members unusually receptive. "Our hearings," he reported back to the CTR Standing Committee, "went off unexpectedly well, with the Committee members evidencing great interest in the Soviet results, inquiring about our ability to compete in the demonstration of scientific feasibility, and asking the unheard-of question concerning the level at which we think the CTR program should be supported!"[40]

A meeting of the Standing Committee was scheduled for late June at Buchsbaum's home company, the Sandia Corporation, in Albuquerque, New Mexico. Bishop determined to use the occasion to resolve the tokamak issue. He sent out a call for proposals to be presented at the open meeting; they would be evaluated at the closed, decision-making executive session that would follow.

One of the tokamak proposals that Bishop decided to have considered at Albuquerque actually predated Artsimovich's visit and Bishop's call for proposals. Tihiro Ohkawa, of General Atomic, had submitted a proposal in March 1969 for a tokamak with a kidney-shaped cross section in place of the usual circular cross section. Ohkawa's interest in noncircular cross sections dated back to the early sixties, when he had proposed improving the General Atomic pinch by shaping its plasma into a noncircular form. The idea had then been put aside in favor of initiating a multipole program. In 1967, however, Ohkawa had decided to combine the multipole idea with his old suggestion and build what he called a "plasma current multipole" (PCM). The field that is ordinarily produced in a multipole by two metallic rings would be generated in Ohkawa's PCM by currents in a double-lobed plasma. The PCM would thus retain the stability of a multipole configuration while avoiding all the problems connected with trying to encase metal conductors within a hot plasma. A small model had been built in-house in 1968 with about $100,000 of company funds. Ohkawa had had a chance to discuss it informally with the Soviets at Novosibirsk, and after Novosibirsk he had rechristened it the Doublet for "double tokamak." The Doublet had operated successfully in early 1969. Ohkawa now wanted a much larger Doublet II. The asset of these machines, as he saw it, was that the multipolelike stability would make it possible to confine plasmas with values of β a factor of 10 greater than could be held in tokamaks of comparable size with circular cross sections.[41]

When the members of the Standing Committee assembled in Albuquerque in late June, they had, as one of them later recalled, "tokamak proposals coming out of our ears."[42] Besides General Atomic's Doublet II, Oak Ridge was submitting Ormak, MIT was submitting its high-field machine, and Texas was proposing its turbulent-heating tokamak. The one proposal not included was the one that Bishop most wanted: The Princeton Model C was precisely the right size to be converted quickly and cheaply into a tokamak.[43] Princeton, moreover, had the equipment and expertise to do a fast job of confirming or challenging the Russian claims.

Princeton, however, had persuasive reasons for not proposing a tokamak, even disregarding its scientists' suspicion of the Russian data. First of all, the Princeton group felt that the stellarator was a better experimental machine than the tokamak. In the tokamak, the large plasma current produces both an essential part of the confinement field and the heating of the plasma. Thus it is impossible to study confinement in isolation from the heating process and from any other effect due to the plasma current. The stellarator permits this kind of experiment. Second, a stellarator could be run as a steady-state reactor. The tokamak, by contrast, would have to be run in pulses, its plasma alternately heated and cooled. Many scientists believed that only a steady-state machine could make a successful reactor. Third, there were some theoretical reasons for believing that pumpout, whatever its cause, would markedly diminish in toroidal plasmas brought to high enough temperatures. Thus, the stellarator might do as well as the tokamak, if it only could be made to break through to a higher temperature regime. Fourth, the Model C had never been tried with a really substantial shear field, and the high-shear coil for such an experiment had just been finished.

Eighteen years of experimental and theoretical work on stellarators had elevated the Princeton team to the world's most expert team on this machine. Some of this expertise would inevitably be jettisoned in the switch. The very success of the tokamak was a cautionary example, for it represented the fruits of years of effort on a device, the toroidal z pinch, which the fashions of the late fifties had made unpopular. A machine, moreover, develops its own loyalties among those who work most intimately with it.

On the other side of the argument, if the Soviets were correct, the tokamak plasmas were closer in their properties to a reactor plasma and were therefore more "interesting," and more worthy of study, than the stellarator plasmas. To be sure, the consensus in 1969 was that the invention of a fusion reactor must rest upon knowledge of plasmas. But of what kind of plasmas? Most fusion scientists agreed that the knowledge that was wanted was of reactor-quality plasmas in machines that are similar to actual reactor cores.

The Standing Committee convened in Albuquerque with a report by Fred Ribe on an evaluation of the tokamak that he had undertaken for the Washington Office. Ribe concluded, "The evidence is convincing that the reported Tokamak ion temperatures are reliable. . . . There seems little question that the Tokamak results represent a break-through. . . . [There is] no comparable success in the U.S. low-β toroidal program."[44] Wolfgang Stodiek and Donald Grove from Princeton then took the floor to present their alternative view that the Russian data were better explained on the hypothesis of runaway electrons. A long and bitter discussion followed with no resolution of the scientific issues. The critical datum, an unambiguous measurement of the tokamak's electron temperature, was not yet at hand. At that point, Gottlieb came to a decision. As early as the beginning of 1969 he had suggested to his Princeton colleagues that the tokamak was worth taking seriously. Now in Albuquerque, at the meetings, he was being subjected to the pressure of other members of the Standing Committee who wanted Princeton to volunteer its Model C.[45] One would also guess that Gottlieb's position as administrative leader gave him a special perspective. It may have provided him a measure of detachment, since he was not himself engaged in experimental work on the stellarators. It surely meant that he was particularly exposed to the criticisms of the Model C and the doubts about the multipoles. Moreover, Gottlieb's participation on the Standing Committee may have given him a heightened sensitivity to the needs of the national fusion effort. In any event, as the morning session drew to a close, Gottlieb resolved that the debate had to end. A reliable measurement of the electron temperature had to be made. This would be a much more valuable contribution to the program than continuing to look for the cause of Bohm diffusion in the stellarator. It was lunchtime, and he went out with Harold Furth. "We took a swim in the pool at the meeting break, and I told Harold of my decision, expecting a big fight, and Harold said, 'well, maybe you're right.' "[46] When the meeting resumed after lunch, Oak Ridge, Texas, General Atomic, and MIT made their presentations—and Princeton offered preliminary plans for a "stellarator-tokamak."

The Standing Committee members, meeting in executive session in the following days, faced two constraints. They needed a quick and accurate check on the Kurchatov data and they were short on funds. They chose the Princeton conversion of the Model C, which was eventually to be called the Symmetric Tokamak, and which Princeton estimated would take 6 months, and the Ormak, which was estimated to come into operation in a year.[47]

Shortly thereafter, the Soviet interpretation of the tokamak data was resoundingly substantiated. In a manifestation of the international cooperation which more and more characterized fusion, Artsimovich and R. Sebastian Pease had arranged for a team of Britons from Pease's Culham Laboratory, specialists in the technique of measuring plasma properties by the use of scattered laser light, to go to the Kurchatov Institute. The team had brought its laser equipment to Moscow in the spring and had set it up on the T-3 tokamak. By August the British had obtained preliminary, reproducible results. They telephoned their laboratory in England, and the British, in turn, put through a confidential call to Washington. The Soviets had been correct; the bulk of the electrons were at temperatures of up to 10 million degrees centigrade. The measured densities also corresponded to the Russian values. There was no evidence of the population of runaway electrons that Furth and Stodiek had speculated to be the cause of the Soviet results; rather, the velocities of the T-3 electrons obeyed a Maxwellian distribution.[48] It was an emotional moment in the office of the Washington Branch. Staff member Eastlund danced on the table.

Some years later, the Princeton interpretation was quietly substantiated. Further Soviet research on the other tokamak that had been described at Novosibirsk, the TM-3, showed that for significant stretches of the data, the electron distribution was not Maxwellian. Princeton scientists like to speculate about what might have happened if the British laser team had made their measurements on the TM-3, instead of the T-3, in 1969.[49]

The tokamak bandwagon

The results achieved by the British laser team at Moscow had the inevitable result of making the tokamaks planned by Oak Ridge and

Princeton less interesting. The three other proposals that had been made in June 1969 to the Standing Committee all offered machines that, unlike the Ormak I and Symmetric Tokamak, would advance beyond existing Soviet machines. They had been rejected, above all, because lack of funds seemed to preclude building more than two devices. In late September the British team made a definitive presentation of its results to the international fusion community at a conference in Dubna, near Moscow. When the Standing Committee met again in October 1969 to rehearse what it had learned at this meeting, the members were worried about the planned replication of the Russian devices.[50]

The committee members were anxious to go beyond the Russian results and, in particular, to strike out in directions that were different from those the Soviets were taking. Bishop saw two alternatives: either "formulate a strong and imaginative program on tokamaks which would permit us, in a relatively few facilities, to go well beyond the Soviet program" or give support to the General Atomic, MIT, and Texas proposals as supplements to those of Princeton and Oak Ridge. The second alternative would enable a test, in parallel, of a diverse array of tokamak features. Bishop advocated the second alternative and after some discussion the committee accepted it.[51]

Congress was astonished. Joint Committee Chairman Holifield asked, "We want to know . . . why you are going to five devices simultaneously. Is this a new romance of these laboratories? . . . Is there a duplication involved. . . . What is going on here?"[52] Los Alamos was dismayed. Departing President Lyndon Johnson's lame-duck budget for fiscal year 1970 had been purposely designed as a lean one. President Nixon, nevertheless, had decided on still more austerity, and had sent out a call to the departments and agencies to cut their 1970 requests further. No one had much hope for supplemental funding in the budgetary climate of mid-1969; tokamaks were obviously going to be financed by cutting other projects. "We have now been put on the spot," Taschek wrote to Washington, "by having the Joint Committee *expect* a major breakthrough from this Tokamak emphasis since, after all, if everyone in the CTR community jumps madly on the Tokamak bandwagon, it clearly must be going to pay off! What a shambles! . . . Somehow, all the painstaking effort to get a balanced U.S. program has gone with the wind."[53] Even the Standing Committee members who supported the tokamak wondered whether they had gone too far in the direction of this new device in the enthusiasm generated by the Dubna meeting. At issue, aside from any special-interest pleading, was this question: Was it better to push the most promising

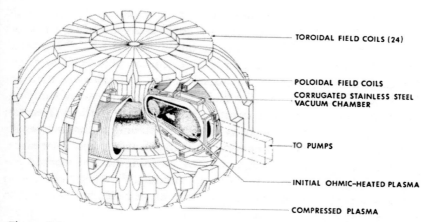

TOROIDAL FIELD COILS (24)

POLOIDAL FIELD COILS

CORRUGATED STAINLESS STEEL
VACUUM CHAMBER

TO PUMPS

INITIAL OHMIC-HEATED PLASMA

COMPRESSED PLASMA

Figure 9.2
The Princeton Adiabatic Toroidal Compressor (ATC). [From *JCAE, The Current Status of the Thermonuclear Research Program, 1971*]

existing approach to achieving a reactor-grade plasma or to pursue many approaches in order to obtain a more many-sided knowledge of plasmas and to foster a climate of inventiveness?

For Oak Ridge, the dream of a tokamak monopoly was now, of course, completely shattered. The competition that worried Postma most was that of Princeton. Each of the two laboratories was formulating an ambitious program of tokamak research culminating in a scientific feasibility experiment. Oak Ridge envisioned progressing from Ormak I to Ormak II to the feasibility test.[54] Princeton had two proposals before the Standing Committee. One was the Adiabatic Toroidal Compressor (ATC), which would test what Princeton then thought was the most promising method of postohmic heating for tokamaks. Princeton's ATC was to heat by a compression effected by forcing a ring of ohmically heated plasma inward, thus simultaneously decreasing its circumference and squeezing its cross section (see figure 9.2). The other Princeton proposal was for a tokamak called the Proto-Large Torus. As its name implied, the Proto-Large Torus was conceived to be a half-way station to the full, feasibility-sized torus.[55] Clearly, it was not likely that two laboratories would both be deputized to carry out the tokamak feasibility test.

In May 1970 the Princetonians brought their Symmetric Tokamak into operation, only 4 months after the dismemberment of the Model C stellarator. By the time of the July Standing Committee meeting at Princeton, their data, although preliminary, were considerable. The Symmetric Tokamak team had been able to set up their experiment with online computer analysis of the data, a capability the Soviets did not have. The rapidity with which it could analyze and repeat shots put the team in a position to catch up quickly with the T-3.[56] Princeton got confinement times and temperatures comparable to those of the Soviets. Oak Ridge had been scooped. It was an unpleasant moment for the Oak Ridge team, which had already put in a year of studying, designing, procuring, and constructing. "George [Kelley], Herman [Postma], and I [Roberts] were in Princeton listening to the first results of the first U.S. tokamak experiments with stiff upper lips. That night, George spent a sleepless night, and by morning he had worked out a scheme to put us once again in the running."[57] What Kelley, who by now was setting Ormak design policy as team leader, resolved to do was to discard altogether the mammoth and unusual single-piece copper coil that had been so carefully machined for the Ormak. Instead, the larger transformer doughnut was to be relieved of its iron core and itself be made into the plasma chamber.

There were several reasons behind Kelley's move. First, the Princeton work had eliminated the need for mere duplication of the Soviet results, which had been part of the rationale for Ormak I. What was now needed was an advanced tokamak that could go beyond the Russian machines. Further, given the Oak Ridge-Princeton competition, this advanced machine had to be constructed with a minimum of time and money and a maximum of effect. The conversion of the transformer to a plasma vessel would give a much fatter plasma cross section, and therefore a smaller aspect ratio than the original Ormak II would have had. (Princeton's Symmetric Tokamak, by contrast, had an aspect ratio even larger than those of the Soviet tokamaks.)

Second, while at Princeton, Kelley had been viewing the assembly of the ATC. This machine had toroidal coils large enough to allow installation of a plasma chamber with a larger cross section than the Ormak would have had on the original plan. Of course, the Princeton ATC team was not intending to made use of the full circular coil cross section. But were they to choose to do so, that aspect of the Ormak experiment that was to test the effect of a larger minor cross section could also be scooped. Finally, there was a technical problem that had emerged during the construction of Ormak. The leads that would carry the current from the transformer to the copper sheath had to feed the current in a completely uniform manner, in order to maintain the planned symmetry. But Kelley and the Ormak team had a difficult time engineering the crucial feature of uniformity. The new design would simply eliminate that problem. The drawback to Kelley's new Ormak was that it would sacrifice the high magnetic fields that had been incorporated into the plan for the first Ormak—and that had been one of its strong selling points.[58]

Oak Ridge was now faced with the necessity of mounting a heroic effort to redesign and construct the new machine. The Thermonuclear Division, however, did have a second string to play in the meanwhile; this was neutral-beam heating. In the summer of 1969, some time after the first discouraging calculations on tokamak heating by neutral beams had been made by Kelley and Postma, the Oak Ridge Thermonuclear Division was host to a visit of several weeks by R. Sebastian Pease, head of the British Culham Laboratory. At a picnic lunch one day, Pease chided the Oak Ridge scientists for not doing more to develop neutral beams for tokamak experiments. Oak Ridge, after all, was a world leader in beam development and injection techniques. It was perfectly clear to the Oak Ridge physicists that the adaptation of beams to tokamaks would be a sensible undertaking for them. Neutral-beam heating, if it worked, would be an important contribution. And if Oak Ridge, by providing the answer to the tokamak heating problem, could establish itself as an important part of the US program, the thermonuclear group would go a long way toward providing itself with the security it needed. Postma decided that the division should take another look at the neutral-beam problem. A team headed by beam expert O. B. Morgan was set up under the overall supervision of Ormak chief Kelley.

Pease had pointed out that they were reckoning in terms of beams with too high a value of particle energy; less energetic particles would actually deliver higher temperatures by sharing their energy more effectively with the plasma ions.[59] The difficulty the group faced was that it was hard to make beams with low particle energy but sufficient power. To retain the same power at 40 kilovolts that they had had with the 600-kilovolt DCX beam, the group needed very intense beams. As they worked, Kelley, Morgan, and their team at Oak Ridge began to see the first steps toward a neutral beam sufficiently intense to make a significant improvement in tokamak temperatures. The ion sources that had been used on the DCX experiments had provided streams of 600-kilovolt ions through electrodes provided with wide, single apertures. For the 40 kilovolts they now contemplated, however, the current from a single aperture would be insufficient. The first change the scientists made was to go to multiple apertures. They soon found, however, that their ion source gave a very uneven density of plasma across the larger effective cross section of the multiaperture accelerating electrodes. A second innovation, forced by the shift to multiple apertures, was a new design of the ion source to produce a more uniform beam cross section. By the fall of 1971 these changes would enable Oak Ridge to boost the current of the beam from the 200 milliamperes that had been used on the DCXs to 4 amperes.[60] The team also decided to eliminate the electromagnet previously used to focus the ions and instead allow the beam to pass directly to the gas cell. Focusing would now be done by very careful design of the holes in the accelerating electrode grids. The ion source could now be brought closer to the port of access to the plasma chamber. This simplification was eventually to prove crucial for boosting the beams to give tens and hundreds of amperes.[61]

The years from 1968 to 1970 saw a rebirth of optimism among the members of the American fusion community after a decade of despondency. The optimism was due in some degree to the improved performance of the stellarators. It was due to a greater degree to the multipoles. These had broken the Bohm barrier on toroidal containment. By 1968 the most advanced multipoles were giving confinement times of about 100 Bohm times. Above all, however, it was due to the tokamak, which permitted toroidal devices to break through into entirely new regimes of plasma parameters.

The scientific importance of the new regime is hard to overemphasize. From beginning to end, one of the guiding rules in CTR strategy was the achievement of plasma conditions as close as possible to reactor conditions; *these* were the plasmas that had to be understood. Yet, part of the glee must surely have had political wellsprings. The program had been under pressure from Congress for years to get higher parameters. The Joint Committee had continually asked Bishop and his predecessors about improvements in density, temperature, and confinement time. On the one hand, increasing parameters was one clearly understandable indication of progress in a field whose complexity baffled congressional attempts to monitor it. On the other hand, the Joint Committee members were suspicious that the fusion scientists were attempting to bootleg pure research into a program that the committee was funding as technology, and whatever their problems with the science, the committee members understood well that higher parameters meant that the program was closer to the practical goal. Thus the tokamak results must also have given the fusion community a sense of relief.

This is not to say that the tokamak plasmas were understood. It is a mark of the state of fusion science that two of the foremost groups in toroidal research, the tokamak group at the Kurchatov Institute and the stellarator group at Princeton, could disagree so long and so vehemently about the interpretation of the experimental data. Furthermore, the way in which tokamaks might behave as their features were varied—the so-called scaling laws relating plasma properties to increasing toroidal magnetic field, larger diameter, lower aspect ratio, differently shaped cross section, and the like—were completely uncertain. For this reason, all the advanced experiments that the Standing Committee had authorized by the spring of 1970—the modified Ormak, the Texas Turbulent Tokamak, the Alcator, and the Doublet II—were truly exploratory.

The fusion community had at least one reason to look forward to these experiments with confidence: Great improvements had occurred in diagnostic techniques. In this respect, the work of the British laser team was more than just an episode in the narrative; it also exemplifies a new kind of capability. The Culham team was able to get temperatures and densities for the T-3 plasmas both at successive moments of the same discharge and also at differing points along the radius of the plasma column. Such "time-resolved" and "space-resolved" measurements were now coming to replace the previous "global" results in which only average temperatures and densities could be obtained. The ability to achieve parameter "profiles" in time and space gave a new purchase to the theorists in their attempt to forge some unity between experiment and prediction. They would now have experimental points with which to compare detailed theories.

The story of the American switch to the tokamak illustrates two other features of the history of the program's strategy. It clearly exhibits the change in the locus of strategic decision making away from the laboratories. It is hard to believe that Princeton would have proposed the Symmetric Tokamak if the laboratory had been left entirely to its own devices. The pressure of the Standing Committee on Princeton was an important factor in Gottlieb's decision. In turn, the presence of the four outside members, Buchsbaum, Brueckner, H. B. Crane, and William A. Fowler, was one cause of such pressure. Because of them the Standing Committee was guaranteed a larger vision; decisions could not represent merely an accommodation among the laboratories. It is conceivable that without the outside members only the Ormak would have been commissioned. Richard Taschek of Los Alamos was opposed to tokamaks, Van Atta was indifferent, Gottlieb had the weight of some of his leading scientists against him, and Postma saw no advantage for his laboratory in a Princeton tokamak. The result of the Albuquerque meeting was one more token that the Standing Committee was evolving into an effective governing body. The fusion program had entered on the second stage in what was to be a continuing process of centralization.

Finally, it was still the case in the late sixties that most of the factors that determined program strategy were internal to the community. This can be seen by examining the political determinants. The politics of institutions played an obvious part in fixing the steps to be taken; Oak Ridge's need for a new "backbone" project is one outstanding example. The institutions involved, however, were almost all within the little world of fusion science, and the AEC leadership, the government, the public at large were little concerned. This situation was soon to change.

Fusion Enters the Energy Marketplace

Concern over damage to the environment increased sharply in the middle and late sixties and reached a vivid culmination in the nationwide, grass-roots celebration of Earth Day on April 22, 1970. Congress responded with the National Environmental Policy Act of 1969 and the Clean Air Act of 1970, and in late 1970 President Nixon gathered federal pollution-control agencies into the new, vigorous Environmental Protection Agency.[1] The accelerating environmental movement, however, coincided with a perceived growing shortage of energy. A serious blackout had afflicted the Northeast in 1965, and a succession of smaller blackouts and "brownouts," or reduced supplies of electricity, had continued through the rest of the sixties. The need to clean up the atmosphere and the waterways seemed to be in conflict with the need for more electricity.[2]

The Nixon administration responded by pronouncing the production of unlimited supplies of energy from technologies with minimum environmental impact a national goal of the highest priority. The foremost candidate for such a technology in the minds of the Atomic Energy Commission, the Joint Committee on Atomic Energy, and, by early 1971, the Nixon administration, was the fission breeder reactor. Spokesmen from these groups argued that the breeder had environmental advantages. It would produce 50 to 70% less thermal pollution than did the current generation of light-water fission reactors. Nixon and his science advisor, Edward E. David, Jr., took the position that the light-water reactors already polluted the air far less than fossil-fuel plants and had excellent safety systems. Breeders, they claimed, would preserve these features. They judged that the problem of safe storage of radioactive wastes was real and difficult for all types of fission-power technology, but they believed it was solvable.[3]

Those citizens who were already alarmed over light-water nuclear plants, however, were not at all convinced. Instead, in May 1971, the

Scientists' Institute for Public Information (SIPI) brought suit under the National Environmental Policy Act to require the AEC to prepare an environmental impact statement for the breeder. SIPI characterized the breeder's plutonium fuel as a toxic material that was also highly radioactive. Thereafter, the protests of environmentalists were increasingly directed against the breeder technology that the Joint Committee and the administration were promoting.[4]

In this atmosphere, with segments of the public actively seeking more benign energy sources, the small and hitherto obscure fusion program could go out into the energy marketplace with its news of recent tokamak successes and be assured of an audience. The fusion community had qualitative reasons for claiming that fusion promised less noxious effects upon the environment than fission reactor technology. First, the only radioactive component of either the fuel or the reaction products in fusion is tritium, and tritium's half-life of only 12 years is far less than the half-lives of uranium, plutonium, or most of their fission products. Second, although fusion neutrons would react with the atoms in the reactor structure to create radioactive isotopes, the scientists presumed that reactor designers could choose structural materials that would result in a minimum amount of short-lived radioisotopes. Third, fusion reactors are not subject to "nuclear excursions," situations in which the rate of energy release "runs away." A nuclear excursion leading to a dangerous meltdown is possible in fission reactors. Finally, Sherwood scientists were confident that second- and later-generation fusion plants could make use of "advanced fuels," such as deuterium-deuterium or even deuterium-helium mixtures. In these cases both the tritium and the activation problems could be much diminished.

The Washington CTR Branch now began to stress fusion's environmental advantages to government officials and news reporters.[5] Fusion was even elevated into something of panacea for modern industrial society's ills by the exuberant suggestion of two AEC staff members, Bernard I. Eastlund and William C. Gough. They sketched the physics and technology of a "fusion torch" whereby the high temperatures of the fusion plasma would be applied for both industrial heat and the dissociation of waste materials into their constituents for recycling. They unrolled thereby "the vision of large cities, operated electrically by clean, safe fusion reactors that eliminate the city's waste products and generate the city's raw materials."[6]

With the widespread interest in environmental problems, the claims of the fusion community were picked up by the press, government, and public. *New York Times* journalists told their readers that this form of energy "produces little or no radioactive by-products and [is] virtually foolproof against runaway reactions." Its editorialists urged Nixon to "move constructively to meet [the energy dilemma] by substantially increasing the amount now being spent on fusion power research and by initiating serious efforts to develop solar power plants."' The *Washington Post* wrote, "Fusion power . . . is a far cleaner and safer power source. It won't have the troublesome waste heat discharge that atomic power plants have. . . . It would produce a fraction of the radioactive byproducts. . . . Assuming disaster—a war, an earthquake or a collision of some kind—a fusion plant would be thousands of times safer than an atomic power plant."⁸ The *Cleveland Plain Dealer* editorialized, "The United States should be investing much more." Victory is "predictable" and would be the "greatest technological triumph ever achieved."⁹

In Congress, Senator Mike Gravel of Alaska sent a letter to a number of distinguished plasma and fusion physicists in which he suggested that they lobby Congress: "The only chance to win increasing funds for fusion seems to lie in Congressional debates, which will not take place without strong support and even agitation from the scientific community."¹⁰ Elsewhere in government, Professor Rolf Eliassen, a Stanford University environmentalist and a member of the AEC's General Advisory Committee, told reporters that fusion was virtually free of pollution and called the program's modest funding "just heartbreaking."¹¹ AEC files and the mailboxes of members of Congress swelled with letters from constituents advocating a crash program.¹² Representative Chet Holifield, chairman of the Joint Committee, complained that fusion advocates "are convincing, unfortunately, a great body of emotional environmentalists, who are insisting that we immediately put our eggs in the fusion basket."¹³

A quantitative, as opposed to merely qualitative, estimate of the safety of fusion reactors in comparison with fission reactors was not possible without specific designs for fusion generating plants. The biological hazard posed by radioactive tritium, for example, depends upon the total inventory of tritium held within all parts of the fusion reactor and the pathways by which it can be released into the atmosphere or the groundwater around the plant. The tritium inventory can vary a hundredfold, however, according to the choice made for the various reactor subsystems. Tritium is also extremely mobile and diffuses easily through solid structures. The difficulty of containing it could be considerable, and this burden cannot be assessed without detailed reactor studies.¹⁴ We must therefore go back to see how far the study of fusion reactors had progressed by the end of the 1960s.

Fusion reactor design studies

The question whether to carry out fusion reactor design studies had been vehemently debated within the fusion community in 1964 and 1965. They were championed, above all, by MIT's David Rose. He wrote in June 1965, "These questions are just as important as those of plasma confinement. What is the good of plasma confinement unless we have a fair idea that the [reactor] system would be feasible?" In his graduate courses, and as a thesis advisor, Rose had set his students tasks like calculating the amount of heat that would be deposited in the innermost or "vacuum" wall of a reactor or designing a dual-purpose blanket that could absorb fusion energy, transfer it to a steam system, and simultaneously regenerate tritium from lithium for reinjection into the plasma. He concluded that solutions to such tasks were possible, but that "the engineering will be a vastly more difficult task than is generally recognized [and that] many problems need attention now, not later."[15] Without reactor studies to complement plasma studies, Rose maintained, the whole CTR program was seriously out of balance.

Up to that time there had been a handful of reactor studies. Following the pioneering Model D study of 1954, Robert G. Mills of Princeton had treated some of the engineering aspects of closed-system confinement devices. Richard Post's 1960 assessment of mirror reactors[16] had called forth a number of articles on open-system reactors. These had shown that mirror plasmas possess an "ambipolar potential," a positive electrostatic potential that drives positive ions out of the plasma and, even more serious, enhances instabilities. The effect of the ambipolar potential was to degrade the power balance from Post's original estimate of 5:1 down to about 1:1. All in all, however, reactor studies had been a minuscule percentage of Sherwood's efforts.[17]

Rose's view found little support within the fusion community, which had just reached a degree of consensus that the major task was to understand plasma properties, and above all plasma stability. Furth and Post declared, "Arguments about the eventual feasibility and economic potential of any of the particular present approaches to thermonuclear power cannot be settled objectively at the present time, and . . . such arguments have no real bearing on the question of the directions in which the present course of fusion research should be steered. What we do consider important from a scientific standpoint is the relative ability of the present approaches to answer specific immediate questions in physics, and this is the basis on which we believe they should be judged."[18] Ruark himself was strongly opposed to studies of integrated systems. To be sure, there were isolated engineering questions that he believed needed attention. He tried in particular to interest Sherwood scientists in the development of materials for the vacuum wall.[19] (He himself had investigated materials for fission reactors during and immediately after World War II.) But for Ruark the central issue was still whether the laws of physics permitted the creation of a well-behaved plasma. "As important as [the engineering] problems are," he and Hilliard Roderick wrote, "they are not the main issue in deciding whether a fusion reactor will be feasible or not. . . . We believe the crucial issue is whether plasma of suitable density can be confined long enough to provide net power output."[20] Nor did there seem to be enough money to do both plasma physics and reactor studies. Ruark was afraid that studies aimed at designing a whole system could become "a bottomless pit" for resources as they raised one problem after another calling for still further studies.[21]

In 1967 Rose, who had been visiting Oak Ridge for shorter periods since the early sixties, decided to take the first part of his 1967–1968 sabbatical year there. He wanted to devote some time to the question whether fusion reactors were feasible from an engineering viewpoint. Oak Ridge had a strong group in fission reactor work and useful capabilities in the chemistry of reactors. Herman Postma, who was then just moving from associate director to director of the Thermonuclear Division, was strongly interested in technological questions. Rose assembled a small group straddling the thermonuclear and fission reactor divisions, using money Postma took from the discretionary funds of the Thermonuclear Division. In addition to himself and Postma, there were two men from the Reactor Division. One, Art Fraas, had been trained as a mechanical engineer, but had acquired a broad nuclear background in his job as associate director of the Reactor Division. The other, Don Steiner, had recently received his degree in nuclear engineering from MIT, with a major in fission reactors.[22]

Fraas and Steiner illustrated the kind of engineering expertise that was relevant to fusion reactor studies. The core of a fusion reactor, with its plasma chamber and complex magnets, is very different from the core of a fission reactor. But outside the core, many of the components of fusion and fission electrical generating plants are similar. The people who were trained to design fission reactors—that is, the nuclear engineers—were equally competent to design the outer structures of fusion plants. Now they were becoming needed. Rose was particularly anxious to promote "reference designs." These designs of whole systems, from plasma core to electrical generators, were not blueprints for any actual reactor, but rather mechanisms for revealing where problems lay.

In late 1967 Bishop began to weigh the usefulness of convening an ad hoc panel on reactor engineering problems. Until then he had held mixed opinions on encouraging efforts in reactor technology. He wondered, as did other fusion leaders, whether CTR was at a sufficiently mature stage of development. In 1967, however, Commissioner Gerald F. Tape had called Bishop into his office and pressed upon him the importance of determining fusion's engineering feasibility. Pioneering reactor studies being issued by British scientists at Culham had led Tape's counterpart, John Adams of the UK Atomic Energy Authority, to serious misgivings about the practicality of fusion reactors. Tape felt that it was vital to clarify the point. Furthermore, it was getting harder and harder for the program to win funding with the justification that it yielded scientific knowledge, in a period when the Vietnam War was squeezing the federal budget and President Lyndon Johnson's program for a Great Society was placing emphasis on practical goals. It was prudent, rather, to stress the program's technological aim. For that purpose, it would be helpful to show plausible plans for entire fusion reactor systems.[23]

To evaluate the desirability of an ad hoc panel, the first, small, programwide fusion technology meeting took place in Washington in February 1968.[24] It was organized by CTR staff member William C. Gough, who had a master's degree in engineering from Princeton, and who had always been deeply concerned with the program's practical ends. Rose, who attended, wrote back to Postma from the meeting, "The ORNL study now underway is more comprehensive than others elsewhere. Thus, ORNL, provided it pursues these studies with diligence, will be the 'first among equals' in guiding the program."[25] Both he and Postma hoped that Oak Ridge's Thermonuclear Division could find a niche for itself, within the US program, in technological studies.

An ad hoc panel was not constituted, but the February meeting was nevertheless an important step toward institutionalizing fusion reactor studies. The first allocation of funds from the CTR Branch was now made—a total of $240,000 for fiscal year 1968, or something under 1% of the budget for the year.[26] A regular series of national fusion technology meetings was inaugurated. And the other laboratories began to emulate Oak Ridge's lead.

Livermore's CTR Director Chester Van Atta was the next to set up a team within his division. The team leader, Richard W. Werner, had been trained as a mechanical engineer but, like Fraas at Oak Ridge, had been working on fission reactors. Post, who with Christofilos had strongly supported its formation despite Livermore's tight budget, now enlisted the group to help him defend his mirror concept. The mirror was under severe attack as a reactor candidate because of its low ratio of power produced to power consumed. The ratio of 5:1 that Post had estimated in 1960 was marginal; 1:1 would be out of the question.[27]

There was, however, a unique technological possibility in mirrors that arose from the very particle loss that was causing concern. This was "direct conversion," in which the charged particles emerging from the mirror's ends are used directly to generate electric current, thereby eliminating the step assumed thus far in every detailed fusion reactor design of changing particle energy into heat to boil water for powering a conventional steam turbine. The idea of direct conversion had been noted from the early fifties as one of the exciting possibilities of fusion. Post worked with Werner's group to develop a concrete scheme for direct conversion. It was a scheme that had problems; for one thing, the system would be about 100 meters in radius, as large as a football field. Nevertheless, it was politically effective for it made mirror reactors look more plausible at a moment when they were especially vulnerable.[28]

The month of September 1969 is traditionally taken by the fusion community as marking the beginning of sustained and serious interest in fusion reactor design studies. In that month the first international conference on fusion reactors was held at Britain's Culham Laboratory, organized by Rose and Culham's scientists. The papers were overwhelmingly British and American, but the audience was drawn from 18 nations.[29] From that time onward, the US reactor studies program grew steadily. Starting at less than 1% of the program monies in fiscal year 1968, fusion reactor technology was 1.4% in fiscal year 1969 ($365,000), 1.7% in fiscal year 1970 ($475,000), and 2.4% in fiscal year 1971 ($665,000).[30]

A formal reactor study group was set up at Princeton in 1971 under Robert Mills.[31] Fred Ribe organized a special development and technology group within the Los Alamos CTR division in 1972.[32] Gough and Bishop also wanted at least one program of reactor studies that was not tied to any particular laboratory or type of containment device. They were aided by the fact that University of Wisconsin Engineering Professor Harold Forsen had been collaborating in some of the work at Oak Ridge and was anxious to start a fusion group at his university. Wisconsin utilities were to undertake part of the financing. Such financing was the more attractive to Washington in that it meant that the branch could limit itself to committing a sum too small to come under the scrutiny of the Standing Committee. Inevitably, that committee was bound to raise the vexed question of in-house versus outside efforts in a time of tight budgets. Forsen started off with $400,000, with equal parts contributed by the government and the local power industry.[33]

As reactor engineering studies expanded, a new group started entering the fusion community, with a different background and novel perspectives. Of course, engineers had participated in Sherwood from the start. But the bulk of their work had involved the design of plasma experiments. They had developed vacuum systems, designed electric circuitry and magnet systems, and constructed support structures. They tended, in line with these duties, to have degrees and experience in mechanical and electrical engineering, power engineering, and electronics. These engineers had influenced research strategy, to be sure, but their influence was limited to the choice of research apparatus. For example, one desideratum in research is to minimize the time that elapses between the conception of an experiment and its execution. Thus one kind of guidance that engineers gave to physicists was in the direction of simplicity, relying upon what is known and what is quick to do and avoiding the need for engineering development.

By contrast, the engineers engaged in reactor studies came from outside the program. There were exceptions, to be sure; Robert G. Mills, the ranking reactor expert at Princeton, is a notable example. He received his doctorate in physics, did his first work in Project Matterhorn on the design of the Model C, and continued to work on the design of experiments, with only occasional forays into reactor problems, until the seventies. Mills, however, went outside the fusion community for part of his team, hiring from among nuclear engineers. Los Alamos's development and technology group was organized around Keith Thomassen, an MIT engineer, and Robert Krakowski, a nuclear engineer who had been in the Rover nuclear reactor propulsion project until it was closed down. Forsen recruited, among others, Gerald L. Kulcinski, a chemical engineer who had worked on fission at Battelle Laboratories, and Robert W. Conn, an applied physicist who had been studying the effects of fission neutrons and safeguards against nuclear proliferation.[34] Fraas and Steiner of Oak Ridge and Werner of Livermore all came from outside the program.

This world of nuclear engineers had by and large been unconcerned with fusion.[35] Now, as they turned their attention to fusion, they saw that a host of power-plant engineering problems had scarcely been tackled. From their viewpoint, fusion was an inviting frontier, a virgin field where a diligent and gifted worker could easily make a major contribution.[36]

These fusion reactor engineers saw themselves as set apart quite distinctly from both the fusion physicists and the fusion hardware engineers. The hardware engineers had organized themselves in 1965 by beginning a regular series of symposia on engineering problems of fusion research. Sponsorship of the symposia had been taken up by the Institute of Electrical and Electronics Engineers (IEEE) in the late sixties, and ultimately the group crystallized into a separate society within the IEEE.[37] The fusion reactor engineers formed a professional organization of their own, the Technical Group for Controlled Nuclear Fusion, within the American Nuclear Society, and it mushroomed.[38] Meanwhile, graduate training was stimulated; the fusion engineering program at Wisconsin, for example, had 6 graduate students in 1971; it was to have 30 in 1975 and 44 in 1979.[39]

That reactor studies were bound to have an effect on the strategy of the whole fusion program follows from the intimate connection between the properties of a plasma and the design of the engineering systems that have to surround it. Before the Culham reactor conference, it had been tacitly assumed by most in the community that the engineers would work with whatever plasma the physicists would be able to give them. Now, however, as the reactor engineers came to grips with the limitations placed upon their part of the enterprise by, for example, the strength and properties of materials, they would begin to make demands upon the physicists. In the years from 1969 to 1972, however, this development was still in the future. Systematic work on reactor designs was at its beginning.

Thus, an air of unreality hung about many of the environmental claims that fusion advocates were making. The fact was that no one could yet give a quantitative comparison of the environmental effects of fusion and fission or translate the journalists' phrase "virtually pollution-free" into a numerical estimate of biological hazard.

Laser fusion

Fusion also captured the public imagination at the end of the 1960s because of the emergence of laser fusion. The laser's highly concentrated light made possible a method of fusion-energy production that eliminated altogether the use of costly magnets. A tiny pellet, for example, solid deuterium and tritium, could be placed in a container of exceptional strength and subjected to a momentary pulse of intense laser light. The pulse would last less than 1/1,000,000,000 of a second (1 nanosecond).[40] The pellet would implode, changing into a plasma of enormous temperature and density. A burst of fusion reactions

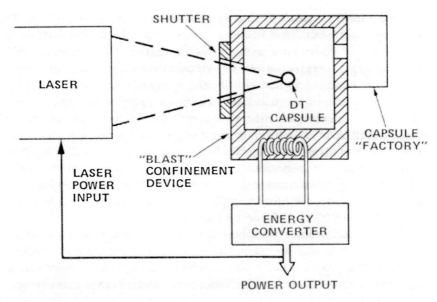

Figure 10.1
Model of a laser-fusion power generator. [From *JCAE, The Current Status of the Thermonuclear Research Program, 1971*, part 2, 386]

would occur, and then the plasma would explode, delivering its pulse of energy to the container (see figure 10.1). The energy of the fusion reaction products would be taken up and converted to thermal energy in the usual way.

This possibility had been apparent to Ruark shortly after the discovery of lasers in the early sixties, and the CTR Branch had initiated in 1964 a study of laser fusion. The focus of their study was the question of how to produce lasers of adequate energy; beams with the requisite short pulse times were available, but their energies measured only in the tens of joules.[41] The techniques at hand for achieving higher energies did not look sufficiently promising. The amount of energy calculations showed to be needed, 10^7–10^9 joules, was close to being unmanageably large.

In 1968, however, scientists in the Soviet Union and, independently, in France, succeeded in producing neutrons from deuterium targets irradiated by lasers.[42] At about the same time, new calculations showed that laser energies of 10^5–10^6 joules, that is, a factor of 100–1,000 less than originally proposed, should be adequate for fusion. Available laser energies were still only about 100 joules. Laser fusion also faced the problem that the efficiencies of the glass lasers then in use were unacceptably low for any application in power generation; the ratio of energy produced as laser light to energy supplied to the laser as electricity was less than 1%.[43] Nevertheless, the new results rekindled the interest of the CTR Branch, and the Standing Committee requested Keith A. Brueckner to chair an ad hoc committee to look into the matter. Brueckner, a distinguished theoretical physicist, had been on the periphery of the program since the fifties, and in early 1968 he had become one of the outside members of the Standing Committee. After an extensive review, he and his committee concluded that laser fusion had promise and that some of the resources of the branch should be allocated to it.[44] The budget was tight, however, and the leadership decided it could not afford this new approach.

Laser pellet fusion also had potential in weapons research. The military needed to test the effects of hydrogen bomb explosions, with their release of particles and radiation, on equipment like satellites and nuclear warheads. Microexplosions in deuterium-tritium pellets could give a mix of electromagnetic and particle radiations more closely approximating a bomb than did the radiation sources then in use. In addition, laser pellet fusion could be extremely useful in weapons design.[45] A small program in laser fusion had been proceeding at the Livermore weapons laboratory since 1963 under the AEC's Division of Military Application.[46] The Soviet and French results stimulated interest in the Livermore project. In addition, Keith Boyer at Los Alamos started an independent project, also funded by the Division of Military Application (DMA). The growth of laser fusion under this division set the stage for tension between laser and magnetic fusion. The military importance of laser fusion was to give DMA strong arguments against relinquishing this research program to the CTR Branch. But the continuing existence of the two forms of fusion under two different managements inevitably led to competition for funding.[47]

Of more immediate importance to the CTR Branch, however, was the fact that an organization outside the AEC was also becoming interested. Brueckner had devised a method of his own for a laser fusion device. In late 1969, when he was serving as director of the Technical Center of KMS, Inc., a private firm, he brought his scheme to the attention of the firm's president, Keeve M. Siegel. Siegel contacted the AEC and telegraphed President Nixon. Laser fusion was classified as "Secret-Restricted Data," and Siegel needed permission to carry out experiments to test Brueckner's ideas. While Brueckner was optimistic, Siegel arrived at a rare point of enthusiasm. He wrote to AEC Chairman Glenn Seaborg that, once Brueckner got under way, he would be able to prove scientific feasibility within 18 months. KMS could then bring "efficient fusion power" into "availability in the next few years."[48]

Siegel's request evoked a violent response. Aside from security questions, there was the issue of KMS's desire to secure patents. Brueckner had not only been a consultant to the CTR Branch but had also been an advisor to the Department of Defense. The commission, and particularly Commissioner Ramey, wanted to know how many of Brueckner's alleged innovations could be traced to the privileged information he had received from the national laboratories. The Joint Committee was also concerned. Its chairman, Chet Holifield, wrote for information, remarking, "I and the other members of the Joint Committee have supported and obtained the authorization of hundreds of millions of dollars for controlled fusion research over the years. It is, at the very least, distressing to contemplate the entire CTR discipline being put in a position of economic disadvantage relative to an individual or group whose main source of information has been from research funded by the United States Government."[49]

For its part, the management of KMS was irritated by the delays that the commission was causing in its program. Siegel was mortgaging his enterprises and enlisting other private companies to secure funding for KMS fusion. In KMS's eyes, the Atomic Energy Commission, with its rigid classification policies and its unwarranted possessiveness, was thwarting the free enterprise process and denying the inventor the fruits of his labor.[50]

The KMS affair stimulated public support for all fusion; the fact that private industry was confident enough to put up its own money could not but make the fusion option more credible. On the other hand, it also hurt the magnetic fusion program because of Siegel's astonishing estimates of the speed with which KMS could commercialize a reactor. No one with expertise in the field believed the KMS timetables; nevertheless, these numbers put pressure upon the leaders of the program to get dramatic results quickly.[51]

Just as magnetic confinement fusion began to face a rival in the glamorous new technology of laser fusion, so too the fission reactor program began to find fusion an increasingly serious rival. Tuck wrote to Seaborg, "Taschek tells me that the fission reactor people, both inside and outside the division, are highly jittery about any mention of the relative advantages of fusion over fission and especially about any mention of reactor hazards."[52] Chet Holifield exclaimed, "I don't want this fusion thing, this pie-in-the-sky deal to distract either the AEC or the Congress from going ahead and doing [the fission breeder] that we have been preparing to do for so long, and we have such a great stake both in investment of money and in our hopes for the future for energy."[53] James T. Ramey, the leading advocate of the breeder among the commissioners, concluded his reply to a pro-fusion letter "I also question the tactics (and sometimes the ethics) of those scientists who go around rapping the fast breeder effort in the mistaken belief that money saved there could be applied to the fusion program."[54]

Fusion workers, for their own part, were not innocent of a sense of rivalry. They understood the obvious fact that fusion and fission both aimed to supply the same good, an energy source with an almost inexhaustible fuel supply, and that breeders might preempt the world market if they were commercialized too much in advance of fusion reactors.[55] They realized as well that an emphasis on breeders diminished industry's willingness to invest in research on controlled fusion.[56]

Mechanisms within the AEC prevented the public expression of fusion-fission rivalry. A system of concurrences ensured that public statements originating with one division, for example, the Division of Research, had to be approved by other divisions, like the Division of Reactor Development. Articles written at the laboratories could be held up by headquarters. Furthermore, Research Director McDaniel was alert to the dangers of pitting fusion against the breeder and took pains to avoid comparisons.[57] In addition, the commissioners smoothed over the antagonisms with the official position that the two programs were "complementary." Breeders represented an intermediate power source; they were at the development stage, and they would be ready for demonstration in the eighties. Fusion by contrast was an ultimate source; it was in a research stage, and if all went well, it would contribute to the energy supply in the twenty-first century.[58] As Commissioner Clarence E. Larson explained it, "The question often arises as to the relative priorities of the breeder reactor program and the CTR program. I believe that this is due in part to a general misunderstanding. . . . It is . . . vitally important to continue our very vigorous support of the fast-breeder program. . . . Fusion power . . . is yet to be demonstrated as technically feasible. . . . By pursuing these two complementary programs, we can make steady progress toward the . . . energy sources of the more distant future while developing and improving those energy sources that will be needed in the near term."[59] Whatever the validity of the arguments, however, these pronouncements at the top merely papered over a strong sense of competition among the practitioners at the grass roots.

Robert Louis Hirsch

The political opportunities provided the magnetic fusion program by the new emphasis on energy and the environment were of special interest to CTR staff member Robert Louis Hirsch. Hirsch had joined the Washington office in 1968, at the age of 33. In that year, the Fort Wayne, Indiana, laboratory of International Telephone and Telegraph, where Hirsch had been supervising an exotic electrostatic confinement device designed by television inventor Philo T. Farnsworth, had submitted a proposal to the CTR Branch of the AEC Research Division. Bishop had rejected the proposal, but he had recruited the project's energetic young leader.

Hirsch had come in time to join in the trip to the international conference in Novosibirsk. The tokamak had immediately aroused his interest, and he saw the results that the Russians were getting as a major breakthrough. He had expected a strong, positive reaction among Sherwood personnel returning from the conference. Instead, to his mind, Bishop had been overcautious and too willing to be ruled by Princeton, while Princeton's response was partly "sour grapes" in the face of the stellarator's difficulties with Bohm diffusion.

Hirsch was buoyed by the positive conclusions of Fred Ribe's evaluation of the tokamak in the spring of 1969 and was gradually coming to the conclusion that in the tokamak, fusion scientists finally had a device that could serve as the core of a working reactor. By the fall of 1969 he had done battle not only for the Ormak and Princeton tokamak but also for General Atomic's Doublet II, MIT's Alcator, and the Texas Turbulent Tokamak.

Hirsch's own view was that the fusion community could do far more than it thought it could. As he saw the history of the program, the early wild optimism of the fifties had been succeeded by a period of exaggerated prudence, underselling of the program's achievements, and an unwillingness to turn its successes into political and financial capital. Hirsch was also distressed at what he perceived as the parochialism and selfishness of the major fusion laboratories. He thought that some of them attempted to freeze out industry and university programs in order to preserve their budgets. In his view it was essential to take control of the program away from the laboratories and vest it in the Washington office.[60] He believed the fusion program was finally ready to be put on a road that could lead it to the stage of development. At times he spoke of a crash program on the order of the Apollo moon program; whether or not he really wished to go that far, he did believe that fusion had to be moved out of the rut of pure physics in which he saw it sitting. But the program budgets were incompatible with the needs of a development program. Actual fusion energy, Hirsch thought, was simply not to be had at $30 million or even $40 million a year. By 1970 he had formulated a strategy for raising the budget. It incorporated going out to other government agencies, Congress, and industry, and attempting to mobilize their support. This effort, Hirsch felt, had to be combined with work within the program to create the enthusiasm, courage, and purposiveness that he thought indispensable for selling the program to the outside world.[61]

Hirsch's hopes that his point of view could prevail rose in 1970. Amasa Bishop was becoming increasingly interested in environmental problems and accepted a position with the United Nations' Geneva-based division for environment and human settlements. Hirsch had a number of qualities to suggest him as Bishop's successor. He was aggressive, self-confident, and engaging and had a talent for administration. The central issue, however, was whether the program should move as quickly as Hirsch wanted it to. In the upshot, Bishop elected to bypass his young staff member and recommend instead Professor Roy W. Gould of the California Institute of Technology. Gould, then in his early forties and just completing 15 years at Cal Tech, had not yet had administrative experience. He saw the CTR post as an opportunity for exploring the congeniality of a career as an administrator and government official. Gould was, in turn, interesting to the CTR community because he was a first-rate plasma physicist who was equally comfortable in science and engineering. It was also important that Gould was unsullied by any ties to a particular device or laboratory. Gould offered the program respected neutrality.

Gould, who took over in the spring of 1970, had a conception of the role of the assistant director for controlled thermonuclear research very different from Hirsch's. Gould saw his job to be, not managing the program from Washington, but reacting sensibly to proposals put forward by the laboratories. Gould was conscious of the considerable rivalry that existed among the laboratories, and he also felt that an important part of his work would be impartial arbitration. As someone utterly new to administration, Gould chose to keep as close as possible to the pace, programs, and forms of organization instituted by Bishop. But he did recognize Hirsch's considerable political abilities. Hirsch was a good salesman—perhaps occasionally too good—and he got along well with the "outside world" of congressmen, administrators, and business executives. Gould, who was innocent of the politics of protecting and promoting government programs, here gave Hirsch free rein.[62] Hirsch used this freedom to take advantage of the increasing interest in fusion by working actively with outside groups to translate their goodwill into concrete program support.

The utilities industry was one such group.[63] Up to the late sixties the interest of the electric utilities in fusion research had been relatively modest. Only the Texas Atomic Energy Research Foundation, by then representing 10 Texas utilities, had put money into controlled fusion investigations, first at General Atomic and later at the University of Texas. After 1969, however, the utilities' concern, while still small, swung markedly upward.[64] At this time utilities looked forward to a continuation of the unusually rapid expansion in the use of electricity of the 1960s. Public Service Electric and Gas Company (PSE&G) of Newark, for example, projected the need for a sevenfold increase in its electric generating capacity between 1970 and 2000. In light of this "prodigious growth," reliance on fossil fuels was seen as difficult. Not only was their exhaustibility a problem, but the cost of transporting and storing such large quantities of fuel, and controlling pollution from commensurate quantities of effluent, could be staggering. Given this prospect, both fission breeders and fusion appeared promising.

Fission breeder technology, however, was coming under public attack for its toxic wastes and the danger of plant accidents. These same problems were evident to utility leaders, who saw that waste disposal was expensive, while the potential for accident made it necessary to site fission reactors too far from population centers. Robert A. Baker, of PSE&G of Newark, was led to evaluate fusion, therefore, as definitely promising more. "In comparing . . . fission vs. fusion, we are impressed with the potential advantages of fusion, namely: negligible radioactive wastes, higher efficiencies (especially with direct conversion), simplified thermal effluent disposal, fuel availability, the inherent safety of such a facility, and the flexibility in siting."[65] Utility leaders were also sensitive to the issue of public acceptability. Public distrust of fission breeders could raise obstacles to the construction of new generating facilities.

As a reflection of the rise in interest, more research money was forthcoming for fusion. By early 1971 a total of 33 utilities, together with the industry trade association, the Edison Electric Institute, were giving the combined sum of over $1 million a year to university research. The beneficiaries were the Universities of Wisconsin and Texas, Cornell, and Princeton.[66]

Hirsch's opportunity to work with the utilities came when the utility-sponsored Electric Research Council organized an R&D task force in the fall of 1970. The assignment of the task force was to review the research needs of the industry. The utilities feared that the manufacturers of electrical generating equipment, motivated by a desire for "salable products and a profit within a relatively short time-span," were offering piecemeal and short-sighted solutions to the problem of the need for vast increases in the supply of electricity.[67] The task force was to make its own evaluation of electrical generation, transmission, utilization, and probable environmental effects, in a survey that was "the first coordinated effort by the entire electric utility industry to arrive at a position with respect to research and development."[68] The task force leadership was strongly interested in fusion. Its chairman, Raymond A. Huse, manager of research and development for PSE&G of Newark, had got to know Sherwood at first hand in the late fifties when he had been assigned as a consultant to Project Matterhorn to aid in the design of the power supply for the Model C. Huse had subsequently developed into one of the utilities' leading champions of the fusion option. The head of the Energy Conservation Task Group, charged with evaluating and assigning priorities to the several advanced concepts for power generation, was Howard R. Drew of Texas Utilities Services Company, the vice president of the TAERF.[69] The task force was clearly going to give the fusion program an enthusiastic hearing.

Some members of the fusion community were aghast at the entrance of the utilities. They believed that their research was at far too early a stage. "Utilities do not understand fusion and do not appreciate [the] difficulties," Los Alamos's James Phillips reported back to Richard Taschek. "When all is said and done it is simply the optimism of the CTR labs that encourages the utilities to support fusion."[70] Hirsch, however, was delighted with the utilities' interest. They were, after all, fusion's ultimate customer. Moreover, their involvement presented an excellent way to bring more outside attention to the program. Hirsch sought out Huse, gave him materials, and actively aided the task force with its report. "I would love to see us help them with this first phase projection," he noted to Gould, " . . . and proceed so as to produce a roadmap (or program plan or program projection) that we can distribute to the Commission and elsewhere."[71]

The formulation of long-range program plans was something that Hirsch thought vital. There had been some interest; McDaniel was encouraging the CTR Branch to look more systematically into the stages past plasma research. But by and large, Hirsch believed, planning had been grossly neglected. Hirsch saw planning as a tool to sharpen the program's focus and help minimize research that was merely physics and irrelevant to the energy goal. He thought success was more likely if the scientists could be mobilized behind concrete goals.

Planning would also answer demands that were continually being made from without. "We would like to be informed," Joint Committee member Craig Hosmer had told Gould, in one such instance, "about the probable dates of three important milestones: The first is the proof of scientific feasibility. . . . The second milestone would be the demonstration of technical feasibility by a fusion energy source with a net output. The final milestone would be . . . the date when [a] commercial fusion energy source might go on the line." Hirsch perceived correctly that the preparation of specific program schedules was one of the prerequisites for winning external support and funding.[72]

There was also interest in fusion within the White House, and Hirsch turned this to advantage, using it both as another prod to push fusion leaders to think about long-term schedules and as a way of publicizing the program. In the spring of 1971 Edward E. David, Jr., recently appointed presidential science advisor and chairman of the White House Office of Science and Technology (OST), decided to institute a thoroughgoing review of energy research and development in order to be in a position to affect the fiscal year 1973 budget. Some of the OST staff members whom Hirsch knew, notably J. Frederick Weinhold and Richard E. Balzhiser, were already favorably disposed toward fusion. It was natural for Hirsch to cultivate their interest. In light of his conviction that the program budget had to be increased, it was also an obvious move for Hirsch to suggest to them that, as a part of David's review, the OST ask the Atomic Energy Commission to prepare alternative plans embodying a more aggressive fusion program. "In discussions with the staff we formulated questions that I later helped to answer." By having the questions posed and then answered, Hirsch hoped to get an audience for fusion among all the legislators and administrators who would read the OST review.[73]

In early May, David sent a request to AEC Chairman Glenn Seaborg to develop an outline for the two traditional cases, a "significantly expanded" program and "an 'all out' program that seeks to develop fusion energy in the shortest feasible time." The CTR Branch was instructed to take the second case "quite seriously."[74] The laboratory leaders gathered in Washington with the AEC fusion staff on May 14 to discuss the response they wished to submit. It is worth looking at the deliberations of that day in a little detail because they reveal two conflicts that lay at the heart of the program's development.[75]

The first conflict was between what fusion leaders wanted to buy with the sharply augmented funding that they felt they needed and what they thought the outside world—groups such as the Congress and the Office of Management and Budget—wanted to pay for. The fusion community wanted to explore plasmas with higher, more reactorlike, values of temperature, density, and confinement time. To get the new plasma regimes, they thought, they must build larger machines that would be an order of magnitude more expensive than the set of tokamaks that had just been funded. The tokamaks had been budgeted originally at under $1 million, and were actually costing several million dollars apiece. The Scyllac, in comparison, was costing $8.5 million for the buildings and power supply and $1.5 million for the device itself. The Proto-Large Torus, now being called the Princeton Large Torus (PLT), was also being estimated to require $10 million. PLT was designed to be a significant step up in magnitude; it was being planned for a plasma current of 1.6 million amperes as against the T-3's 50,000 amperes, and for 500 milliseconds of operation and 20–30 million degrees centigrade as against the Soviet 10 milliseconds and about 5 million degrees centigrade.[76] Beyond the Scyllac and PLT, the fusion leaders wanted three feasibility experiments in which the plasma would be brought to the full temperature and density of reactor operating conditions in each of the three main devices of θ pinch, tokamak, and mirror. Each was estimated to cost $30 million or more. These new machines would be highly visible budget items, and the fusion leaders anticipated that Congress and the administration would therefore expect them to perform as predicted. Yet it was just because they could not trust the theoretical predictions that the fusion scientists wanted the experiments in the first place. It was precisely the unexpected that they wanted to look for. In these unfamiliar regimes of different geometries and higher parameters, wholly new phenomena might manifest themselves, and the scientists felt that it was essential for the program to discover these phenomena, and learn to deal with them, as soon as possible.

The second conflict was between the demand made by Congress and administrators for detailed program plans and the state of fusion research in 1971. Planning in any R&D program is provisional. In one that depends as decisively upon new scientific insights as fusion energy did, detailed planning can be foolhardy. The crucial question was whether any existing device was good enough to serve as a frame around which to drape plans. If not, planning now would at best be superfluous. Better, as one participant phrased it, a slow, steady, relatively undirected research program, over half a century, that wiggled out like an amoeba to encompass new questions as they arose.[77] Neither plasma data nor engineering studies had progressed to the point at which anything more than a hunch could be offered whether any of the devices could be a reactor. Perhaps the fusion reactor of the twenty-first century would be based on a scheme not yet invented.

After discussing this problem, the program leaders nevertheless proceeded to formulate plans. In part, they felt a sense of responsibility to the government that was supporting them, and a sympathetic understanding of the government's need for information in its decision making. In part, they were aware of the political dangers of failing to comply. Were they to admit that they could not predict dates past the first stages, the administration and Congress might choose to delay the increased funding until those stages were completed. Further, none of the scientists believed that an amoebalike half-century of research had any political chance at all.

The first milestone they laid down—scientific feasibility—had already accreted a set of semi-intuitive meanings. Their task was to give the term a more rigorous definition. They decided that the scientific feasibility experiments would "attempt to reach 'break-even' fusion plasma conditions (threshold reactor values of density, temperature and plasma confinement time) in laboratory configurations which lend themselves to development into net power-producing systems." This test was to be carried out under conditions of minimum radioactivity and therefore without tritium. To this end, the leaders introduced the idea of "equivalent-energy breakeven": "test plasma conditions such that if a fusion fuel were substituted for the test plasma gas, the resultant thermonuclear energy production would equal the energy originally invested in the plasma. . . . This demonstration is envisioned without the actual production of a large quantity of neutrons or the handling of significant amounts of tritium, both of which would require significant additional expense and necessitate a time delay."[78]

Where further milestones were concerned, the fusion leaders fell back upon analogies to the development of fission technology. One or more experimental reactors, using fusion fuels and producing net energy, would comprise the next stage. A demonstration reactor would follow; it should include "all of the elements of a commercial power plant," and its successful operation "would be a prelude to commercial sales."[79] They set 1980 as the target for the first proof of scientific feasibility. The group was reluctant to make any statement at all about the dates for the stages that would follow.[80] In the end, it settled on admirably cautious language: "Estimates of the time and costs . . . to carry through . . . experimental fusion reactors and demonstration fusion reactors are very difficult to make" for the case of significant expansion. "It is even more difficult" for the all-out program plan. A demonstration plant in the midnineties for the first case, and the late eighties for the second, was suggested by "a rough estimate." They were, in fact, not at all in favor of an "all-out" scenario for their program.[81]

At this time, the internal AEC projections for fusion funding for operating expenses were

fiscal 1972: $28.0 million,
fiscal 1973: $38.9 million,
fiscal 1974: $39.6 million,
fiscal 1975: $37.0 million,
fiscal 1976: $34.6 million.[82]

The fusion leaders' budget for the case of a significant expansion was

fiscal 1973: $42 million,
fiscal 1974: $54 million,
fiscal 1975: $60 million,
fiscal 1976: $70 million.

These amounts would be swollen by major construction sums of $10 million in 1974, 1975, and 1976 for feasibility experiments. By 1980 the budget was projected to level off at about $90 million.[83]

The utilities' task force, simultaneously preparing their own figures, were arriving at something much more consonant with an "all-out" case. Their report, incorporating both federal and private R&D funds, and representing total expenditures (operating plus equipment plus construction), called for

1972: $ 50 million,
1973: $ 60 million,
1974: $ 95 million,
1975: $110 million,
1976: $135 million.

From there, the task force's budgets soared upward to reach $245 million by 1980.[84]

Even while the CTR Branch was preparing its response for the White House Office of Science and Technology, the budgetary straitjacket was finally beginning to loosen. The original presidential submission to Congress for the fiscal year 1972 magnetic fusion operating expenses had been $28 million, down $400,000 from the budget for fiscal 1971. The decrease corresponded to the Nixon policy of transferring the funding for basic research out of "mission agencies" like the Atomic Energy Commission and into the National Science Foundation. For controlled fusion, it was the university research on basic plasma problems that was slated to be moved to the NSF.[85] In May, however, for totally unrelated reasons, the AEC requested an add-on to the fiscal 1972 authorization bill of $48.3 million. Fusion, as one of the odds and ends, came in for $1.8 million of this amount.[86] Then, when the commission's budget passed through the Senate, Gravel proposed a successful amendment which gave the program another $1.2 million.[87]

In July 1971 circumstances became still more favorable through a dramatic change in the AEC leadership. The AEC had been under increasingly severe criticism from environmentalists. They accused it of being too cosy with the nuclear power industry and neglecting issues involving safety and environmental effects of nuclear plants. In the summer of 1971 Kennedy appointee Glenn T. Seaborg, who had not been on close or even comfortable terms with the Nixon White House, resigned. Nixon chose James R. Schlesinger, Jr., a former professor of economics and at that time assistant director of the Office of Management and Budget, to succeed Seaborg. Schlesinger intended to alter the AEC's image as an advocate for nuclear interests and instead present it as a mediator between the nuclear industry and public interests. Moreover, he wanted to expand the objectives of the agency to include nonnuclear energy, in the spirit of the comprehensive Department of Natural Resources that Nixon was unsuccessfully urging on Congress. In addition, Schlesinger planned to revise the AEC's structure, both to streamline its organization and to change the weighting among its various objectives.[88]

One organizational change that Schlesinger's consultants urged was that CTR be extracted from the Research Division and elevated into a division of its own. It had always been anticipated that fusion energy would some day leave the Division of Research. Eventually, it must graduate from a research into a development project. Hirsch was actively pressing to have the change effected. As Gould's chief lieutenant, he hoped to be the next CTR head, and he wanted to be able to preside over a unit with greater autonomy and weight. There were good reasons why the time was ripe for the new division. First, the program budgets finally seemed destined to start growing. It would be uncomfortable, and politically inexpedient, for the Research Division to shelter a subunit with swelling funding. Second, Schlesinger's purposes of giving the commission the image of pursuing multiple energy options could also be served. Fusion energy is not, strictly speaking, nonnuclear; nonetheless, it was attractive to the antinuclear factions. In December 1971, therefore, as part of Schlesinger's reorganization, Gould's tiny branch of five technical people and five secretaries became the Division of Controlled Thermonuclear Research.[89] The submission for fiscal 1973 funding remained, as had been planned, at about $39 million in operating funds.

Gould operated his division through the first half of 1972 in the same manner as he had run the branch when it was lodged within the Division of Research. By this time, however, he had decided against the life of a Washington science administrator and was preparing to return to the California Institute of Technology. Again, Hirsch set his cap for the post. Gould, like Bishop, disagreed with Hirsch about the pace at which the program should proceed. This time, however, it was not in the CTR chief's hand to determine his own successor. The decision was instead made by General Manager Robert E. Hollingsworth and Assistant General Manager for Research and Development Spofford G. English. Hirsch was the clear choice, if the next CTR chief were to be chosen from within the Washington office. He was not, as Gould had been, a prominent fusion scientist, but he was a dynamic manager and an excellent promoter. Reviewing the choices open to them, Hollingsworth and English nominated Robert Hirsch to be the new division head, to take office in August 1972.[90]

Although Hirsch was not picked by Schlesinger, it was easy to sell him to the chairman. Like Schlesinger, Hirsch was articulate and keenly interested in planning and effective management procedures. Hirsch also matched Schlesinger in another regard; he was determined that research should not be an end in itself, but a stepping stone to solutions to the nation's energy problems.

11

The Transition to Big Science

In the summer of 1972 many of the ingredients were in place for a transition from the concentration on plasma physics research that had marked the sixties to a new phase. There was, first, the tokamak. It was widely felt that toroidal magnetic configurations offered the best hope for reactors. The toroidal tokamak had burst through the low ceilings on temperature and confinement time that had oppressed many of the toroidal machines of the sixties, and had awakened new hopes for reactor-level toroidal plasmas. Second, there was a new level of public support. Environmental issues had finally become a matter of wide concern, and fusion was perceived as "clean energy." Energy shortages had already been appearing, and an energy crisis was soon to make an even stronger claim on public attention. There was every reason to expect that forceful managers could translate public interest into program dollars—dollars that could underwrite a more expensive stage of the program. The new management of the Atomic Energy Commission was signaling a desire to move beyond research. The establishment of the independent Division of Controlled Thermonuclear Research was one sign. Budget projections were another; between January 1971 and January 1972 the level at which the commission projected supporting CTR in the future doubled.[1] A final ingredient was added in August when Robert Hirsch was made acting director. Hirsch had a deep commitment to advancing rapidly to the phase of development and commercialization: "My primary personal goal is to get something practical accomplished."[2]

As soon as he was appointed, Hirsch began to act. Within his first 2 years in office, he expanded the Washington staff to provide a greater pool of technical expertise at the center and simultaneously diminished laboratory autonomy. He asserted Washington's control by terminating several projects that he believed drained resources without showing much potential for meeting their objectives. He altered the plans for the tokamak feasibility test to make it a deuterium-tritium, rather than a deuterium, experiment. Hirsch thereby led the program into what German historian Guenter Kueppers has called the "reactor-relevant" phase.

Kueppers maintains that the transition to this third stage of fusion history must necessarily engender resistance on the part of scientists. Like any other discipline, plasma physics has its own autonomous tendencies; new goals arise spontaneously out of previous research. The greater degree of external direction that is embodied in a reactor-relevant phase, as opposed to the research phase, entails a greater clash over research directions between management and the people at the laboratory bench. Added to this, Kueppers points out, is resistance on the part of the scientists whose projects are cut out since the new phase requires greater concentration around a smaller number of concepts.[3] Kueppers focused on the history of West German fusion, but these things have also happened in the US program. After more than a decade of emphasis on research, fusion scientists were enmeshed in a social system in which many of their rewards derived from doing plasma physics, rather than reactor development.

What had also happened, however, and what inevitably showed up most in the documents, was resistance based upon disagreements over tactics. The years from 1972 through 1974 were marked by an undercurrent of opposition to Hirsch on two grounds. First, it was claimed that the knowledge of plasma physics was simply not adequate for the steps Hirsch wanted to take and, in particular, for a deuterium-tritium experiment. Second, it was argued that since it was by no means certain that any of the machines in existence could make an adequate reactor, the best strategy was to pursue many approaches in parallel.

Hirsch therefore took up the director's post at a time of unprecedented opportunities. But the addition of this vigorous new leader to the complex fusion scene also created the potential for greater tensions and contradictions within the program.

Hirsch in office

In the fall of 1972, upon assuming the leadership of the division, Hirsch launched a major reorganization. He divided program responsibilities among the staff by creating assistant directors for confinement systems, research, and development and technology. For the research group, he recruited physicist Alvin W. Trivelpiece from the

University of Maryland. For confinement systems, he chose veteran CTR staff member Stephen O. Dean, who had returned to the division in February after completing the work for his doctorate at the Naval Research Laboratory. For the important area of development and technology, he picked Robert W. Bussard in 1973.[4] He also began to hire more staff. There were 5 scientists and engineers in the division in June 1972; 1 year later there were 16; by the end of 1974 there were to be 27 technical staff members and 5 administrative workers, while the supporting secretarial staff was to swell to 14; and by October 1975 there were to be 50 technical members in a staff of 75.[5]

The laboratory leaders watched the moves of their new man in Washington with apprehension. Gould's style of leadership had supported the laboratories. He had seen his role as spokesman for their needs and arbiter of their conflicts. Hirsch, on the contrary, by the nature of his reforms, was moving the center of control and policy-making away from the field and toward Washington. His decision to create three separate sections within his office meant that the laboratories would now receive three separate budgets earmarked for research, confinement systems, and development and technology, respectively, which would restrict the flexibility with which the laboratory leaders could apportion their funds. It also meant that the lines of communication had been altered. Laboratories now reported directly to the assistant directors, and the laboratory heads wondered to what extent they would still have the ear of the division director. They noted that Hirsch was in the process of surrounding himself with new, technically expert personnel, and asked how much attention Hirsch intended to pay to technical advice from the field. They liked the Standing Committee Bishop had created and worried that it might be emasculated. The laboratory leaders did not trust Hirsch completely, and they judged that Hirsch did not entirely trust them.[6]

Hirsch had already taken the planning process yet a step further than the blueprints for a significant expansion that the leaders had submitted in 1971. In the spring of 1972, while Gould was still division director, Hirsch had chaired a panel of project leaders, scientists, and engineers that had prepared a statement on fusion's potential and requirements, in response to a request from the White House's Office of Science and Technology and the Federal Council on Science and Technology. It was the most detailed statement of milestones that had yet emerged from the fusion program. It articulated the steps that lay between the proof of scientific feasibility and the demonstration reactor plant that was to be the prelude to the phase of commercialization. Hirsch had also convened a subpanel on technology, which had devoted its attention to the concrete technological advances that would be needed in order to attain both scientific feasibility and the later milestones. The panel had selected the period 1980–1982 for the first scientific feasibility experiment, and it had tentatively suggested that a demonstration plant, might be in operation by the year 2000.[7]

In October 1972 Hirsch took steps to implement this program plan. He went to the Standing Committee for a formal endorsement of the scientific feasibility goal that he could carry to the commissioners as a fixed program objective. Hirsch laid out the steps that would be required to meet this benchmark. The laboratory leaders were somewhat taken aback by the pace that would be necessary; some feared that an 8- to 10-year schedule was inadequate. Nevertheless, Hirsch was able to get their agreement and to report to the commission that "this statement . . . represents a commitment on the part of the project directors. . . . The Division of Controlled Thermonuclear Research plans to adopt the achievement of this goal in the indicated time period as its formal policy and planning strategy."[8]

The transfer of control from the field to the Washington office was symbolized by a dramatic event at the end of 1972—the termination of Astron. Christofilos's program was one of three that Hirsch wanted to eliminate when he took up the director's post in August 1972. Hirsch wanted to terminate Livermore's superconducting Levitron. It duplicated the Princeton FM-1, which had a larger staff and better diagnostics, and it lacked support in the fusion community. Hirsch thought that the California laboratory should be concentrating instead on making mirrors work and, of still greater importance, the technology of neutral-beam injection. He also wanted to end the IMP program at Oak Ridge. And he was determined to close down the Astron.[9]

For years the Astron had been controversial. It was tantalizingly attractive; if it worked, it would provide toroidal confinement in mirror geometry with an external coil that, because of its simplicity, would be both flexible for experiments and economical for reactors. Moreover, the interaction of high-energy particle beams with fusion plasmas had become increasingly interesting to the CTR scientists. Nick Christofilos's Astron, if it functioned, would be a valuable source of information on the physics of plasma-beam interaction.[10] The plain fact was, however, that Astron had not been made to work. After successfully fashioning his innovative accelerator in late 1963, Christofilos had not been able to achieve the necessary second step of building an electron layer to a density sufficient to produce field reversal. Thus, he had not brought Astron to the point where it could confine a plasma, so that the basic questions of plasma confinement, stability, and heating had not yet even been addressed.

In 1964 Christofilos had pledged to achieve field reversal by the program review of 1965 or close the Astron down. He did not build up a layer dense enough to reverse the field. The electron layer was stable, however, and the Herb-Allison committee had recommended he be allowed to proceed to the next benchmark.[11] In 1967 Bishop had constituted an ad hoc Astron review panel, chaired by Keith Brueckner. The panel returned a "guardedly negative" report to the Standing Committee in 1968.[12] Even more than the failure to achieve reversed-field magnetic configuration, the panel was troubled by the fact that physics helpful to the rest of the program was not coming out of the Astron. They saw this as a result of both Christofilos's management style and his research style. On the one hand, he did not easily share authority, and on the other, he himself was more an inventor than a physicist, a vestige of fusion's earlier predilection for trial-and-error methods. "In the Astron, there has been virtually no experimental program in the sense of testing theory against experiment."[13]

Although scarcely enthusiastic, the Standing Committee had nevertheless decided to continue to support the Astron. First, there was as yet no proof that the scheme was unworkable. Second, research with relativistic electron beams was worth doing, and Astron was the only relativistic beam experiment in the United States. Finally, the committee members were faced with Christofilos's forceful personality, his ability to muster support outside the agency, and his deep dedication to his project: "To try to turn off Nick Christofilos's experiment would be like trying to turn off Ernest O. Lawrence's accelerator or J. Robert Oppenheimer's atom bomb project."[14] The Standing Committee did insist that a second review be carried out soon.

A new panel was convened in the summer of 1971. Then, in late 1971, field reversal was finally achieved—but not by Astron! Hans H. Fleischmann and his colleagues at Cornell University had mounted the Relativistic Electron Coil Experiment (RECE), using a type of accelerator radically different from Christofilos's to create the relativistic electron beam. The Cornell group used a pulse with a current of tens of thousands of amperes and electrons of about 0.5 MeV energy, as opposed to Christofilos's current of 500 to 800 amperes and electron energy of 5–6 MeV. It also relied upon a single pulse, as opposed to Christofilos's scheme of stacking a series of pulses. Cornell reached complete reversal by the end of 1971.[15] Christofilos's goal was evidently achievable, but was it to be had by Christofilos's methods?

The review panel, led by Louis D. Smullin of MIT, did not give much ground for hope. The panel again attacked Christofilos's methods. Acknowledging that he had a strong physical intuition, energy, and enthusiasm, the members nevertheless characterized his ideas as ad hoc and improvisational. "The approach has been to look for ingenious ways to avoid or circumvent difficulties rather than to understand them."[16] They criticized Christofilos for concentrating too much of his attention on the accelerator itself and too little on the events in the Astron vacuum tank; they wanted more physics and less engineering. Astron, they said, despite its large size, was still being run as a one-man show. There was no theoretical way to predict whether pulse stacking was physically possible, and hence whether the Astron method would succeed or fail. They noted, however, that Christofilos had been at work now for 15 years and had spent $25 million of AEC funds. He had driven field reversal up to 15%, had conquered a serious "precessional" instability that had affected the E layer, and had just designed a much improved accelerator. But he had not achieved field reversal, and that was telling.[17] An independent report issued by the General Advisory Committee in August 1971 buttressed the Smullin panel's conclusions.

In the spring of 1972 Gould, armed with the Smullin report, devised a plan. The Astron group was to establish precise goals and timetables and to submit monthly reports on their fulfillment. Another review and a decision on termination was to be scheduled for spring of 1973. The GAC report, which was circulated outside the agency, would help the CTR leadership sustain this plan against the pressure of Christofilos's allies.[18]

Hirsch, taking over from Gould, had a strong weapon in the monthly Astron reports; as the Washington staff had expected, the schedule of milestones was not being met. At Livermore, T. Kenneth Fowler had recently replaced Chester Van Atta as head of the CTR program. Fowler was far from pleased with the turn of events. He was not so happy with mirror physics as to feel that the mirror was the single basket for his eggs. Moreover, the Astron, if it worked, appeared to him to offer a much better product than any other reactor idea that had been brought forward.[19] On the other hand, it was hard to argue against the fact of the unmet schedules. Gould had given Christofilos until spring 1973 to make good. The deadline was chosen in part to allow him to determine how much improvement he could get with his new accelerator, and Christofilos, a congenital workaholic, was laboring at his installation around the clock. Hirsch, however, came to a firm decision to terminate both Astron and Levitron in early

September 1972. Then, on September 24, Christofilos died suddenly of a heart attack, at the age of 55. In December the decision on the Astron and Levitron was formally announced to the commissioners. "It was a historic moment. . . . It was the first time that CTR programs had been turned off because *the Washington management* had decided that they didn't want to support those experiments any more."[20] The Levitron was terminated at the same time. In April 1973 Hirsch formally closed down the Oak Ridge IMP, the last of the line of that laboratory's mirror experiments.[21]

Tokamak Fusion Test Reactor (TFTR)

One central task Hirsch set for himself was to redefine the milestone of scientific feasibility. As they had been described in the July 1972 program plan, the feasibility experiments, estimated to cost $30–50 million each, were to run on fuels of hydrogen or deuterium. Hirsch wanted to run the feasibility experiments with a deuterium-tritium (DT) fusion fuel instead.

Hirsch saw four advantages to a deuterium-tritium experiment. First of all, there was a scientific advantage. The DT reaction is

$$H_1^2 + H_1^3 = n \text{ (14.5 MeV)} + He_2^4 \text{ (3.5 MeV)}.$$

The charged helium nucleus (or alpha particle), whose energy is roughly 300 times the approximately 10 keV of the fuel ions, would be captured by the magnetic fields and, remaining in the plasma, would substantially alter its properties. Hirsch felt that it was essential to confront such new physical situations since they would typify a functioning fusion reactor. He himself had used a DT mix in his work in the sixties at International Telephone and Telegraph.

Closely allied to the scientific benefit was a psychological advantage. Hirsch thought that too many of the scientists in his program were there simply because of an interest in plasma physics research. He wanted to start them thinking in a radically different way. The idea of a DT experiment was then nebulous and consigned to some point in the future; bringing it down to earth would be a wrenching change in tactics that would, Hirsch believed, awaken fusion scientists from their scientific slumbers into the practical world of energy generation. "It was very obvious for some time that we were playing hydrogen [i.e., H_1^1 and H_1^2] plasma physics," Hirsch recalled later. "Every part of me and the way I think says you've got to get into the dirt and slop of making the real thing happen before you know what your real problems are."[22]

A third advantage would be the opportunity that a DT experiment would give for the investigation of engineering problems. One of Hirsch's goals was to increase the attention paid to the engineering, as against the physics, aspects of a fusion reactor. "It is my firm position that a significant part of the justification for a [DT tokamak, or DTT] will be based upon engineering considerations. For instance, the DTT will likely serve as the focus for toroidal superconducting magnet, neutral beam, high current switching, and tritium handling development in the near term."[23]

Finally, there would be an important political advantage. A DT experiment would provide actual power. Hirsch believed that the goal of "equivalent break-even" was too esoteric to bring home the importance of fusion to the public and the politicians. He wanted, metaphorically speaking, to light a bulb. More exactly, he wanted to demonstrate the production of actual power in order to dramatize the proposition that fusion offered a real option.

Hirsch's suggestion was received by the fusion community with widespread dismay.[24] Both the engineers who had been designing CTR devices and the physicists had objections. A DT tokamak was precisely the kind of experiment that engineers abhorred. Far from being simple and flexible and offering the shortest possible lead time between the idea and the data, a DT experiment would be plagued with all the problems of radioactive structures and hazardous neutron fluxes. Much of the work would need to be done by remote control. The scientists, for their part, questioned whether the physics of tokamaks was sufficiently well in hand. There was a consensus that the outlook was good for macroscopic stability in tokamaks.[25] There was also optimism on the chances of finding heating methods; the technique of compression had been demonstrated in 1972 on the Adiabatic Toroidal Compressor, and the technique of neutral-beam heating was in the first stages of being tested there and on Ormak.[26] Both methods looked promising. Two other aspects of tokamak behavior, however, inspired less confidence: impurity control and loss by diffusion.

Impurities looked like the worst tokamak problem in mid-1973. The walls and other structures within the vacuum vessel were at that time made of, or coated with, heavy substances such as steel, gold, and tungsten. Atoms of these substances would sputter into the plasma under the bombardment of energetic plasma ions diffusing outward toward the walls. Heavy atoms have too many electrons to be completely stripped as they enter the plasma, and the remaining attached electrons can radiate an enormous amount of energy. All the data pointed to the conclusion that the amount of sputtering would increase sharply as plasma temperatures climbed from the 3–6 million degrees centigrade the tokamaks of 1973 were recording to the 100 million degrees centigrade needed for a reactor.[27]

The precise extent to which particles and energy would be lost by diffusion as tokamak temperatures increased was unknown in the early seventies, but theoretical studies by the Soviets B. B. Kadomtsev and O. P. Pogutse had uncovered a troublesome possibility. As temperatures rise, more and more plasma particles go into what fusion scientists call "banana orbits." Kadomtsev and Pogutse had predicted that if a sizable population were trapped in these banana orbits, both the number of particles diffusing out of the plasma and the quantity of heat carried away by conduction would increase greatly. The effect would be to sink reactor temperature tokamaks back into the mire of Bohm diffusion. There were no data available to confirm or falsify the prediction since the required temperatures had never been reached.[28]

Rehearsing the uncertainties surrounding impurities and diffusion, Marshall Rosenbluth explained his own assessment to Hirsch: "I think you can now understand my negative feelings about the DT feasibility proposal. . . . To deal with our problems we should retain the maximum flexibility in large experiments. A setup compatible with radioactivity problems would seriously compromise this."[29]

The two major tokamak laboratories took diametrically opposite positions to Hirsch's suggestion. At Oak Ridge, the tokamak team supported Hirsch. Oak Ridge was not afraid of radioactivity; on the contrary, it sold itself as the rational site for radioactive experiments. It viewed itself, first of all, as an applied, reactor laboratory. Then, it had the acreage to allow for remote installations and a community already accustomed to acquiesce to radioactive experiments. "Because Oak Ridge, with its reactor experience, its experience in radiation damage and safety, and its remote site, stands in a unique position amongst CTR sites, it seems rational that such experiments be done here."[30]

It had been the Ormak scientists who had first opened Hirsch's eyes to the possibility of inaugurating DT experiments at a time earlier than had been anticipated. They had promised him in the first months of 1972 that if the Kadomtsev-Pogutse prediction were false, they could build him a DT tokamak experiment at reasonable size and expense.[31] Now, in 1973, Ormak leaders Mike Roberts and John Clarke actually went further than Hirsch. They plumped for a machine that would reach not only break-even but ignition, the state in which burning is so vigorous that the energy of the reaction products is adequate to sustain the plasma temperature. Roberts and Clarke argued this position as a matter of both principle and institutional politics. On grounds of principle, they held that the program step beyond the Princeton Large Torus, then under construction, would be so expensive that it could only be justified if it could advance the program by a significant increment. On political grounds, they recognized that an ignition to-kamak would require superconducting magnets. Oak Ridge felt that it had the engineering capability to develop large-volume, high-field superconducting magnets. Thus, while Clarke and Roberts believed that superconducting magnets were so important to the program that it was sensible to begin a development program early, it was also common sense to promote a machine that would make use of their own laboratory's capabilities.

The Ormak group's interest in a burning plasma experiment intensified the split within the laboratory between the (mainly) young enthusiasts of the Ormak group and the (mainly) older scientists working on IMP and Elmo, the experiments of the mirror and hot-electron plasma effort. Gareth Guest, leading theorist for the mirror group, had helped to put together, in 1972, Oak Ridge's earliest exploration of a tokamak feasibility experiment. He had been convinced by the exercise that a 1980 tokamak feasibility experiment was unwise, since a simple consideration of staff size showed that the effort needed to do the tokamak experiment at Oak Ridge would squeeze out both the mirror and "alternate approaches." Guest thought that it would be a mistake to abandon other approaches at a time when no one could be certain that the tokamak would be successful. Rather, an attack in parallel on several concepts would provide alternatives if one line were checked, and would "foster a climate for invention and discovery."[32]

In 1971 and much of 1972 the Ormak team had been at a disadvantage with respect to the mirror team. Ormak plasmas were poor. The laboratory's lack of experience with toroidal machines had shown up in design features that made Ormak repair difficult, and in slowness in getting diagnostic equipment in place. But in late 1972 the relative standing of the two teams reversed. First of all, Ormak had begun to give good data for its first, ohmic-heated, phase of operation. Second, Hirsch, who had just come into office, did not like the IMP experiment and wanted to close it down.[33] Finally, Ormak was operating in a universe of increasing rivalry and therefore needed more of the division's resources. "World-wide tokamak construction and research has . . . been progressing . . . [and] competition is about to stiffen considerably. . . . The pace of physics research on Ormak must accelerate sharply if the ORNL toroidal effort is to prove competitive."[34]

Herman Postma, as division head, was cast into the arbitrator's role. Postma was at that time coming to believe more and more that the future of Oak Ridge lay in research on and development of the technological aspects of fusion. An advanced machine like the DT ignition tokamak would fit in nicely because it would define some of the technological requirements for fusion. At the same time, Postma did not have great confidence in IMP; he and Ray Dandl, moreover, did not get along. Postma therefore chose to line up with Roberts and Clarke, cooperating with Hirsch to phase out IMP.[35]

At Princeton, on the contrary, the laboratory leaders opposed the DT burning experiment. They did not dispute that radioactive fuel would put the program face to face with engineering problems of reactors, but they disputed that the scientific problems DT burning would bring were the important ones. Princeton scientists maintained that the behavior of a plasma containing α particles could be relatively reliably predicted by theory. Where it was necessary to supplement the theories, they claimed, cheap experiments could be mounted in which the effects of the α particles could be simulated by injecting very energetic particle beams into the plasma. Far more important than α particle effects, the Princeton leaders believed, were the two crucial phenomena of impurities and the variation of particle loss with temperature. The best way to attack these problems was an experiment without burning.[36] For all these reasons, Princeton, along with many others in the fusion community, viewed Hirsch's proposal as more of a publicity stunt than an orderly forward step on the road to a reactor.[37]

Their scientific judgment fit neatly with the Princeton scientists' institutional goals. Princeton was as eager as Oak Ridge to house the tokamak feasibility experiment. The laboratory, which had secured final approval for the Princeton Large Torus in the spring of 1972, was given to calling the coming tokamak feasibility device the PLT II. But the laboratory most emphatically did not want to complicate its experiments with problems of radioactive materials. Just as Oak Ridge saw itself as a reactor laboratory, so Princeton conceived of itself as a plasma physics institute.[38] The day that Princeton had to get into the radioactivity business was the day that Princeton would bow out.[39] Until then, it believed that the expertise that it had acquired made it the natural low-β toroidal avant-garde. Rosenbluth wrote to Hirsch, "The physics program up to and including the hydrogen feasibility experiment should be done here at Princeton Plasma Physics Laboratory, although with considerable help from ORNL. . . . It seems to me that the problems are so difficult, and the options so numerous, that it is essential to focus the greatest possible expertise on the matter.

I think there is simply no question that the plasma physics knowledge and experience are much greater here, we have had far more success with diagnostics, . . . and even in the electrical engineering of big machines we have the best track record."[40]

The scientific battle between Oak Ridge and Princeton was joined in a committee set up by Stephen Dean to produce an authoritative manual on the status and prospects of tokamak research.[41] There were at that point two chief predictions of the way in which particle diffusion would vary with rising temperature. Kadomtsev and Pogutse had derived the prediction that trapped-particle effects would cause diffusion to increase dramatically with temperature. Artsimovich had formulated a different and optimistic law called "pseudoclassical scaling," which was essentially empirical.[42] Artsimovich's law had the quality of a principle of confinement, increasing as the square root of the temperature, rather than decreasing with temperature. Harold Furth, representing Princeton on the tokamak committee, pushed Kadomtsev-Pogutse diffusion, while John Clarke, representing Oak Ridge, pushed a modified form of Artsimovich's law. Other scientific conclusions also had implications for the future of the laboratories. The committee members jockeyed to frame technical conclusions in a way that was both scientifically justifiable and politically consonant with their institutional interests.[43] "It was peer review of the most intense kind. . . . If you came down one way, you would favor one laboratory's program. If you came down another way, you would favor [a different] laboratory."[44]

While controversy thus raged within the fusion community on the scientific merits of mounting a DT experiment, the tactic began to look more and more advantageous from a political viewpoint. The harsh winter of 1973 had been marked by disrupting fuel shortages. In June President Nixon responded to public and congressional pressure for more energy research and development. He called for an additional $100 million to supplement the $772 million he had previously proposed as the federal energy R&D budget for fiscal year 1974. He also announced that he intended to ask for $10 billion to be spent over the 5 years starting in fiscal year 1975, almost double the $6.1 billion previously envisioned.[45] Nixon entrusted the job of deciding who should get the extra funds to Dixy Lee Ray. He had appointed Ray, a marine biologist, to the AEC in August 1972 to demonstrate the commission's increased concern with the environment. She had become chairman in February 1973, after James Schlesinger had left to take over the Central Intelligence Agency.[46]

Hirsch must surely have wanted a piece of the pie that Dixy Lee Ray was about to carve up, for he deeply believed that the program needed much larger annual budgets than the roughly $45 million it was slated to get in fiscal year 1974. One way to qualify for additional funds was to show that the program was proceeding at a pace faster than had previously been anticipated. In June 1973 Hirsch circulated to the project leaders a "conceivable strategy," which banked on a successful outcome of new experiments with neutral-beam heating on Ormak. The Ormak results would be announced as a "major breakthrough," and the division would follow them by "requesting support for a larger new DT burning tokamak facility . . . beginning in FY 1976, [and] projected for completion in 1979." This would constitute an obvious acceleration; previously a DT tokamak had been scheduled to start operation in fiscal 1987. It would also require a change in financial projections. A DT machine would cost another $50 million above the $50 million that had previously been planned for the tokamak feasibility phase. But Hirsch pointed out that his plan could decrease the total cost since the old separation of "scientific feasibility" from actual fusion "burning" involved an additional and, in his new view, unnecessary machine.[47]

To provide a proper setting for consideration of the radical reorientation of the program he was proposing, Hirsch arranged for a Standing Committee meeting to be held in a luxurious hotel on the island of Key Biscayne, Florida, close to the grounds of Nixon's mansion. Hirsch was determined to take control of the program, but at the same time he thought it a mistake to ride roughshod over its scientists. The first Washington magnetic fusion leader to view the field scientists as a breed apart from himself, he nonetheless respected them as knowledgeable and capable people. And like an impresario presiding over his divas, he understood that they were the ones who would make the thing happen. The fact that lack of support among field scientists can undermine a program would surely not have been lost on Hirsch, given the difficulties Milton Shaw, director of the Division of Reactor Development, had had. The Key Biscayne meeting would be the forum for full discussion and democratic acceptance of his proposal.[48]

"We could have had it in the salt mines in Kansas as far as I was concerned," one participant said later. "It was a nice place to stay, but God, it was a stressful time."[49] Not one of the four project leaders wholeheartedly wanted the DT experiment. The experiment, first of all, would exacerbate the competition among the three major approaches. A tokamak DT experiment would cost $100 million instead of $50 million. It was clear, therefore, whatever the rhetoric, that the program would now be able to afford only two feasibility experiments, instead of the three envisioned in the 1972 program plan. Hirsch had already indicated the way he thought that the competition would go. In circulating his suggestion to the project leaders, he had also proposed that the θ pinch would be the principal "alternate concept" and that "a large DT burning theta pinch experiment . . . could be initiated any time from FY 1977 to 1979, depending upon scientific progress." As for the mirror, he thought that it was unsuitable for a practical reactor, but could perhaps be well used in the core of an engineering test facility to create the neutrons and other accoutrements of a fusion environment for the test of power plant components.[50]

This put Kenneth Fowler from Livermore in an extremely unpleasant position. But Fred Ribe from Los Alamos also had cause to worry about the inroads a DT tokamak might make on his program. A very expensive tokamak might commit the program de facto to tokamaks as the single approach to a fusion reactor. This was the more to be feared in that Hirsch, like every Washington leader from the time of the McCone commission onward, was under pressure to decrease the number of approaches.[51] Mel Gottlieb from Princeton, as we have seen, did not want to house a radioactive experiment and, equally, did not want to lose the feasibility experiment. Even Herman Postma from Oak Ridge feared that Hirsch was going too fast. His reasoning, which he shared with others, was that if the DT experiment should fail, it could be a disaster for the entire program.

Hirsch did win the support, however, of both the outside members of the Standing Committee who were present, Solomon J. Buchsbaum and Edward C. Creutz. He also had leverage in his control of laboratory budget allocations, and he did not refrain from using it. The result was a compromise. A provision was incorporated specifying that "appropriate new hydrogen experiments should be planned to answer those plasma and reactor-related questions, which are more simply and economically examined in non-burning plasmas." Simultaneously, a cautiously worded endorsement, in principle, of DT burning was included: "The program should seriously plan for DT burning experiments at a time earlier than previously anticipated."[52] Hirsch now had what he needed. Returning from Florida, he alerted Ray and the Congress that "the decision should result in a saving of a number of years in the time required to develop commercial fusion power."[53]

Hirsch had support for the acceleration he wanted both within the AEC and in Congress. Ray liked the work of the CTR Division. John P. Abbadessa, the commission's controller and one of its most powerful figures, thought highly of Hirsch's management, and also believed that fusion could provide the optimal long-range energy source. The program was also profiting from the fission breeder's troubles. Milton Shaw at the Division of Reactor Development was plagued by cost overruns, public opposition, and the antagonism of field scientists to the liquid metal fast breeder reactor program. As commission leaders saw fusion progressing well and the breeder badly, they were coming to feel that it might be wise to place more of their bets on fusion. This attitude gained ground as the breeder lost its two principal advocates. Shaw had left the AEC in June 1973, and in August President Nixon replaced Commissioner James T. Ramey with former astronaut William Anders.[54]

In Congress, the program had a supporter in Mike McCormack, representative from the state of Washington. McCormack, himself a nuclear chemist at the AEC facility at Hanford, had taken energy legislation as his special province almost immediately upon his arrival in Congress in January 1971. McCormack thought that energy supplies should be increased by regulation and, above all, research. He was enthusiastic about fusion and happy to cooperate with Hirsch. He had been maneuvering even before the Key Biscayne meeting to push the CTR budget from the president's submitted figure of $44.5 million for fiscal year 1974 past the $50 million mark.[55]

The late summer and early fall of 1973 was an unpleasant time for the Princeton Plasma Physics Laboratory. During Gould's time in Washington, the conventional wisdom had been that the first DT experiment would be sited at Oak Ridge. Hirsch also favored Oak Ridge. He was very encouraged by the recent Ormak performance and, as we have seen, he had high hopes for the coming neutral-beam work. Moreover, the technological orientation of Oak Ridge was more to his taste than Princeton's penchant for physics. Princeton scientists nervously sounded out their Oak Ridge colleagues about job prospects in Tennessee.[56]

During the fall months, however, Washington's enthusiasm for Oak Ridge faded. Hirsch had anticipated a doubling or tripling of ion temperatures once neutral beams had been added to Ormak, but that goal was proving elusive.[57] Once again Oak Ridge was scooped as Princeton became the first US laboratory to succeed with neutral-beam heating; two beams with a combined power of about 60 kilowatts were raising the temperature of the Adiabatic Toroidal Compressor plasmas by about one third.[58] Moreover, doubts were emerging about the DT burning experiment that John Clarke and Michael Roberts were designing at Oak Ridge. First, cost estimates were beginning to rise to several times the $100 million the Washington office was allotting for tokamak feasibility. Second, the Roberts-Clarke design included superconducting magnets. Everyone agreed that a fusion reactor would have to use such magnets to reduce the cost of the consumed power to acceptable levels. But there was concern over whether an intermediate machine so vital to the fusion program as the first DT burner should be based upon a technology that was as yet untested.[59]

In December 1973 Hirsch summoned his leading scientists to headquarters to discuss plans for the DT experiment. The tactic Princeton took at the meeting was to snipe at the Oak Ridge proposal. Princeton scientists pointed out that the Oak Ridge machine called for temperatures 10 times larger than anything achieved until then, a very large step. They presented calculations to show that the Oak Ridge machine would cost 4 times the $100 million budgeted for a DT burner. They pointed out that if, as Hirsch intended, the machine were begun in fiscal year 1976, its design would need to be fixed before the results from the intermediate-sized Princeton Large Torus became available.[60]

On the second morning, Ken Fowler asked Herman Postma if Postma really thought Oak Ridge could meet a target date of fiscal year 1976. Postma said that he wasn't sure. Hirsch exploded in anger, and the meeting adjourned for lunch. When it resumed, Harold Furth of Princeton took a step that, in effect, placed Princeton in a new position vis-à-vis the Oak Ridge proposal. He stepped up to the blackboard and began sketching an idea that he, John M. Dawson, and Fred H. Tenney had published in early 1971.[61] At that time, their motivation had been to design a machine that could provide small power outputs of 10–300 megawatts and that would be easier to achieve than a standard tokamak reactor.[62]

They had pointed out that there was an alternative to raising a plasma to such a high temperature that it could ignite and burn by virtue of the fusion reactions among its ions. One could instead take advantage of the emerging neutral-beam technology. High-energy beams, for example, of deuterium, entering a tokamak are ionized and gradually slowed to the temperature of the "cold" plasma around them. Before they slow down, however, a certain proportion of them undergo fusion with the background plasma (which could, for example, be made of tritium) and produce a modest multiple, perhaps up to two or three times, of the energy that was used to inject them. It was this energy produced by the interaction of the hot, injected deuteron component on the cold, background triton component that Dawson, Furth, and Tenney had proposed as the source of electricity for a reactor. The advantage of the scheme was not only that one could operate with a relatively low-temperature background plasma, but also that the particle density and the confinement time did not have to be as high as for a conventional one-component reactor plasma. The more modest two-component reactor they proposed was called, informally, a "wet-wood burner."

By 1973 the wet-wood burner no longer looked like an attractive reactor concept. The recent progress in reactor engineering studies had shown that an energy multiplication factor of two to three was inadequate, when the totality of systems necessary for power generation was considered. If what was wanted, however, was a cheap DT experiment with $Q = 1$ for break-even power production, then, Furth pointed out at the December meeting, the two-component tokamak was a way to do it.[63]

Hirsch was extremely interested and asked Princeton to take a closer look at the two-component idea. Princeton was unenthusiastic. It was still leary of taking on a radioactive experiment and still believed that the proper next step for the tokamak program was larger hydrogen experiments. On the other hand, Princeton scientists were relieved that the machine proposed by Oak Ridge might now be dropped. They regarded it as a behemoth; its cost was enormous and its objective of reaching a quality of confinement of 10^{14} cm^{-3} sec seemed to them to be a monstrously large step. At this time, tokamaks actually in existence were only achieving 1/100th of that product of density and confinement time.[64]

Princeton, furthermore, was facing an uncertain future. Stephen Dean was pointing out to them that there was going to be a DT experiment of one kind or another and that if Princeton refused it, it was hard to see what tasks the laboratory might be assigned in its place. Princeton was also beginning to feel limited by the 200-megawatt power source it was using, which had first been installed in the early sixties for the Model C. If they took on the DT experiment, they would be in line to buy a much larger power source and hence to acquire the facility to do still other large experiments.[65]

At this point, therefore, a competition between Oak Ridge and Princeton began in earnest. At the Oak Ridge Thermonuclear Division—now led by John Clarke because Herman Postma had been appointed director of the entire laboratory—Michael Roberts put together a design team and began to guide them toward a formal proposal for a Fusion/Burning Experiment (F/BX). Princeton enlisted the aid of Westinghouse and started on the design of Two-Component Torus-Fusion Test Reactor (TCT-FTR).[66]

The Princeton and Oak Ridge proposals were aired at a large open meeting in July 1974 at Germantown, Maryland. The new Fusion

Power Coordinating Committee (FPCC), just constituted by Hirsch to replace the Standing Committee, was given the responsibility for deciding whether the division should request the DT machine for fiscal 1976 or fiscal 1977. The decision whether to choose Princeton or Oak Ridge, however, Hirsch reserved for himself. There were two major issues. First was the fear of failure, which was widespread among the program's scientists. Hirsch had summarized this worry in a list of "Fears and Problems": "If it fails to meet its objectives, CTR will be dealt a death blow."[67] Second, there was the need to choose the more able laboratory.

Of the two designs, Princeton's was the more conservative. Princeton had chosen ordinary magnets, while the Oak Ridge design used superconducting magnets. Princeton had selected a pulse length of tenths of seconds, to avoid the difficult problem of refueling a burning plasma, and to allow each shot to end before impurities could evolve in such numbers as to choke off the discharge. The Princeton design was limited to two-component operation. Most of the fusions would occur as the very energetic deuterons of the beam hit the cool "bulk-plasma" tritons. If the beam were turned off, the reactor would necessarily stop producing power; it was a "beam-driven" generator. In 1974 no one was thinking of this kind of tokamak reactor as a candidate for commercialization. The consensus was that a tokamak would function in the "ignition mode"; that is, that beams would be sent in only for heating in the initial phase of the pulse, until the plasma itself could ignite, maintaining the requisite temperature with the energy of its reaction products. Oak Ridge had scaled down its original design and built into it the capability for a two-component operation. But Clarke and Roberts saw no future in an experimental reactor that could not be extrapolated into an ignition device. Consequently, they designed F/BX so that it could also run at or near ignition, and they insisted on retaining the feature of superconducting magnets.[68] Hirsch responded to the two considerations by opting for the conservatism of the Princeton-Westinghouse design and for the more than 20 years of Princeton experience in toroidal confinement systems.[69]

Hirsch had now got his deuterium-tritium machine through the fusion community. In the process, the outlooks of the two tokamak laboratories had dramatically reversed in the course of a single year. In September 1973 Oak Ridge was favored to house the DT burning experiment, and Princetonians were wondering anxiously about the future of their jobs. In July 1974, on the contrary, Princeton was given what was now being called the Tokamak Fusion Test Reactor (TFTR) and was thereby put on the path of a major expansion.

Hirsch had also won funding increases to support an accelerated program. In her report on "The Nation's Energy Future," submitted to the president in December 1973, Chairman Ray had outlined a 5-year energy R&D budget about twice the size of previous projections. In particular, the plan, which Hirsch had helped develop, recommended $135 million for magnetic confinement fusion in fiscal year 1975, growing to $376 million for fiscal year 1979. This is to be compared with the $150 million Hirsch had given out as AEC policy only a year before for fiscal year 1980.[70] Of course, the actual dollar amounts finally authorized for any particular year would depend upon a complicated budgetary process involving the commission, the Office of Management and Budget, the White House, and Congress. By July 1974, however, when the TFTR was ratified within the fusion program, Hirsch expected about $100 million to be authorized for fiscal year 1975, an increase of about 40% over the fiscal year 1974 budget, and the CTR Division was asking for another, comparable jump for fiscal year 1976.[71]

By 1974, with its high budgets and its plans for the Tokamak Fusion Test Reactor, the magnetic fusion program had ascended to the level of big science. How much had it moved into a development phase and how much was it still in a research phase? We note, first, that the institutional structure had changed. A large number of technically trained persons were brought into the Washington office. These men and women began to assume important roles on the various peer-review committees. Some of them fell into the habit of consulting field personnel less, and directing them more, than had been customary. Robert Bussard, for example, ran his Development and Technology Branch tightly from Washington, despite considerable field resistance. The Washington staff thus began to take over decision-making functions that had formerly been the prerogative of the laboratories and advisory functions that had formerly been vested in the laboratory leaders. Hirsch personally took more control of and responsibility for the program, thus stripping the Standing Committee of much of the governing role it had had under Bishop and Gould. These changes were instituted precisely in order to substitute the infrastructure of a development program for the old research arrangements in which "the conduct of the research was left in the hands, and minds, of the senior scientists involved."[72]

The elaboration of program plans and the articulation of milestones and schedules were also part of a conscious adoption of the procedures of a development phase. As one by-product, the plans brought about a change in the way fusion scientists thought about the program goals. Before Hirsch, the demonstration of scientific feasibility, a research experiment, was the one clear point on the path to the future, and everything beyond was shadow. By mid-1974 a development aim, the demonstration reactor, had replaced feasibility experiments as the sharply defined end point.[73]

All this constitutes the trappings or the psychology of development. The actual work being done in 1974, however, and the work projected for the immediate future, was predominantly research. Roughly 90% of the fiscal 1974 budget supported research. In the 5-year plan that the CTR Division had formulated in 1974, about 85% of the total projected funding was to support research.[74] Further, Hirsch and the Washington leaders were well aware that the research on confinement systems formed the indispensable portion of the program; when budgets tightened, it was the advanced development elements that they cut.

Thus, fusion had not yet become a development program. Instead, it had changed from a loosely structured research program to one that was more tightly focused and "relevant." The research being done was more carefully governed by the end goal. The Confinement Systems Branch, for example, tended more and more to judge the devices coming before them by the touchstone of whether they could extrapolate into reactors. The staff discussed how to choose among basic research topics in terms of the probable requirements of the development phase. To some extent there was even a return to the old spirit of trial and error that had dominated the fifties. Washington's milestones put pressure on the laboratories to achieve better parameters, and at times this could be done only by delaying or eliminating the work that would yield understanding. And the very successes of the tokamaks led in the same direction. Since tokamaks were behaving well, and the goal was energy, not understanding, the obvious strategy was to keep going until something went wrong.[75] Only in two respects, then, had magnetic fusion moved out beyond the reactor-relevant research phase. It had taken on an institutional structure characteristic of development, and a "development consciousness" had begun to arise among its practitioners.

Choosing among Approaches

By mid-1974 the magnetic fusion leadership had committed itself to a major new tokamak experiment designed to reach energy break-even. The estimated price of the Princeton TFTR, in July 1974, stood at $100 million. In addition to the tokamak, Hirsch wanted to put significant resources behind one or two other confinement schemes that had the potentiality of being scaled up into practical reactors. One of this small group of competing designs would then be served up to the nation as a demonstration reactor ready for commercialization, a short 20 years in the future.

In fact, given the large commitment of funds being made to the TFTR, most fusion leaders believed that the program could not really afford more than one additional feasibility experiment, and therefore could not sustain more than one competing concept. The implication for the θ pinch at Los Alamos Scientific Laboratory and the mirror at Lawrence Livermore Laboratory was obvious. Both the Los Alamos and the Livermore teams already faced experimental problems of their own in the second half of 1974. The TFTR decision threw upon them the additional weight of a rivalry for the remaining, second place.

It will be necessary to go back in time in order to understand the roots of the physics difficulties confronted by the mirror and pinch machines. Then it will be possible to trace how the competition was resolved in the mirror's favor, so that, by late 1976, the program had narrowed down to two reactor candidates, the tokamak and the mirror. Strategy, however, does not remain static in an ongoing program, especially one that, like fusion energy, functions in a political world. Following the story two years into the Carter administration, to late 1978, we shall discover that even as Scyllac was being phased out, forces were gathering that would enlarge the field of active reactor contenders again.

Livermore: The mirror in extremis

The termination of the Astron and superconducting Levitron in late 1972 left the mirror to become, willy-nilly, the mainstay of Livermore's plasma-confinement investigations. And both the main experimental mirror sequences, the 2X series under Frederic Coensgen and the Baseball series under Charles Damm, were facing difficulties.

By 1972 the 2X experiment, which employed an ordinary mirror coil in combination with a set of Ioffe bars, was giving way to the 2XII, which used a pair of curved magnets, called "yin-yang" coils because of their shape, instead of Ioffe bars to create the magnetic well. The machine had been put into operation in 1971, and reliable data were forthcoming in 1972. The 2XII team got high densities, of the order of 2×10^{13} particles per cubic centimeter, but a low confinement time of about 0.4 millisecond. The instability that was curtailing confinement could not be identified. The reverse side of this problem was equally perplexing. In 1965 Richard Post and Marshall Rosenbluth had done a theoretical analysis of mirror systems and had identified a microinstability they called the drift-cyclotron loss-cone (DCLC) mode. The 2X and 2XII plasmas should show this instability according to the theory, yet Coensgen and his coworkers could find no trace of it.[1]

Baseball II had also been put into operation. It preserved the baseball-seam configuration of Baseball I for its magnetic coil, but its magnet was slightly larger and was made of superconducting niobium-titanium wire. Baseball II produced a magnetic field twice as intense as that of Baseball I, and one that could be kept at a more precise value over much longer time intervals. The vacuum system had also been upgraded. As experiments began in 1971, however, and despite the new magnet and exceptionally good vacuums, Damm's group found that instabilities were setting in far sooner than was expected from either theory or previous experience with Baseball I. Thus, the Baseball group was confronting a density barrier of 3×10^9 particles per cubic centimeter, which they were unable to transcend and unable to explain.[2]

The Washington office differed with the Livermore fusion leadership on the relative merits of 2XII and Baseball II. Livermore's Fowler believed that the Baseball could be extrapolated more easily into a reactor since it used a relatively steady current in its magnetic coils. 2XII, on the contrary, heated its plasma just as its predecessors had back to the days of Toy Top, by compression in a rapidly rising magnetic field. A reactor based on 2XII would therefore face all the problems that came with rapidly fluctuating currents, including the need for fast energy switching. Baseball also used superconducting magnets and thus gave emphasis to a technology whose development would be vital for reactors. Damm's experiment was therefore enjoying greater resources at the laboratory, with 56 workers to Coensgen's 35. The Washington staff, however, was impressed by the fact that 2XII was, quite simply, turning in a better performance in terms of plasma parameters than was Baseball. Hirsch, Dean, and the others worried that Baseball had no clear procedure for transcending its density barrier.

To ready a mirror feasibility experiment by the eighties would require budgetary authorization some time between 1974 and 1976. The mirror program thus had only a few years to reach a base level suitable for a credible experimental design. In November 1972 Dean convened a panel to consider how the program could get there. The US mirror program comprised the Livermore mirrors, the IMP at Oak Ridge, and a project at United Aircraft under John Luce, but Washington's focus was on Livermore.

The Dean panel recommended that Baseball continue but that the highest priority be given to 2XII. It also proposed specific milestones. The 2XII team was given 2 to 3 years to increase plasma confinement time a factor of 20, to raise the ion temperature, and to reach an understanding of the physical processes in the plasma. To support this effort, the panel recommended that "increased funds should, if possible, be provided to the 2XII program" in fiscal year 1973.[3]

The panel's recommendations, duly forwarded to the Office of Management and Budget, were now on record as the official "objectives" for the mirror program. The Livermore scientists were taken aback by the whole process. They vividly remembered the CTR program's first decades, and how the breaking in the sixties of brash promises made in the fifties had endangered the whole enterprise. Nevertheless, they buckled down to the task that had been assigned them.

Theory suggested that the way to meet the objectives for the 2XII was to increase its size considerably. The physicists reasoned that the DCLC instability might, in fact, be the culprit for the machine's untimely plasma loss. Their theoretical analysis of the DCLC instability implied that the instability, which depends upon a gradient in velocity space, was extremely serious for small machines, but was milder for large ones. A large machine was out of the question, however, because there was no hope of getting money for it. The Livermore group decided instead to improve the 2XII plasma parameters and then to inject neutral beams into the upgraded plasma. A neutral beam equivalent to some 20–120 amperes would, they expected, sustain an improved plasma against the unexplained particle loss. A new magnet would be needed, and the device resulting would be called 2XIIB, where B signified "beams."

As 1973 wore on, however, 2XII plasma improvments were too slow in coming. The Livermore leaders felt themselves under extraordinary pressure. They noted that work on the neutral-beam units, which was being done at the Berkeley Radiation Laboratory, was going extremely well, and they decided to forget about improving the target plasma and instead to hit what they already had with an unprecedented total of 12 neutral-beam modules. In taking this route, they were falling back on an alternative plan they had already weighed early in the year. It capitalized on observations made in 2XII experiments that plasma containment improved at higher densities. Neutral-beam currents from 200 to 600 amperes, the 2XII team calculated, would sustain a high-density plasma by this "brute-force" feeding method. Twelve modules at 50 amperes per module would provide 600 amperes, enough to allow for unfavorable contingencies.[4]

Fowler had already struck a bargain with Hirsch when the Astron and Levitron were cut, specifying that the money released would be plowed back into the Livermore mirrors. Now, throughout 1973, he pressed Washington for funds for the new 2XIIB program. Washington pressed back to have Livermore downgrade Baseball II. In a virtuoso piece of management, Hirsch and Dean monitored, cajoled, and threatened while at the same time encouraging the mirror group and creating an atmosphere for frank communication with the laboratory. And they promptly and unstintingly supplied the money that Livermore needed.[5]

At this point, while the Livermore scientists were struggling to meet the milestones set for the 2XIIB, they also came into stark confrontation with the difficulties involved in transforming a mirror into a practical reactor. Post's 1969 elaboration of a mirror reactor using direct conversion had been suggestive but not definitive. Two of his assumptions, in particular, were open to question. He had assumed efficiencies for the direct conversion subsystem that corresponded to ion energies above 700 keV. Actual ion energies in mirror reactors were more likely to be in the range of 100–200 keV, and this might lower subsystem efficiency to as little as 60 or 70%. Post had also assumed that beams of particles could be injected into the mirror with efficiencies up to 95%, but this was predicated upon a considerable advance over the neutral-beam technology then available. Using the more reasonable value of 85 to 90% for beam efficiencies, and lower values for the efficiency of the direct conversion system, the power balance, of $Q =$

5 and more, that Post had calculated for the direct-conversion mirror reactor could not stand up.[6] Fowler had therefore charged Richard Werner and his reactor design study group to delineate the best possible reactor that could be designed around a Q of about 1.2. In 1974 the group returned its answer: On the most optimistic possible assumptions, such a reactor would be unsatisfactory. Some method would have to be found to enhance the Q.[7]

In July 1974 came the program decision to request authorization of the TFTR. That decision could not fail to be daunting to Livermore. Hirsch, who doubted that the mirror as he then knew it could ever be developed into an economic reactor, had nevertheless suggested, in 1973, that it be chosen as the core of an engineering facility for testing reactor components.[8] But the Livermore scientists were fearful that Hirsch's commitment to a DT burning tokamak—whose estimated total cost had doubled by late 1974 to $200 million—would freeze out *any* large mirror installation, an engineering facility equally with a mirror feasibility experiment. The gloom at the laboratory was unmistakable.[9] The Astron and Levitron had been terminated. Now funding for the next round of larger mirror experiments was in doubt. Livermore was in real danger of losing its last remaining confinement program.

Los Alamos: Scyllac in decline

The Achilles' heel of Scyllac physics was to be plasma stabilization. The original scheme for Scyllac used a Meyer-Schmitt configuration of magnetic fields. This configuration seemed capable of sustaining the plasma in equilibrium along the geometric axis of the torus, but it was clear that the equilibrium would be unstable, in that any excursions from the equilibrium would amplify. The "dynamic stabilization" being planned to overcome this problem was generally regarded as flawed, both at Los Alamos and elsewhere.

Harold Grad of NYU had been a member of the ad hoc panel that had reviewed the Scyllac proposal in 1966. The occasion had given him a chance to survey the accumulated information on the θ pinch. His conclusion was that the magnetic windings needed to produce the corrugated Meyer-Schmitt fields were excessively complicated, and he had a strong hunch that properly chosen helical fields, of the type that was used on the stellarator, might be a better way to produce equilibrium. Moreover, he suspected that such "high-β stellarator" equilibria would be stable. Was Ribe sufficiently interested to make it worthwhile for the group at NYU to undertake a detailed analysis?

Los Alamos definitely wanted to hear more. Grad and a colleague, Harold Weitzner, worked up the idea as a paper for the international plasma physics conference to be held at Novosibirsk in August 1968. By July they had extensive results that looked very good. Ribe wrote, "I have just received your letter with preprints of the Novosibirsk papers and am quite excited at the possibilities opened up by your free-boundary solution which may allow a stable high-β torus without dynamic stabilization. We shall soon start looking at parameters with a view to engineering some of the helical conductors."[10]

In the theoretical studies of equilibrium and stability that followed, it became clear that the Grad-Weitzner equilibrium was not entirely stable. The computations indicated, however, that the perturbations would grow too slowly to be serious. Another discovery was more disturbing. When the theory, which had been developed at NYU for a linear θ pinch, was applied to a toroidal one, the equilibrium position of the plasma shifted well outward from the geometric center toward the wall. Ribe proposed, therefore, that the stellarator fields be made slightly more complex than those envisioned in Grad's original proposal, so that the plasma could be brought back to center. How much would that change enhance instability? No one was quite sure in early 1969. It was clear, however, that some mechanism for providing stabilization would have to be added to the Grad-Weitzner high-β stellarator configuration, as amended by Ribe. Although the configuration was certainly better than the Meyer-Schmitt fields, the hopes that Grad and Ribe had held for an equilibrium that was also stable could not be sustained.[11]

In 1969 the weaknesses of dynamic stabilization—the procedure that had originally been chosen for plasma stabilization on the Scyllac—were becoming ever more apparent. The Scyllac team therefore turned to a different scheme, one that was being talked about rather generally in the late sixties. In this method of "feedback stabilization," delicate optical sensors follow the position of the luminous plasma. If that position shifts, the sensors activate power modules which send current through auxiliary coils, producing corrective magnetic fields to force the plasma back into place. Feedback stabilization was written into the Scyllac program in 1970, as the new means by which instabilities could be arrested. Studies of the details of a workable feedback system for the Scyllac revealed that the crucial condition for success was the speed with which the feedback system worked. Theory suggested that the rate of growth of the instability needed to be kept lower than the time between the sensing of the plasma column's displacement and full application of the corrective field. The criterion was expressed as

$T\gamma < 1$, where γ is inversely proportional to the rate of instability growth and T is the time for the response to the plasma displacement.[12]

Meanwhile, plans for the 15-meter linear machine were being whittled down. The ad hoc panel that had reviewed the Scyllac proposal in 1966 had been emphatic in its judgment that the 15-meter Scylla experiment should precede the Scyllac. There were doubts about some of the assumptions that underlay the Scyllac proposal. There were also difficult scientific and technological problems that could best be attacked by work on the simpler, linear system. The pros and cons of the 15-meter experiment had to be considered, however, not merely in terms of what would constitute a well-founded experimental sequence, but also in terms of interlaboratory politics. Scyllac now had a competitor in the tokamak, and the competition was getting stiffer. During the winter and spring of 1969, the rivalry was between the United States and the Soviets. Los Alamos leaders could use this situation to counsel high budgets for Scyllac. Thus Division Director Richard F. Taschek wrote to Bishop in May 1969, "If you ask yourself 'what are the best U.S. horses on which to place my bets in competition with the Tokamak horse' then it is inescapable that they are Scyllac and 2X! . . . *They are better than anything else we have in the U.S.*"[13] But after June 1969, when the United States established its own tokamak program, Scyllac moved into the more difficult position of having to strive against programs feeding at the same budgetary trough.

Two courses of action seemed reasonable. One was to use the long linear machine to demonstrate quickly that θ pinches could attain still higher values of $n\tau$. As Taschek wrote in mid-1970, "What might otherwise have been a fairly straightforward course of research has been strongly affected by the success of the Tokamak and the competitive position of Scyllac relative to the USSR approach. Furthermore, the IAEA sponsored international CTR conference to be held in Madison in June of 1971 is likely to become a sounding board for successes or advances in this field. Thus . . . there may be some real tactical and impact merit in noting that a linear theta pinch . . . would provide a major contribution to the $n\tau$ derby which now seems to have arisen on a short time scale."[14]

The other course of action was to use the available money and manpower to move to the circular Scyllac as rapidly as possible so as to establish it as a bona fide reactor candidate. This was the path Ribe chose. In February 1969 Ribe proposed to the Washington office that Los Alamos fabricate a 120-degree toroidal sector concurrently with a 10-meter linear pinch, instead of doing the full 15-meter linear machine and subsequently converting its sections into a torus. In 1970

Ribe decided to shorten the linear device still more in the interests of starting experimentation on it at an earlier date. He decided on a 5-meter central section stoppered by 2-meter-long magnetic mirrors; the total length was now less than half that of the machine originally envisioned.[15]

With the promotion of Hirsch to division director in 1972, the story of the Scyllac program started a new chapter. Under Hirsch, the meeting of milestones, and the question whether a machine did or did not have a high potential for being made into a reactor, took on more importance than they had under Bishop and Gould. The game began to be played with changed rules, and in the play Los Alamos was to make one fatal move. The division there, under pressure to meet its milestones and suffering from a shortage of personnel and money, was to bring its program for feedback stabilization to a successful end too late to meet the political exigencies.

The toroidal sector had first been brought into operation in April 1971. No feedback systems were incorporated yet; the intent was to test the validity of the Grad-Weitzner-Ribe scheme for a well-centered equilibrium. Although instability soon intervened to disrupt the plasma, the equilibrium seemed indeed to be established. Ribe and the CTR group threw a big celebration.[16]

The next step was to attack stabilization. Separate testing of the feedback systems for Scyllac had begun in 1971. The crucial parameter for the power modules, T, the time needed for a full corrective magnetic field to be established in response to a sensing of a plasma displacement, was measured to be 0.9 microsecond. By this time engineering professor Keith Thomassen had come to Los Alamos from MIT to assume the direction of the technological side of Scyllac. His calculations showed that with this value for T, the condition $\gamma T < 1$ would not be met, given the current design for the torus. In late 1972 Ribe decided to attack the problem by changing from a planned torus of 4.8-meter diameter and 15-meter circumference to one of 8-meter diameter and 25-meter circumference. The rationale for this change can be grasped by remembering that it was the departure from linearity that was at the root of Scyllac's disequilibrium and instability. Increasing the radius increases the "linearity." (As a circle's radius grows, a small piece of arc approaches a straight line.) A 25-meter torus would therefore have a slower growth rate for instabilities, and hence a smaller γ.[17]

The feedback system was now redesigned to fit the larger torus. At the same time, Robin F. Gribble, the physicist who had had principal responsibility till then in the feedback program, was assigned, because of the lack of personnel, to spend much of his time on another experiment. This may have contributed to what one obsever has called a "loss of collective memory." The crucial figure T was believed to have retained its original value of 0.9 microsecond. In actuality, the figure had increased as the result of two separate developments. One was a series of modifications to improve the power modules. Each individual step caused only a small change; taken together, they became significant. The other development was a doubling of the number of coils each power module would drive in the 25-meter, as opposed to the 15-meter, torus. This change had been ordered because power modules were expensive and money was short. Unnoticed by anyone in 1973, T was now 1.5 microseconds.[18]

The original plans for the feedback program had called for testing the systems on both the old, linear Scylla IV and the 5-meter toroidal sector that had come into operation in early 1971. Under the pressure of the schedule of milestones, the Scylla IV experiments were never completed and the 5-meter sector experiments were never begun. Ribe and his staff instead hurried on to the construction of the full torus.[19]

The 25-meter Scyllac was dedicated in April 1974. By October the Los Alamos scientists discovered that feedback was not working. They had at last recognized the true value of T. To make matters still worse, they had discovered that their criterion was in error; better theory showed that γT had to be kept below 1/2 instead of the more liberal limit of 1. Finally, although they did not yet know this, the equilibrium that they had celebrated in 1971 had not been a static equilibrium, and this fact had implications for the stabilization procedure. The equilibrium plasma was dynamically displacing itself to and fro, and such feedback forces as there were were kept busy correcting this displacement and had no potency left over to attack instabilities.[20]

On top of their experimental troubles, Scyllac's leaders faced the competition of the tokamak. The July 1974 decision to build the TFTR posed the same problem for Los Alamos as for Livermore, that funds for new nontokamak experiments were going to be difficult to find.[21] Scyllac shared yet another liability with the mirror program: There were doubts abroad as to its engineering feasibility.

As a reactor core, a machine like Scyllac had two outstanding advantages. The first, which it shared with every high-β fusion device, was the economical use it made of its magnetic fields. As a consequence, there was no need to worry, as tokamak scientists must, about developing the complex new technology of high-field superconducting magnetism to a point where it could be used for fusion energy. Ordinary copper coils would suffice. Its second advantage was that θ pinches were already running at thermonuclear temperatures. For tokamaks, in contrast, the need for auxiliary heating systems had been a problem from the start. On the other side, Scyllac reactors had several major engineering problems. Two of them were bound up with the fact that it would be a fast-pulsed machine, with its magnetic fields rising and falling in intervals of under half a second. This meant, first of all, that energy had to be transferred back and forth between the magnetic coils and the energy storage system at a rapid rate. Further, the amount of energy in the energized coils was of the same order of magnitude as the amount to be delivered to the power line. Thus, in order to avoid a situation in which the losses in the energy transfer system would eat all up the power generated, the transfer and storage system would require efficiencies of the order of 95%. Second, the pulsed nature of Scyllac meant that the wall of the vacuum chamber was alternately heating and cooling at a rapid rate, and this put it under severe stress.[22]

The Washington staff was deeply concerned with Scyllac's engineering troubles. James M. Williams, who had been appointed assistant director of the Development and Technology Branch in the spring of 1974, was convinced from his own reactor experience and his conversations with people at the Electric Power Research Institute that the utilities would never buy a highly pulsed device. Hirsch, whose doctorate was in nuclear engineering, had made his own reactor study and concluded that finding practical solutions to Scyllac's problems would be difficult. And he was listening closely to the nuclear engineers at MIT. They had always been critical of the idea of a pulsed reactor, and particularly of the difficulties it would pose for the first wall.[23] The last months of 1974 were therefore as dark for Scyllac as they were for the mirror program.

The struggle for second place

1975 opened with changes in Washington. Congress had abolished the Atomic Energy Commission in 1974, and its divisions had been absorbed in January 1975 by the Energy Research and Development Administration (ERDA), a new agency with a wider energy mission. Hirsch was also making an important change in his own staff. He had begun to feel the need for a deputy director, and he chose Edwin E. Kintner, whom he had met in the early seventies when they both participated in an AEC trip to the Soviet Union. Kintner was an experienced engineer, and thus precisely the sort of person Hirsch wanted to help him continue to shift the focus of the fusion community toward practical applications. There was another attraction in Kintner. There were two men in whose management policies Hirsch took the keenest interest. One was Hyman Rickover, the maverick admiral who had created a nuclear submarine fleet. The other was Milton Shaw, whose push toward a commercial breeder reactor was being thwarted. Hirsch wanted to know why Rickover had accomplished his objectives and why Shaw had failed, and Kintner was in an excellent position to tell him. Kintner had worked in the Navy's nuclear propulsion program from 1950 to 1963, and he had subsequently spent 10 years with Shaw in the AEC's Division of Reactor Research and Development.[24]

Despite the changes, the fusion program experienced little disruption. ERDA's incoming administrator, Robert C. Seamans, Jr., had received an extremely favorable report on magnetic fusion from the members of the Office of Management and Budget. The program, they had told him, was one of the best conceived, and best managed, within the AEC. Seamans was himself convinced, in any event, that fusion, together with the breeder and solar energy, constituted the only possible long-term energy source that had yet been discerned. Continuity for the program was further reinforced by the fact that John M. Teem, the new assistant administrator for solar, geothermal and advanced energy systems, had also been Hirsch's boss in the AEC.[25]

Meanwhile, neutral-beam lines were being connected to Livermore's 2XIIB. By June 1975 the laboratory engineers had solved their major problem—how to shield the beam lines from each other so that a malfunction in one of them would not disturb the circuitry of the others[26]—and had brought the neutral-beam lines to the point where 300 amperes of beam current, out of a theoretical maximum of 600 amperes, were available.[27] Experiments now began. At first, the 2XIIB team found that confinement time, far from improving, was actually getting worse. The first unequivocal signs of the drift-cyclotron loss-cone instability predicted in the 1965 calculations of Post and Rosenbluth could also be identified. In 1967 Post had proposed a cure for the DCLC mode, should it ever be found experimentally: the passing of a stream of lukewarm plasma through the main body of hot plasma. Coensgen decided to try warm-stream stabilization on the recalcitrant 2XIIB experiment. By July the results were dramatic: Confinement time increased 10-fold, and ion temperature doubled. The 2XIIB group was jubilant. The machine was close to meeting the milestones that the Washington office had set in November 1972.[28]

Warm-stream stabilization brought an unexpected bonus. The 2XIIB scientists discovered that they could use the steady stream of warm plasma as a seed upon which the incoming neutral beams could ionize to form a very hot plasma. It was therefore possible to dispense altogether with the old methods of injecting and heating the plasma. The scientists were no longer constrained to use rapidly pulsed, magnetic fields to effect heating by compression and could go over to direct-current magnetic field coils. The original division of labor between the 2XIIB sequence of experiments and the Baseball sequence called for the Baseball to be the steady-state experiment. With warm-plasma stabilization, however, the 2XIIB had solved the Baseball's problem, and in so doing had acquired for itself all the attractiveness as a reactor candidate that steady-state fields bestow.

Hirsch at first remained cautious. He was worried by the fact that Q, the ratio of power output to power fed into the neutral beams, was still limited to a maximum of about 1 in mirror machines. Livermore Division Director Fowler, however, radiated an impressive confidence. The warm-plasma stabilization experiments had been coupled with new work by the Livermore theorists, and the whole had added up to an enhanced understanding of the physics of mirror plasmas and, in particular, of the DCLC instability. Fowler made the case to Hirsch that the new level of knowledge to which Livermore had ascended would give the laboratory an excellent basis for inventing a higher-Q machine. Rosenbluth, whom Hirsch sent to Livermore, supported Fowler, and Hirsch grew increasingly optimistic. "I found great encouragement in the fact that Marshall Rosenbluth agrees with the Livermore position," he wrote in his diary for October 16, 1975. "I think that we are very close to having another concept that we can say will produce net fusion power."[29]

As a concomitant to this change in appreciation of the mirror program, Washington began to send ever clearer signals to Los Alamos: If the laboratory is to remain one of the major CTR sites, it must drop Scyllac and pick up a more promising approach. The physics of Scyllac, the Washington staff felt, was being elucidated at too slow a pace, and a Scyllac-based reactor appeared altogether too complicated. Hirsch asked Ribe and his men to "take a serious look at other approaches, and if their reactor technology looks feasible, then work on the physics."[30]

A number of senior people at Los Alamos agreed with the Washington position. James Phillips and Joseph DiMarco, for example, had resurrected a modest z-pinch program in the mid-1960s. At that time, Phillips had been mulling over the question of why the θ pinch was able to produce hot plasmas and thermonuclear neutrons, whereas the z pinches had never attained temperatures over 500 eV and had only produced spurious neutrons. Working with DiMarco and Louis

Burkhardt, Phillips had pinpointed the problem as the slow rate of rise of the magnetic field in the z pinch. They had proceeded to build a linear pinch with a faster power source and then, in 1970, a toroidal pinch, the ZT-1. Computer analyses suggested that their pinch might be stable at β values up to a high 40%. Furthermore, the analyses indicated that, unlike the tokamak, which cannot be run above the Kruskal-Shafronov current limit without exciting macroinstabilities, there was no current limit to stability in the z pinch. The group, now under Don A. Baker, was therefore hopeful that a z-pinch reactor might be brought to ignition temperatures by high-current ohmic heating alone.[31] Others, like George A. Sawyer, thought that the linear θ pinch could be superior to the toroidal Scyllac. Some means would have to be found to prevent the loss of plasma from the ends, but the Scyllac's problems with equilibrium and the enhanced instabilities created by the equilibrium fields would be absent.

Ribe, however, felt that the stakes were too high to back down. Scyllac was the only substantial machine in the country embodying a high-β, θ-pinch approach to fusion. The other important project of its type, the German ISAR in the Garching Laboratory near Munich, could easily be knocked out by an American show of no-confidence in the θ pinch. Ribe also believed that in evaluating Scyllac, Hirsch and Dean were dominated by the compulsion they felt to reduce the total number of concepts in the program. On this view, complaints about Scyllac's reactor potential were mere convenient supports for a position dictated at bottom by political needs. Ribe himself thought that Scyllac's prospects were good, and he was determined to go to the wire for his experiment.[32]

The struggles of Livermore and Los Alamos were played out against a background of tokamak successes. The most important of these was Alcator. The MIT tokamak had hardly had an auspicious beginning. The Alcator team had had an exceptionally difficult time assembling the machine, and when it was finally ready in late 1972, the plasma was so badly contaminated that the discharge was often quenched by the impurities. Such troubles were hardly unusual at the start-up of a major fusion experiment, but some of the MIT scientists new to CTR were shaken by the poor runs and unhappy at jokes at Alcator's expense that were circulating at fusion meetings. Bruno Coppi and Bruce Montgomery, however, stood unwaveringly by Alcator, and so did two younger men. One was a heterodox but creative and resourceful experimentalist and engineer named Robert Taylor. The other was Ronald R. Parker, an assistant professor from the Department of Electrical Engineering, who had originally joined the team as a part-time specialist in techniques of auxiliary heating.[33]

Alcator's journey to success started in early 1973. A short circuit in the magnet produced stresses that collapsed the vacuum vessel. The MIT research leadership had to decide whether to cut their losses and get out gracefully or rebuild. They decided to rebuild, and to put Parker in charge. Hirsch, who had always been supportive of Alcator, agreed to continue its funding to give the MIT group a second chance. Parker redesigned the mechanical connections between the vacuum chamber and the magnet coils and instituted a zealous simplification and improvement of the vacuum system, incorporating every advanced "clean-room" technique that he could. Meanwhile, Taylor, working with a few thousand dollars and spare parts from the Alcator, invented the Versator, a classroom-sized tokamak, and on it discovered a new method for decontaminating plasma by discharge cleaning. When the rebuilt Alcator went into operation in January 1974 with its improved vacuum system and its new technique of Taylor discharge cleaning, it produced, from the start, a deuterium plasma of unparalleled purity.

It was with this clean plasma that Alcator achieved its important results; it is an interesting illustration of the relation between fusion theory and fusion experiment that obtained in the early seventies, however, that several significant results were entirely unexpected. Instead of a plasma with an anomalously high resistivity and, in consequence, an exceptionally high temperature, two distinct regimes were found: a low-density regime in which resistivity was anomalous, and a high density regime in which it was not.[34] Then, using a Princeton technique of raising the density by puffing in additional gas, the Alcator team found a highly important new effect. The confinement time was roughly rising with density whereas analogy suggested that it would fall off with the increasing density. The quality of confinement, measured by the product of density and confinement time, was therefore rising as the square of the density: $n\tau \sim n^2$. By late November 1974, the team had obtained an $n\tau$ of 3.7×10^{12} cm^{-3} sec in their best run.[35]

In the summer of 1975 a delighted Hirsch asked the Alcator leaders to return the favor he had done them in supporting their machine and drive the $n\tau$ product as high as possible by Thanksgiving. Parker phoned Hirsch at home on October 27 to tell him that their preliminary data showed a quality of confinement of 10^{13} cm^{-3} sec at ion temperatures of about 10 million degrees centigrade. "[Parker] was relatively calm about it but obviously was extremely excited."[36] Alcator had been entered in the temperature sweepstakes, but was instead winning the confinement sweepstakes.[37]

The Oak Ridge Ormak team was also getting good plasma parameters and important scientific results. Like MIT, Oak Ridge had gone through a period of adversity. It had been scooped in 1970 when Princeton's Symmetric Tokamak had been the first US machine to confirm the Soviet tokamak data, and again in 1973 when Princeton's Adiabatic Toroidal Compressor team had published the first US documentation that neutral-beam heating could significantly raise the plasma temperature in tokamaks. Oak Ridge had lost to Princeton also on the competition to house the Tokamak Fusion Test Reactor. Although the laboratory had been awarded the contract to develop the neutral-beam systems for the Princeton Large Torus, it was losing out to Berkeley on the contract to provide neutral beams for TFTR. Division Director Clarke and Ormak leaders Lee A. Berry and O. B. Morgan were therefore both relieved and delighted as they finally began to get significant data from neutral-beam heating on Ormak. The neutral beams tripled the ion temperature, pushing it to a world tokamak record of close to 20 million degrees centigrade, more than double the previous high.[38]

Still to come were the results on the Princeton Large Torus. First plasmas were expected in November or December 1975. This machine was the largest in the American tokamak program, with a confinement chamber radius of 40 centimeters. Only the Soviet T-10, with a 37-centimeter radius, was comparable. PLT was designed to show how confinement time varied ("scaled") as ion temperature mounted above the values reached by Ormak. In September Hirsch dictated for his diary, "If the scaling is poor, the size of a tokamak reactor quickly becomes prohibitively large. This is a period in which these facts, which we have recognized for many, many years, really have penetrated deep into our conscious being. . . . It means that the PLT experiment and its performance is really very critical to the longer term future of the program."[39] On Friday, December 19, 1975, Mel Gottlieb of Princeton telephoned Hirsch at 2 A.M. Princeton had gotten its first PLT plasma, and Gottlieb and a group of his associates in the control room were opening a bottle of champagne.[40] It would take a very large step, however, to achieve a decisive determination of the law for temperature dependence that was so eagerly sought.

At this juncture, Edwin Kintner took over the post of director of the CTR Division from Robert Hirsch. Hirsch had taken the step, in December 1975, of turning the day-to-day affairs of the division over to Kintner, and secluding himself to draft a revision of the program plan issued in early 1974. He never came back. Before he could complete his draft, John Teem resigned as assistant administrator for solar, geothermal and advanced energy systems, and Hirsch was offered the promotion into Teem's place. As Kintner's superior, Hirsch continued to play an important role in the formation of fusion policy. His concerns, however, now became necessarily wider.[41]

Kintner's first major issue was the mirror program. Basing its case upon the 2XIIB successes, Livermore began once again to sound out Washington about supporting a mirror scientific feasibility experiment, at a cost of some $70–80 million.

The tokamak community of scientists from Princeton, Oak Ridge, MIT, and General Atomic looked with some nervousness at the Livermore request. In late 1975 there were signs that the sharp upward swing of funding for magnetic fusion was nearing a check.[42] There were also technical arguments against the request. Oak Ridge leader John Clarke worried that its low Q made the mirror a dead end, and that it was consequently pointless to pour money into a larger-mirror experiment. On the contrary, Stephen Dean, who was emerging as an increasingly influential voice in the Washington office, supported the Livermore proposal. Dean had just chaired a review committee that had recommended a vigorous program to develop mirrors, both for pure-fusion reactors (via schemes yielding higher Q values) and for other applications. Kintner himself was anxious to use his new position to maintain the program's breadth. As the Dean review committee had pointed out, "Tokamaks could falter. Tokamaks could ultimately be surpassed by other concepts having more attractive features . . . or . . . end-uses could emerge for which non-tokamak concepts are more suitable."[43] Kintner therefore pushed hard at the April 1976 Fusion Power Coordinating Committee meeting. "I got [Clarke] out in the parking lot and convinced him . . . [and I told Fowler], Ken, look, I'm only going to do this on one basis, that we honest-to-God identify ways to fix . . . Q-enhancement."[44] The FPCC ratified a mirror scientific feasibility experiment (eventually named Mirror Fusion Test Facility, MFTF), to begin operation in 1981, stipulating that Livermore make a concerted attack on the problem of Q enhancement.[45] The mirror was now in second place as backup to the tokamak approach. And the systematic search for high-Q mirror machines that the FPCC mandated exemplified the influence that reactor engineers were beginning to exert upon the direction of research. For the Livermore reactor study group, their days of isolation were over, replaced by a vigorous, daily interplay with the people who worked on the plasma physics.[46]

Before Hirsch's promotion, the Washington office had already decided to close down Scyllac. The senior staffs of the Los Alamos CTR Division and the Washington office—just rechristened, on Hirsch's directive, the Division of Magnetic Fusion Energy—met in December 1976 at ERDA's Albuquerque Operations Office to discuss the experiment. The Scyllac scientists had by then "derated" the Scyllac plasma by lowering its temperature and decreasing the strength of the magnetic fields to a point where the feedback system they had in place would be able to stabilize it. They had also worked out a better design for the magnetic field coils that provide equilibrium. Ribe therefore pleaded to be allowed to carry on with the experiment at the minimal operating level of $1–2 million a year.[47] It was too late.

But what role should Los Alamos retain? Dean took the extreme position that it should be entirely cut out. The group, Dean argued, had not come up with an interesting alternative to Scyllac, and there were more urgent uses for the money. The New Mexico congressional delegation would be angry in any event; it seemed better to bite the bullet and suffer congressional wrath for terminating both Scyllac and the Los Alamos fusion program at one time. Kintner, on the contrary, thought the Los Alamos team strengthened the program and preserved its diversity.[48] Scyllac money could profitably be diverted to the toroidal z-pinch, linear θ-pinch experiments and an array of supporting projects in high-β fusion research. In the end, it was decided that Scyllac would be terminated at the end of August 1977. In the additional few months, the Scyllac team was able to verify the ability of the feedback stabilization system to control the derated plasma and to get encouraging, though incomplete, results for their new coil.

The overall result of the changes were, however, as Kintner wrote to Los Alamos, to "... modify what might have been called a 'mainline' effort at LASL to a broader program of research into high-beta alternate concepts for magnetic confinement systems."[49] Put otherwise, Los Alamos had plummeted from a position as one of the program's lead laboratories, with one of the serious reactor contenders as its main machine, to the position of a supporting laboratory.

A new course

A few weeks after the inauguration of President Jimmy Carter in 1977, division head Kintner was in Princeton delivering a pep talk to the hundreds of members of the TFTR work team. When he finished his talk, Mel Gottlieb came up to tell him that he had a telephone call from Washington. It was ERDA Controller Merwin C. Greer. "How would you like," Greer asked Kintner, "to take a $60 million reduction in your program for fiscal 1978?"[50] The Ford administration had left behind it a budget incorporating $370.9 million for magnetic fusion. The new cut now proposed by Carter would leave the division with even less than the $316.3 million they had received, in budget authority, for fiscal year 1977.[51] With inflation, the real cut would be still greater.

The Division of Magnetic Fusion Energy had been looking forward to rising budgets. The updated program plan initiated by Hirsch in December 1975 had been completed by a committee under Dean and issued by the division in July 1976. The plan had offered five possible schedules, corresponding to five levels of effort; it had plumped for Logic III, which scheduled a demonstration reactor for about 1998. Logic III called for budgets of over $400 million in fiscal years 1978, 1979, and 1980, and over $600 million in fiscal year 1981.[52] The Carter administration did not promise to be hospitable to these plans.

Hirsch decided to leave the government. He felt at odds with the new administration's way of thinking and doing things. He was also concerned about the current ceilings on government salaries and about impending "revolving door" rules that would tighten the restrictions governing a transition from government employ to a job with industry. He resigned his post in March and took a position with Exxon Corporation.[53] Kintner stayed on and fought for a congressional add-on to the Carter figure, but he succeeded in winning back only $10 million of the $60 million markdown. In the meantime, Carter had submitted to Congress a plan to abolish ERDA and form a Department of Energy. Kintner and the Division of Magnetic Fusion Energy waited with apprehension through the spring and summer of 1977 for the coming transition to the new department.

Carter had campaigned on a platform of conservation and solar energy. He had criticized the breeder reactor as a dangerous stimulus to nuclear weapons proliferation; instead, he wanted to reduce American dependence on foreign oil by introducing energy technologies based upon the nation's ample coal supplies. The Department of Energy and its first secretary, James R. Schlesinger, Jr., began operation in October 1977 with a pledge to shift emphasis away from nuclear technologies and toward solar energy, coal conversion, and conservation techniques. Schlesinger was also committed to emphasizing short-term energy problems over long-range technologies, and to decreasing the proportion of money spent on the development of central station electricity—then about 70 percent of energy research and development. None of this boded well for fusion.[54]

Schlesinger was also alarmed by the rapidity with which the CTR budget had mushroomed since he had left the AEC in 1973. He wondered whether it would not be wise to trim $100–200 million from the fiscal year 1979 fusion budget for distribution to some of the DOE programs to which he gave higher priorities.[55] As a means of obtaining a broad overview of the research needs of all the energy technologies, Schlesinger established an Office of Energy Research in the department. He recruited John M. Deutch, chairman of the Massachusetts Institute of Technology's Chemistry Department, to head it. Deutch, who had broad contacts in government, and particularly in the Department of Defense, had known Schlesinger for many years. A relatively young man and something of a "whiz kid," he was "aggressive, hard-headed and energetic. [His style is] to come on strong and to move quickly. Rather than wait for problems to come to him, he likes to take hold of a situation, get good people to advise him, and act."[56] One task that Schlesinger gave Deutch immediately was to determine whether it would be possible to trim back the fusion program.

Deutch was not well acquainted with the program when he took office. He did share a common view that fusion had become a playpen into which millions of dollars had been poured so that scientists could amuse themselves. Talking it over with Robert D. Thorne, acting assistant secretary for energy technology, Deutch decided that the first step was to set up another review panel; both men agreed that the panel would probably find the prospects for fusion energy so uncertain that it would be appropriate to cut the program back from a mission-oriented endeavor to the less expensive level of a research program.[57]

Deutch was careful to keep the selection of panel members in his own hands. This transfer of the locus of decision making was crucial. Hirsch and his coworkers in Washington had worked hard to overcome the parochialism of the laboratories and to unite the entire fusion community around a common set of goals. They could not and did not try to transcend the interests of the fusion community itself. Hirsch therefore strove to develop fusion into a practical source of electricity as soon as possible. This achievement would make fusion a real energy option and keep the budgets at the levels Hirsch thought indispensable. It would also get fusion onto the scene soon enough to prevent the breeder from preempting the market. With this goal, it was almost inevitable that Hirsch should concentrate on tokamaks, since no other device in operation was showing such satisfactory plasma behavior. Deutch, on the contrary, had the larger energy picture in mind. Fusion was not going to solve the short-term energy problems with which the Carter administration was grappling. As a long-term power source, it did not make much difference from Deutch's perspective whether fusion would enter the grids in 2020 or 2030 or 2050. He therefore went into the review process with the suspicion that the stress on tokamaks was misplaced.[58]

Since both magnetic and inertial fusion were to come under the group's purview, Deutch's panel included specialists in magnetic, laser, and particle fusion methods.[59] Some of the eight men were academicians, while others, including its chairman, John S. Foster, Jr., had experience in the management of government laboratories and/or industries with a major interest in research and development projects.[60] Many of them were involved in managing or advising on weapons research. They were a strong-willed and forceful group. They did not work independently of Deutch, however; rather, their views and his evolved together through a process of continuing communication.[61]

The panel considered the newly emerging results of the reactor design studies, as well as the data on plasma experiments that the project leaders were briefing them on. Reactor studies had been done for all the major experimental concepts, but the preponderance were based on the tokamak. The tokamak work included a major report from Princeton, directed by Robert Mills, studies from Don Steiner's group at Oak Ridge, and three reactor designs from the Wisconsin group led by Robert Conn and Gerald Kulcinski.[62]

The Wisconsin studies were particularly influential. Aided by a large contingent borrowed from industrial firms, Kulcinski, Conn, and their associates turned out a tokamak study called UWMAK-I in 1973, an improved UWMAK-II in 1975, and UWMAK-III, an unconventional tokamak reactor design featuring a noncircular plasma cross section, in 1976. Their approach in the first two designs was to invoke "as little extrapolation of present-day [technological] capabilities as is possible."[63] They therefore threw away the exotic materials—niobium and vanadium—that physicists had been choosing for the inner wall of future reactors, and substituted stainless steel, a material with well-known properties, for which a manufacturing infrastructure already existed. They also declined to assume any of the "dream cycles," the advanced fuel cycles involving pure deuterium, deuterium and helium, or boron. Nor did the Wisconsin team attempt to optimize the reactor; rather, the goal was to perform the study in sufficient detail to uncover problems. UWMAK-I and UWMAK-II were "reference" designs of the type that Rose and his Oak Ridge colleagues had promoted.

The group did indeed succeed in uncovering problems. The worst concerned the inner wall. This structure was subject to such intense fluxes of neutrons and such extremes of heating and cooling that it was unlikely to last more than 2 years. The Princeton group studying reactor systems came to only a slightly less pessimistic figure.[64] The second severe problem was the large amount of radioactivity induced by the fusion neutrons in the structural material. For the first time, it was evident that for some fusion designs the levels of activity could approach those of fission reactors. This problem would compound the first; replacing the first wall would have to be done by some sort of mechanical robot run by remote control. Such difficulties had never before been exposed in their full severity.[65]

The engineers' conceptualizations of tokamak reactors had led to the criticism within the utilities that tokamaks were not going to meet the needs of central-station electricity generation. The utilities had no quarrel with the choice of the tokamak as a vehicle for demonstrating scientific feasibility. But an optimal device for a demonstration of feasibility would not necessarily be an optimal commercial reactor; the utility executives worried that the tokamak might be frozen into the program as the candidate for commercialization.[66] The projected size of tokamak reactors, producing 500–2,000 megawatts of electricity, was about standard for utility plants expected to come on line during the twenty-first century. For an unproved technology like fusion, however, it was inappropriately large. It would be better, the utilities' critics maintained, to introduce new generating equipment in small units until the dependability of the method had been demonstrated.

The utilities' institutional apparatus for evaluating fusion had improved by this time. After a decade of calls from various quarters, the Electric Power Research Institute (EPRI), dedicated to research problems of central-station electricity generation and transmission, had been established in March 1972.[67] EPRI had in turn set up a Fusion Advisory Committee that included some of the utility people who had been the most involved with fusion; among them were Raymond Huse of PSG&E of Newark, Howard Drew of the Texas Atomic Energy Research Foundation, Clinton P. Ashworth of Pacific Gas and Electric Company (PG&E), and Michael Lotker of Northeast Utilities Service Company of Hartford.

Clinton Ashworth advocated that the program pay serious attention to advanced fuels, like mixtures of deuterium and helium-3. Since these fuels could minimize radioactivity, they would make it possible to site plants close to customers.[68] Ashworth argued that capital costs, estimated in the UWMAK studies to run perhaps 30–50% higher than fission breeder costs,[69] could be cut if advanced fuels were used since shielding could be reduced and the tritium-breeding blanket dispensed with. Maintenance would be simplified, he claimed, and the utilities' political troubles over environmental hazards alleviated. "I see no need for a fusion reactor that is not substantially different [from] and better than, say, a fission reactor. . . . Fusion reactors that are large, complex, and with radioactivity in the hundreds of megacuries would face the same problems that create energy supply doubts in the first place. . . . The advantages to be sought from fusion should be as dramatic as its billion-degree technology is different."[70]

Raymond Huse and Michael Lotker called for more attention to the fusion-fission hybrid. In a hybrid, the blanket surrounding the fusing ions contains thorium-232 or uranium-238, and the fusion neutrons both produce heat for electricity and convert the fertile materials into the fissile substances uranium-233 and plutonium-239 for use in light-water reactors. The advantage from the utilities' perspective was that a mature and proved technology, that of light-water reactors, could be retained, with the addition of a small number, perhaps one in seven, of hybrids.[71] Utility executives were not completely happy with the other alternative, the fission breeder, which would require them to scrap altogether the light-water fission technology with which they were so painfully acquiring experience. For Huse, who was a fusion enthusiast, it was equally important that hybrids might help the fusion program. A hybrid could be built around a fusion core that was producing only about as much power as it was using; such breakeven machines would be coming in the 1980s, with the TFTR. The hybrid would therefore be a relatively near goal and, as such, could win the support of politicians who might otherwise pass over the long-range fusion program in their allocation of R&D money.[72]

The utilities' objections to the tokamak had been brought skillfully before the community of scientists and science policymakers through three knowledgeable articles by William D. Metz in *Science*, the official journal of the American Association for the Advancement of Science, in the summer of 1976.[73] These articles, as well as their own personal contacts, meant that the panel members could not fail to be aware of EPRI's viewpoints.

After a few half-days of briefings and 2 days of deliberations, Deutch's fusion panel jettisoned Hirsch's emphasis on the tokamak. The panel also rejected the goal of a demonstration magnetic fusion plant before the end of the century. Instead, the members substituted, under the code phrase "the highest potential of fusion as a commercial energy source," the goal of optimizing the program so as to achieve the best possible reactor in the shortest possible time. This strategy called for selection of "the best partners in a marriage of scientific and engineering promise." "While tokamaks are currently the most advanced scientifically, they seem on the surface to be the most complex of the possible alternative approaches to fusion from the standpoint of engineering into an energy producer."[4] The group by no means ruled out tokamaks, but they recommended vigorous exploration of the most likely nontokamak devices. In the first instance, this meant giving mirror machines an opportunity to catch up. In the second, it meant selecting the most attractive of the so-called alternate concepts, supporting them for sufficient time and at a sufficient budget to test them, and then either upgrading or dropping them. In addition, the report admonished the Department of Energy to give attention to advanced fuels and to DT devices that might lead to simpler and more easily maintained reactors, "two areas in which significant breakthroughs might be made which could change the whole picture," and to devote some thought to hybrid reactors. "The present fusion strategy . . . emphasizes carrying the front running approach as quickly as possible to a demonstration of commercial viability while maintaining a vigorous backup program. [We recommend a modification.] Pursue vigorously several physics approaches and carry out, in parallel, engineering and materials test programs until at least one potentially economically competitive design is identified. . . ."[5] Clearly, the Foster panel was rejecting major elements of Hirsch's strategy.

When they turned from a strategy for the future to an evaluation of past achievements, however, the Foster panel's findings contradicted Deutch's original expectations. The panel praised the magnetic fusion program's management and scientific successes and vindicated its immediate, as opposed to its long-range, goals. It was impressed with the quality of confinement reached in the Alcator, by this time a quality of confinement of 3×10^{13} cm^{-3} sec and the magnitude of the temperatures reached in Ormak, about 2 kilovolts. The tokamak parameters, it found, "are within modest factors of those required for reactors." The panel members also praised the plasma parameters achieved in mirrors: ion temperatures of 20 kilovolts and β values of about 1. They pointed out that at least two ingenious ideas for improving the mirror Q had been put forth. One of them, the tandem mirror, was soon to be tested in a facility then under construction at Livermore. The tandem mirror would attempt to turn the mirror's ambipolar potential into an advantage, by using a pair of mirrors as end-plugs to a long solenoid. The positive ambipolar potential developed at the end-mirrors was expected to form a barrier to the escape of ions from the central solenoid.[76]

As far as short-term milestones were concerned, the Foster committee judged the goal of achieving break-even in TFTR to be "urgent." It deemed the magnitude and momentum of program effort to be one of the program's greatest assets; by implication, the budget that sustained this momentum was deemed worthy of being maintained.

The committee members also expressed their pleasure at the centralization of program decision making that had been achieved: "Headquarters should be responsible for development and approval of program objectives, strategy and plans. . . . It was clear that such a headquarters group is established and functioning well in the magnetic fusion program."[77] It was a victory for Edwin Kintner and his staff, now reorganized into the Office of Fusion Energy within the new Department of Energy. The favorable evaluation meant that the threat of a $100–200 million cut in budget had been averted.

In September 1978 John Deutch sent Schlesinger a detailed strategy for both magnetic and inertial confinement fusion built on the Foster committee's conclusions. He recommended that the magnetic fusion effort be supported at the same level in real dollars through the eighties that it now had. As a short-term goal, Deutch's strategy gave priority to the aim of demonstrating scientific feasibility in the TFTR, which Hirsch had so skillfully pushed to adoption within the fusion community 4 years before. Deutch also reiterated the Hirsch-Kintner emphasis on reactor studies and on their use as a touchstone of fusion approaches. He called for an acceleration of work on fusion technology. Experiments and computation on technological problems had been urged by Hirsch, and even more strongly by Kintner, but, because of program exigencies, had never really been well funded.

It was on the long-term plans that Deutch, like the Foster committee, deviated from the 1976 program plan. The schedule he laid out was completely different from Hirsch's. Deutch envisioned a demonstration plant in about 2015, in contrast to Hirsch's date of the late nineties, a generating capacity of 3,000–4,000 megawatts in 2020–2030, and "full commercialization" by 2050. The more leisurely timing, in turn, made possible another fundamental difference; echoing the Foster group, Deutch wanted a thorough exploration of alternative types of devices. "When choices are made, we must be confident that they are based on a firm understanding of the significance of all technical alternatives. . . . It is not our object . . . to penalize the leader because it is the leader. . . . [But] tokamaks, for all their promise, appear to possess disadvantages as reactors. . . . [This] suggests that as we try to find methods of overcoming them by improving tokamak design, we vigorously seek new schemes which, by their conceptual nature, avoid the problems altogether."[78]

There were indeed new schemes in the wings. A wide-ranging comparison of 11 conceptions had been carried out by the Washington office in 1977.[79] The most advanced was the Elmo Bumpy Torus at Oak Ridge. The name signified an ingenious juxtaposition of two concepts that had, historically, developed independently. Bumpy tori had long tantalized fusion scientists. They are a hybrid of mirror and torus in which a large number of mirrors are arranged end to end into a closed configuration. A simple mirror has a magnetic field that is weakest in the center and strongest at the two ends. A circular chain of mirrors consequently has a field that alternates periodically between strong and weak; it is this varying field that makes the torus "bumpy." Since the early sixties, it had been clear that this kind of magnetic field would provide a distinct solution, different from stellarators, tokamaks, and multipoles, to the problem of confinement in a closed tube. But it was also clear that a bumpy torus was not magnetohydrodynamically stable, for its field is, in effect, a chain of magnetic hills, whereas MHD stability requires that the reacting plasma lie in magnetic wells. The bumpy torus idea had therefore remained undeveloped.

In creating and investigating hot election plasmas in Elmo and its predecessors, however, Ray Dandl had found that these plasmas formed rotating rings, or annuli. Acting as little, internal, current coils, they set up magnetic fields of their own inside the confinement vessels, fields that oppose those of the external coils and combine with them to form magnetic well regions. Dandl had proposed using this fact as a way of overcoming the MHD instability of the bumpy torus. He suggested building the torus of Elmo mirrors.

In November 1971, Dandl, Elmo theorist Gareth Guest, and a handful of coworkers had submitted a proposal for an Elmo Bumpy Torus (EBT) of 24 mirror segments. By 1974, the EBT team had proved that a warm plasma, threading the torus through the 24 hot-electron annuli, was stable over a useful range of conditions. A few years later, the team confirmed early indications that the EBT showed less diffusion across the magnetic field lines than tokamaks.

An EBT reactor should, in principle, be a steady-state device; whether it would in fact be steady-state would depend upon the solution to the engineering problem of refueling. EBT's modular structure, as a series of iterated mirrors, would simplify construction. Its large aspect ratio, making it a long skinny torus as against the short fat tokamak, would permit easier access and make the design of the systems surrounding the plasma chamber less vexatious. The ions in the EBT then operating were cold, with temperatures in the hundreds of thousands of degrees centigrade, and the product of density and confinement time was orders of magnitude below that of the tokamak. It remained to be seen whether the plasma parameters could be improved to reactor levels. More engineering studies could well turn up new difficulties also.[80] The strategy that Deutch was now initiating meant that EBT would get a serious and sustained look.

Deutch also adopted the Foster committee's suggestion of making an experimental test facility (ETF) the benchmark to follow the TFTR. The ETF's principal function would be to test the technological components of a reactor system. Here, Kintner was in strong agreement. The Hirsch-Dean program plan of 1976 had called for two experimental power reactors (EPR) to be the next steps. It had been Hirsch's intention that the first EPR produce actual electric power. But detailed studies of a tokamak EPR in 1975 had accented the difficulties of so big a step, and Kintner had concluded the program was not ready for it.[81] The ETF was less ambitious, but would nevertheless allow not only tests of individual engineering subsystems, but also answers to the crucial question of how well the subsystems were integrated with each other and with the physical properties of the core.

Kintner had political and institutional reasons for wanting the ETF approved by fiscal year 1981. He had adopted from Rickover's nuclear submarine program the management principle that a healthy research and development effort needs a large project to serve as its backbone. The requirements of the backbone project would force the pace of the other, smaller projects in the program. It would also serve as a measure of their relevance to the overall goals. TFTR was going to be the magnetic fusion program's backbone until the late 1980s. In Kintner's view, the ETF should be in the design and construction stages early enough to take over from TFTR. He recognized that that starting date for ETF meant that it would have to be built around the tokamak. He maintained, however, that this did not entail committing the program to a tokamak. The tokamak approach would, rather, serve simply as a vehicle to carry out the recommendations of the Foster committee for putting an engineering installation into operation in the shortest possible time.[82]

The Washington management believed that the Foster committee had involved itself at one point in a contradiction. On the one hand, it had stressed the need for engineering studies and experiments and for the speedy completion of the engineering test facility. On the other hand, the group insisted that no new tokamak past the TFTR be initiated until the other concepts had the chance to catch up and to have their potentials measured against the front-runner. But, as the committee members themselves said, the TFTR was inadequate for all but the most limited of engineering tests. And no magnetic confinement concept besides the tokamak was far enough along to be the basis for an engineering installation.[83]

Deutch did not agree. He saw the ETF not only as a test facility for engineering subsystems, but as a way of probing for the best among the magnetic fusion schemes. He therefore wished to postpone the time when the program had to commit itself to an ETF until the mirror and the "alternate concepts" had had an opportunity to catch up. He recommended a decision date of 1984 or 1985, with operation of the facility in the early nineties. Indeed, Deutch held that "if at the date a tokamak ETF design seems practicable, mirror results seem very close to achieving similar or better plasma parameters, or EBT [Elmo Bumpy Torus] . . . is approaching the same [point], it may be necessary to delay the decision date for the magnetic ETF one to three years in order to guarantee that we are making the most reasonable choice—the one which, as defined in our goal, will develop the highest potential for fusion energy."[84]

Deutch conceived that one of his missions, as director of the Office of Energy Research, was to strengthen the research base of the energy technologies. An ETF would trail along with it the whole web of "big experiment" problems: basic research would be pared down to the projects most relevant to the ETF, and smaller, more peripheral projects would grow leaner. But Deutch believed that one of magnetic fusion's weaknesses was precisely that some of its existing programs were not yet strong enough. It would be sounder, he held, to put money over the next few years into strengthening these, and into extending the program's base in the universities. It was also important, in Deutch's view, that there be time to get results from the TFTR into the plans for ETF. Kintner and the staff of the Office of Fusion Energy were arguing that an ETF was important to sustain the mission orientation of the magnetic fusion program. Deutch, on the contrary, thought that too strong a mission orientation, too zealous a regard for milestones and schedules, could scuttle the endeavor to obtain the best possible energy technology from fusion.[85]

At this point, and as a kind of ironic punctuation to the new concern with "alternate concepts," the Princeton Large Torus scored a stunning success. The PLT had been running since the end of 1975 without the neutral beams, which were to drive the plasma temperatures up into new and unknown regimes. Kintner, Dean, and Chief of the Tokamak Branch N. Anne Davies urgently wanted the beams running. In their view, the dependence of confinement time upon temperature was the most crucial unknown in magnetic fusion. The MIT Alcator had demonstrated that the confinement time increased proportionally to increasing plasma densities. The scaling of time with size was also favorable, rising as the square of the plasma radius. But if the predicted trapped-particle instabilities were to manifest themselves, the variation of confinement time with temperature might be so bad as to rule out tokamak reactors entirely. Washington had selected Oak Ridge to build the neutral beams for the PLT, and it was driving both Oak Ridge and a harried Princeton to accelerate their neutral-beam work. "The PLT beam project in particular remains the single most important project in the magnetic fusion program."[86]

Then, in the summer of 1978, the complete set of four neutral beams finally was mounted. As the Princeton scientists raised the beam power and thereby moved toward the regime in which trapped-particle effects were expected, they found that they were, indeed, encountering new fluctuations. Whether these fluctuations were evidence of the instabilities that some members of the 1973 tokamak panel had feared could not be determined. What could be determined, however, was that, despite the fluctuations, the plasma temperature rose steadily as the beam power was increased. Whatever was going on, it did not include instabilities that were bad enough to prevent the temperature from climbing.[87]

PTL's power-producing capability was unprecedented. The device used deuterium bulk plasma and deuterium beams; but, had it run on a mix of deuterium and tritium, it would have yielded in fusion power 2% of the energy supplied to it by the beams.[88] The maximum ion temperature attained was nothing short of spectacular: 65 million degrees centigrade. This was an increase of 40 million degrees centigrade over the previous record tokamak temperature. It was more than 10 times the results that had started the tokamak fever in 1969, and nearly twice what Gottlieb and his associates had expected. The electron temperature was also high, over 40 million degrees centigrade. Behind the numbers were two noteworthy achievements: the powerful Oak Ridge neutral beams, delivering a total of more than 2 million watts to the plasma, and the success of the Princeton researchers in suppressing the flow of impurities into the plasma during beam injection.[89]

The PLT results did not affect the properties that engineers thought would make a tokamak reactor cumbersome and difficult to maintain.

But they dramatically raised the confidence that the fusion community felt in its ability to reach break-even in the TFTR.[90]

The rapid growth of the magnetic fusion budget through the mid-seventies had made the program an obvious candidate for a review at agency levels higher than that of the fusion division itself. The formation of the Department of Energy in 1977, under the exceptionally activist James Schlesinger, merely hastened the day on which the review began. Once it did begin, it was inevitable that the program strategy would be judged by considerations different from and broader than those of the fusion community. Two circumstances intensified this difference in outlook. First, the new department came in committed to a shift in emphasis from nuclear energy technology to solar, conservation, and coal, and from long-term concerns to the solution of short-range problems. Second, the person who took the fusion program most directly in hand, John Deutch, was only slightly acquainted with magnetic fusion, and certainly held no brief for it. Understandably, the first impulse of Schlesinger and Deutch was to divert a large proportion of fusion's funds to short-term energy programs.

What came to the rescue of magnetic fusion was its own scientific accomplishments. Above all, there was the progress of the tokamak program; although largely empirical, it was nonetheless impressive. The Alcator had shown that the crucial $n\tau$ parameter improves with increasing density and had reached $n\tau$ values more than a tenth of that needed. PLT had reached temperatures more than halfway toward reactor requirements and had indicated that confinement did not suffer at the new temperatures. At this point, there was every reason to conclude that it would be possible to go on to create reactor conditions in a tokamak plasma. To this could be added the accomplishments of mirrors. Warm-plasma stabilization had given the mirror program both basic understanding and an unprecedented ability to control microinstabilities. New and plausible schemes for enhancing the Q of mirrors existed and were about to be tested. Still other good ideas, like the Elmo Bumpy Torus, for alternative reactor schemes were at hand. An excellent program of fusion reactor systems studies was in place and was guiding research. A beginning had been made on some essential fusion technologies.

The strategy that emerged at the end of 1978 was determined by the combination of the scientific base that had been achieved and the catholic energy concerns of Deutch and his advisors. Because of fusion's successes, its annual budget was to be maintained, so that it would continue to be a major energy program. The TFTR experiment would continue to be vigorously pursued. The program would not be demoted to the status of a purely research program, but rather its engineering and technological elements would be strengthened. Because Deutch's position as director of the Office of Energy Research gave him a concern with the overall energy problem, the fusion program was to be stretched over a period far longer than that Hirsch and his colleagues had planned for in 1976. Deutch estimated that the "inexhaustible" energy sources of solar energy, fission breeders, and fusion would be needed between 2020 and 2050. The longer time span would give the fusion leaders the chance to compare the tokamak with alternative confinement devices that might present a better combination of physical and engineering properties.

13

Which Way to the Future?

Most telling, for both the environment in which magnetic fusion research has developed and the strategies its leaders have adopted, is the fact that practical fusion reactors are not yet clearly visible on the horizon. Because of this, fusion research has remained a government-supported endeavor. General Electric started a fusion program in 1956, but terminated it in 1967 because the research was getting too expensive and the GE management viewed the payoff as too uncertain and too far in the future. General Atomic, which also committed its own funds to fusion in 1957, as well as those of the Texas Atomic Energy Research Foundation, gradually shifted its program to government financing. Today, the General Atomic fusion group receives a larger annual budget from the Department of Energy than does the CTR Division of Los Alamos National Laboratory.[1] Even KMS Industries, which burst on the scene at the turn of the seventies with such exuberant confidence in inertial confinement fusion, has been financially exhausted and reduced to doing its fusion research under a Department of Energy contract.[2]

In any industry-funded research, market considerations assume vital importance. Research into semiconductor electronics furnishes an excellent example. Electronics scientists, engineers, and management alike have worked with an eye to the relation of the expense of the research to the potential profitability of the solid-state device, the possible impact of the device on the firm's financial standing, marketability—including its reliability and capital requirements—and the probable costs of the manufacturing process.[3] Government scientists have considered the market side of fusion, but they have never accorded these considerations the dominant position they have in industry. The contrast between the two habits of thought was illustrated at the time of the Model D study in 1954. The Princeton staff was not unduly worried by the size of the stellarator reactor that emerged from the study. General Electric's Willem Westendorp, on the contrary, returned to tell Kenneth Kingdon, his superior, that a stellarator would be too large and too expensive to merit industry attention.[4] The contrast was illustrated again in 1965, when General Electric's Cook committee judged that fusion was unlikely to become competitive with fission or coal-fired plants. The government-sponsored Herb-Allison committee, on the contrary, concluded that it was a mistake even to attempt to assess the costs of a technology that was at such an early stage.

Since market considerations only weakly influenced fusion program decisions, the gap was filled by a succession of political concerns. In the Strauss era, the US fusion program was pressed into the service of the goal of American prestige. After 1958 this particular pressure ceased. Space technology—and above all the race to the moon— became the vehicle for the struggle for international scientific standing, while fusion energy research metamorphosed into an arena for international cooperation. Fusion remained relatively untouched by political exigencies throughout the sixties. This changed in the seventies. Precisely because fusion research was located within the government, it was able to profit from the increasing impact of environmental and energy issues. Government had to respond, whereas industry would have been far less constrained. The dramatic increase in fusion's resources in the midseventies would simply not have occurred if magnetic fusion energy research had been conducted in industrial laboratories, no matter how politically adroit the leaders of the programs may have been.

The situation is not black and white. It is true that in a democracy government-financed science incorporates a mechanism whereby public needs can influence technical outcomes. But even in industry, where the major determinant is profitability, and sometimes distressingly short-term profitability, public concerns affect decision making to some degree via market forces and government regulations. Moreover, there is no reason to condemn out of hand the steering of technology by politics. On the contrary, I would argue that we should go still further in applying political criteria to the fusion program by adding other broad issues.

Worker-related issues have so far been absent from the discussions, among them the question of the numbers and types of jobs a fusion industry might bring in relation to the nation's employment needs. The question whether a technology as difficult as fusion will be useful outside the highly industrialized countries has hardly been raised. Another relevant issue is that of how to allocate the scarce resource of capital so as to strengthen democracy and increase social justice and the national well-being. Fusion reactors as now conceived are very demanding of capital, and fusion technology would be an appropriate vehicle for bringing the question of the optimal use of capital to the level of a public debate. The results of such a debate could, in turn, affect the technological products that fusion might bring forth.

Yet there is no doubt that the prosecution of large-scale technological projects under government aegis also brings difficulties. To get political support that can be translated into funding, program leaders are tempted to promise short-term results. But the most immediate product a technology can offer is not necessarily its best product. Again, the leaders may be tempted to do the kind of research that leads to easily understood "milestones." But the pursuit of scientific understanding often allows surer progress, even though a progress less visible to the public and the lawmakers. The dilemma is deep because the difficulties are inescapable features of a society governed by elected representatives and in which technocrats are accountable to the citizenry.

Political pressures have been one cause for the curious alternation of periods of crash program development with periods of long-range basic research that marks fusion's history. The historian can seize this phenomenon of alternation to make a historical periodization of the program by the strategy that dominated in each decade. Thus, the fifties were characterized by trial and error, the sixties by basic research, and the seventies by research more closely tied to a final, reactor product. This periodization reflects the fusion community's own view of its history, as refined by historian Gunther Kueppers. The fusion community usually talks of these periods, for the record, in terms of the evolution of the state of their science. They emphasize that trial and error was used before 1958 because the full extent of the physics problems was not yet known, that basic research was vitally necessary in the sixties to achieve an understanding of plasmas, and that the strategies shifted again in the seventies because new knowledge combined with the success of the tokamak to make possible a more result-oriented approach. Clearly, politics has to be factored into this account. Trial and error is a traditional road to invention under the press of time, and the international competition of the fifties was a mighty stimulus to it. The politics of the energy and environmental crises ranks with the tokamak and new scientific understanding in bringing about the goal orientation of the seventies. Relating research more directly to reactor goals was one of the means of building the fusion budget at a time when money for research into clean energy sources was becoming available.

Fusion is not the only technology that has taken shape under political pressure. Space science and technology, for example, were also politicized. But by the time the United States began to accelerate its space program in 1958, there was little doubt that satellites and even manned flight could be made to work. Consequently, a host of government and private interests plunged into the politics of space, among them the three armed services, the National Aeronautics and Space Administration, Presidents Eisenhower and Kennedy, Vice-President Lyndon Johnson, the communications industry, and a variety of congressional committees.[5] Success in fusion, in contrast, has never been viewed as assured. As a result, there have been far fewer groups, either government or private, involved in the politics of fusion.

The Joint Committee on Atomic Energy, which, until its demise in 1977, governed the nuclear enterprise jointly with the Atomic Energy Commission and its successor, the Energy Research and Development Administration, had always been involved.[6] Most members of Congress, however, have had little knowledge of or interest in the fusion program. Representative Mike McCormack, Democrat from Washington, is a notable exception. He championed an expansion of the fusion program almost from the time he arrived at the House in 1971 until he was defeated for reelection in 1980. Fusion formed part of his overarching strategy for solving the energy problem by aggressively increasing energy supplies.

It was McCormack, aided by the fusion community's lobbying, who engineered the passage through Congress of the Magnetic Fusion Energy Engineering Act of 1980. This bill attempted an end-run around Deutch's fusion policy by providing that a "Fusion Engineering Device," a somewhat less demanding version of the Foster committee's Experimental Test Facility, be in operation by 1990. It also called for a reversion to the Hirsch-Dean timetables and budgets by requiring a demonstration fusion plant to be built by about 2000 and a doubling of the fusion research budget, to about $1 million annually, by 1987. The bill was powerless to enforce these goals, however, because the actual authority to change the budget lay with the House and Senate appropriations committees.

In light of the fact that the bill committed Congress to nothing but a sentiment, the circumstance that it passed with an overwhelming majority and almost no debate really testifies to the lack of interest within Congress in the politics of fusion. A handful of pork-barrel issues surfaced, involving the states in which major facilities will be sited. The manufacturers of electrical generating equipment lobbied through the Atomic Industrial Forum, a nuclear industry association, to have more government research contracts placed with industrial, as opposed to government, laboratories. But by and large there were simply too few economic or political issues associated with fusion to give rise to conflicts.[7]

What is true of Congress is equally true of the presidential administrations that have dealt with fusion. Unlike space technology and fission reactor technology, fusion has never risen to the level of presidential decision making. Even at the top administrative levels of the AEC, fusion ceased to be a central interest once Lewis Strauss stepped down in 1958. The military has been interested in inertial confinement fusion, and there has been some pulling and hauling within the Atomic Energy Commission between the Division of Military Application and the Division of Research over this part of fusion. But magnetic confinement fusion research, the subject of this book, has only sporadically attracted military attention.

Nor have university scientists been heavily involved in fusion. Here, it is instructive to compare the program with research in high-energy physics, the branch of science that explores subatomic phenomena by means of gigantic and expensive particle accelerators. High-energy physics has also been supported by government funds, and, because it is fundamental research, it is of even less interest to industry than fusion has been. But the greatest political power within the high-energy physics community has always been wielded by university scientists. The important installations are at Berkeley, Stanford, Brookhaven National Laboratory, which is run by a consortium of eastern universities, and Fermi National Accelerator Laboratory, which was also organized by academic scientists.[8]

In contrast, fusion work involved very few university scientists prior to 1958 because of the secret status of the research. Declassification would have made university research possible; by the early sixties, however, the budget had become so constricted that the laboratory leaders were understandably wary of funding any new group, either industrial or academic, in their struggle to keep their own programs alive. In 1971 fully 85% of the Atomic Energy Commission's magnetic fusion budget still went to the four government laboratories at Livermore, Oak Ridge, Princeton, and Los Alamos, with the remaining 15% spread over about 50 projects at universities, industrial facilities, and the other national laboratories.[9]

The upshot is that whereas solid-state electronics research has been carried out under the commercial pressures of a highly competitive industry, space research has been influenced by a large array of government and private interests, and high-energy physics has reflected the needs of university-based scientists; magnetic fusion research bears above all the imprint of the government laboratories. The laboratories effectively ran the program for its first two decades, albeit in an increasingly coordinated manner after Amasa Bishop took over leadership. Only in the seventies did the laboratories begin to lose power to a steadily more autonomous Washington management.

The laboratories, for their part, have resorted to a handful of characteristic political strategies. Their leaders have repeatedly sought projects that would be unique within the national program. Herbert York, in 1952, gave the new Livermore laboratory the task of exploring a confinement scheme whose open-ended shape would clearly mark it off from Princeton's stellarator and Los Alamos's Perhapsatron. Herman Postma, in 1968, had Oak Ridge look into the tokamak at a moment when only one American laboratory, General Atomic, was in the tokamak business. The Astron's uniqueness was always a potent argument in its defense.

The search for uniqueness is, of course, ubiquitous. Industrial research groups seek to find products that will give their companies a special place within the market. Academic scientists often prefer to stake out research areas for which they have outstanding capabilities, or into which other teams have not yet moved. They are the more drawn to this strategy in that academic teams engaged in large-scale research in the same field and within the same country are likely to be competing for funds from the same government agencies.[10] In the fusion program, the pressures driving laboratory leaders toward uniqueness were even less subtle. Their budgets came straight from a central management overtly, and increasingly, concerned with avoiding duplication.

The leaders also sought to define for their laboratories goals that were central to the national program. In 1957, when the immediate national goal was thermonuclear neutrons, Oak Ridge's leaders reoriented their program around the DCX, which at the time seemed capable of achieving neutrons. This move also served another recurrent strategy: to defend a program by structuring it around a large-scale project representing a substantial financial investment. Similarly, in promoting Scyllac in the midsixties, the Los Alamos leadership was conscious of the importance to the national program of devices that could extrapolate into practical reactors, and it was conscious of the strength the laboratory might gain from a strong backbone project. Oak Ridge and Princeton both struggled, as yet another example, to get the tokamak feasibility experiment, which was clearly going to be the focal experiment of the late seventies and early eighties.

Still another recurrent tactic has been to choose an experiment that, in addition to its scientific merits, would maintain or build the laboratory's capabilities for further research. For example, Shoichi Yoshikawa, one of the scientists who gave direction to Princeton's program in the late sixties, proposed and steered the laboratory's FM-1 multipole. Yoshikawa thought it was vital to compare the multipole's confinement with the stellarator's, but he also wanted to maintain Princeton's engineering capabilities by keeping its engineering group supplied with major projects.[11]

Within this politicized framework, substantial progress has nonetheless been made. Fusion scientists have invented a succession of magnetic bottles—pinch and θ pinch, stellarator, mirror, cusp, tokamak, multipole, Scyllac, Elmo Bumpy Torus, and still others for which there was no space in this book. They have improved them with helical windings, magnetic wells, feedback systems and other devices, to eliminate plasma macroinstabilities, reduce microinstabilities, and adjust equilibria. They have created a variety of methods for placing plasma within the bottles, from Livermore's early titanium washer guns, to the ingenious plasma gun invented in the late fifties by Los Alamos scientist John Marshall, Jr., to the recent use of particle beams on a warm plasma stream in the mirror experiments. When the fusion program began, ion temperatures of a few thousand degrees centigrade could be created in the laboratory. Now, in 1982, temperatures have been raised to tens and hundreds of millions of degrees. Heating has been accomplished by manipulating magnetic fields to compress the plasma, by the high-energy particle beams of Oak Ridge and Lawrence Berkeley Laboratory, and by guiding in power from electromagnetic waves, as in Thomas Stix's ion cyclotron resonance heating or Ray Dandl's hot-electron experiments. It has been aided by great strides toward the elimination of impurities, taken by improving vacuum systems, gettering, and the careful choice of materials. Early methods of measuring plasma properties, such as photographing the luminous plasma or interpreting impurity spectra, have been augmented by a battery of new methods, among which are the analysis of the soft x rays and neutral particles emitted by hot plasmas, the use of scattered laser beams to measure temperatures and densities, and laser holography.[12] Reactor studies have advanced from the pathbreaking, but necessarily broad-brush early treatments to far more detailed, and hence more realistic treatments. The reactor engineers are not yet at the stage of laying down specifications from which a plant might be built. This could not be done today since, for example, materials hardy enough for the first wall of a fusion reactor have not yet been identified. The engineers are at a point, however, where they can expose problems and set conditions on any future plant. Thus, aluminum alloys were suggested as structural materials in the fifties because of their low induced radioactivity,[13] but detailed studies have now shown that aluminum alloys have a high cross section for the production, under neutron bombardment, of helium gas, which remains in the alloy and embrittles it. For another example, experimental tests under temperature cycles simulating the effects of fast-pulsed reactors have enabled engineers to put a floor under the allowable duration of the pulse, a floor based on the cyclic stress that known materials can stand.

Fusion theory cannot yield detailed predictions of what will happen when plasma is brought into new regimes, including the crucial and thus far unattained regime of burning plasmas.[14] Fusion theorists, however, have created a new field—that of collective phenomena in hot, fully ionized, confined plasmas—and have used their results to identify experimental areas for fruitful investigation and to explain the phenomena that have been discovered. On the experimental side, the situation that Artsimovich discerned in 1958, of "a display of ideas . . . only thinly draped with rough and insufficiently verified experimental data" has been replaced by one marked by numerous well-secured results.

But how far has the progress in physics and technology carried us toward a practical fusion reactor? At the very start of the program, the measure of progress was temperature. A threshold for power production, the ideal ignition temperature, was defined as the temperature at which the thermonuclear energy generated was just adequate to balance the *Bremsstrahlung* radiation energy escaping from the plasma. Tuck estimated the threshold temperature for deuterium-tritium fuel, in 1952, as about 50 million degrees centigrade.[15] Progress toward reactors, for a time, was charted in terms of progress toward that temperature.

That this measure was insufficient was clear to everyone by the late fifties. Good confinement, which in turn depended upon stability, was essential. The quality of confinement, measured by the product of density n and confinement time τ was added as a standard by which to judge advance. The finish line of this $n\tau$ derby was the Lawson criterion of $n\tau = 10^{14}$ cm^{-3} sec. In 1968 the tokamak excited the community because its quality of confinement approached 10^{12} while its electron temperature exceeded 10 million degrees centigrade. To extrapolate to a continuing rise of $n\tau$ in tokamaks of the future was an act of faith. There was a real possibility that as temperature was brought still higher, confinement time would collapse. That did not happen, however, and it now (in 1982) appears probable that the TFTR, and comparable foreign machines, will cross the finish line in the 1980s, that is, demonstrate a combination of temperature and $n\tau$ sufficient to give break-even with as much power generated as consumed.

What is next? Beyond the proof of scientific feasibility lies the probing of the engineering feasibility of fusion. There is no consensus yet that it can be proved. Opinion in 1981 runs the same gamut from optimism to pessimism that characterized opinion on scientific feasibility at the Sherwood meeting in Denver in 1952. Optimists point to the way in which one physics problem after another has been solved. Even the stellarator, abandoned in 1969 in the United States but steadily pursued abroad, has finally transcended Bohm diffusion and reemerged as a leading reactor concept. The optimists project an analogous history for the engineering side of fusion. Pessimists point to the formidable nature of the engineering difficulties and the fact that not a single confinement scheme has yet been demonstrated to be technically possible.

When, and if, a technology of fusion does emerge sometime in the next century, the program will face still another question: Will fusion find a niche, in whatever mix of energy sources then obtains, in which it can be economically competitive and environmentally and politically acceptable?

Appendix A   A Glossary of Magnetic Fusion Terms
Prepared by R. S. Granetz

Ambipolar potential
In a magnetically confined plasma, the electrons, which are lighter and faster than the ions, tend to leak out first. This leaves behind a net positive charge in the plasma, giving rise to an "ambipolar potential," which can cause certain types of instabilities.

Ballooning modes
An instability caused by pressure differences that is localized to regions where the magnetic field is curved toward the plasma.

Beta (β)
The ratio of plasma gas pressure to magnetic field pressure (or magnetic energy density). $\beta = 8\pi(n_i kT_i + n_e kT_e)/B^2$ where n_i, n_e are the ion and electron densities, T_i, T_e are the ion and electron temperatures, k is the Boltzmann constant, and B is the magnetic field strength.

Blanket
Structure surrounding the vacuum wall of a fusion reactor whose principal function is to convert the energy released by fusion reactions into heat for use in generating electrical power.

Bohm diffusion
Particle diffusion that exhibits the anomalously poor confinement observed in many early plasma experiments. $D_{Bohm} = (1/16)(kT/eB)$, where e is the electron charge, B the magnetic field, T the absolute temperature, and k is the Boltzmann constant.

Break-even
The point at which the fusion power released in a reactor equals the total input power used to sustain the plasma temperature.

Burnout
The point at which the rate of conversion of neutral particles to ions is exactly equal to the rate of neutral particle influx.

Charge-exchange loss
Loss of hot plasma ions throught the process of recombination with electrons from cold neutral atoms. The resulting hot neutrals are no longer confined by the magnetic field.

Closed systems (toroidal systems)
Magnetic configurations in which the field lines do not leave the vacuum chamber. Examples are the tokamak and the stellarator.

Collective (cooperative) phenomena
A basic characteristic of fusion plasmas in which the motion of plasma particles depends on electric and magnetic fields generated by other particles throughout the plasma volume.

Confinement
Restraint of plasma within a designated volume, especially by electric and/or magnetic fields. Energy confinement time τ_E is a measure of the rate at which energy is lost from the plasma, while particle confinement time τ_p is a measure of the rate at which charged particles are lost.

Cusp geometry
Magnetic configuration in which field lines everywhere curve away from the plasma.

Cyclotron heating
A method of heating charged particles in a magnetic field with radio or microwave power. The radio frequency is synchronized to the rate at which the ions (for ion cyclotron resonant heating) or electrons (for electron cyclotron resonant heating) circle magnetic field lines.

Deuterium
A naturally occurring isotope of hydrogen having a nucleus composed of one proton and one neutron. In a sample of normal hydrogen gas, about 1 atom in 6,000 will be deuterium.

Diagnostics
Experiments that measure basic plasma parameters such as density, temperature, and impurity concentrations.

Direct conversion
Conversion of charged particle kinetic energy directly into electrical energy, bypassing all thermal processes.

Discharge (electric discharge, gas discharge)
An electric current in a gas. In order to conduct current, the gas must be ionized to form a plasma.

Drift-cyclotron loss-cone (DCLC) mode

A microinstability characteristic of magnetic mirrors. It is caused by the deficit of particles with large values of $v_\parallel/v_\perp$, where $v_\parallel$ is the particle velocity parallel to the magnetic field and $v_\perp$ is the velocity perpendicular to the field.

E layer

An integral part of the Astron concept. The E layer is a cylindrical current sheet of high-energy electrons inside a chamber enclosed by a solenoidal coil. A sufficiently dense E layer would generate field reversal.

Feedback stabilization

A method of suppressing instabilities by actively applying electric and/or magnetic fields in proper phase to reduce the perturbations.

Field reversal

A magnetic configuration in which the axial magnetic field in the outer regions of a volume points in the opposite direction to the field near the center (see figure 7.1).

Finite-orbit stabilization

Theoretically predicted stabilization of certain modes (instabilities) when the finite size of particle orbits of gyration is included in the equations used to model a plasma.

Flute instability

An MHD instability caused by a pressure gradient in conjunction with magnetic field curvature. The mode causes the plasma to look like a fluted column.

Gyroradius

In a magnetic field, charged particles gyrate around the field lines in circular orbits having a characteristic radius, called the gyroradius, which depends on the field intensity.

High-β stellarator

A magnetic confinement device formed by bending a high-β ($\beta \approx 1$) θ pinch into a large-radius torus and adding several helical conductors to provide an equilibrium configuration. The Scyllac experiment was an example of such a device.

Hybrids

Reactors that combine functions of both fusion and fission devices. For example, energetic fusion-generated neutrons can be used to produce fission fuels.

Ideal ignition temperature

The temperature at which the fusion power produced in a plasma equals the power lost by the electrons through *Bremsstrahlung* emission (radiation arising from deceleration by coulomb collisions).

Ignition
The point at which a fusion plasma becomes sufficiently self-heated by absorbing the energy of fusion-produced α particles. Once ignition is reached, no external source of power is needed to maintain the plasma temperature.

Independent-particle model
A method of calculating the motion of charged particles in a magnetic field, assuming that interactions between particles can be ignored.

Ioffe bars
A set of current-carrying conductors added to simple magnetic mirror coils in order to make a magnetic well. This enhances plasma stability against flute modes.

Ion
The positively charged residue of an atom that remains after one or more of its electrons have been stripped away.

keV (kiloelectron volts)
A unit of energy commonly used in plasma physics to denote temperature. A particle having a "temperature" of $\sim 11,600°K$ has an energy of 1 eV. A keV is 1,000 eV, or a "temperature" of ~ 11.6 million degrees (°K refers to the Kelvin, or absolute, temperature scale. Temperature in °K = temperature in °C + 273).

Kink instability
An MHD mode caused primarily by large parallel currents in the plasma.

Kruskal-Shafranov limit
A theoretical limit on the current that can be carried by a plasma in order to avoid the worst kink instability.

Lawson criterion
The minimum value of the product of confinement time and density required to have the fusion power offset the plasma losses (break-even). For fusion plasmas having a temperature of 10 keV, the Lawson criterion stipulates that $n\tau > 10^{14}$ cm^{-3} sec.

Line-tying
The shorting together of magnetic field lines at the ends of a linear device by terminating them on a conducting endplate. This allows charged particles to move from one field line to another by traveling along the magnetic lines of force and across the conducting endplate.

Magnetic compression
A method of heating a plasma by increasing the magnetic field strength. If it is done at the proper rate, the plasma will behave like a gas and will be adiabatically compressed and heated.

Magnetic pumping
Method of heating a plasma by alternately compressing and expanding it at certain characteristic frequencies. This can be done by applying a rapidly oscillating magnetic field.

Magnetic shear
A magnetic configuration in which field lines at different distances from the axis have different amounts of twist ("rotational transform") is said to have magnetic shear. In the stellarator, magnetic shear is produced by external, helical coils.

Magnetic well
When the magnetic field increases in every direction outward from a region, that region is in a magnetic well. A magnetic well helps to stabilize certain MHD modes.

Magnetohydrodynamics (MHD)
A model, that is, a set of theoretical equations, describing a perfectly conducting fluid in a magnetic field applicable to the study of hot plasmas. It is a combination of fluid equations and Maxwell's electromagnetic equations.

Microinstabilities
Instabilities that cannot be treated theoretically by the MHD fluid model. Instead, the particulate properties of the plasma must be taken into consideration and kinetic theory must be employed.

Minimum-*B* configuration
Same as "magnetic well."

Neutral beams
High-energy beams of neutral atoms that can penetrate the magnetic field surrounding a fusion reactor and then deposit their energy in the plasma. Neutral beams are therefore useful as an auxiliary heating technique.

Nonlinear theory and phenomena
The study of large-amplitude instabilities that cannot be treated theoretically as small perturbations in the equilibrium.

Ohmic heating
The heating that occurs when an electric current is passed through a resistive medium. It is also called Joule heating.

Open systems
Magnetic configurations in which field lines leave the system. The magnetic mirror is a familiar example.

Plasma
A gas composed of approximately equal numbers of positively and negatively charged particles and a variable fraction of neutral atoms. The charged particles can interact among themselves and with externally applied electric and magnetic fields.

Power balance
The detailed accounting of all the power put into, taken out of, and lost from a system.

Pulse stacking
A technique used to build up the E layer in Astron. Short bursts of relativistic electrons were repeatedly and rapidly injected into the device.

Pumpout
The universally observed and poorly understood decay of plasma density that occurred in early toroidal devices.

Q
The ratio of power produced by a reactor to the power needed to keep it operating. For $Q > 1$ the reactor produces net energy.

Quality of confinement ($n\tau$)
The product of plasma density times the energy confinement time; measured in units of density (particles per cubic centimeter) times units of time (seconds), abbreviated cm^{-3} sec. To achieve break-even this quantity must surpass a critical value (*see* Lawson criterion).

Runaway electrons
Those electrons that have such a large parallel velocity that collisional processes do not slow them down. Consequently they continue to accelerate to very high speeds.

Scaling
The trend in the behavior of a certain parameter as other parameters are varied.

Shielding
A structure interposed to protect reactor components, such as magnetic field coils, from the flux of fusion-produced energetic particles.

Shock heating
A technique for heating with waves moving faster than the speed of sound in the plasma.

Superconductor
A material that becomes a perfect conductor of electric current when cooled to temperatures near absolute zero.

Synchrotron radiation
Electromagnetic energy radiated by charged particles gyrating in circular orbits around magnetic field lines.

Temperature
A measure of the average energy of a distribution of particles. In short-lived plasmas, the ion and electron temperatures usually differ because of insufficient interaction between the two species.

Thermonuclear
Sufficient fusion reactions for net power can only be produced by a distribution of high-energy ions in thermal equilibrium. This precludes the use of high-energy particle beams, for example, as an economical fusion reactor design.

Trapped-particle instabilities
Instabilities due to trapping of a certain fraction of the plasma particles in the magnetic well that exists on the outer side of a toroidal magnetic field.

Tritium
A heavy isotope of hydrogen consisting of one proton and two neutrons. It is radioactive with a half-life of 12.3 years. The fusion reaction between deuterium and tritium is the least difficult one to achieve.

Turbulent heating
A technique for heating a plasma by quickly inducing currents in it. This generates turbulent (that is, random) fluctuations in electron motion and results in anomalously high resistivity and hence anomalously large ohmic heating.

Two-component tokamak
A tokamak consisting of an ohmically heated plasma that is below the ignition temperature and a population of very high energy particles injected by neutral beams. These beam particles produce an extra quantity of fusion reactions in addition to the background thermonuclear yield and therefore lower the Lawson criterion.

Vacuum wall
A chamber (usually metal) that surrounds the plasma. It is evacuated in order to allow formation of the plasma discharge.

Date of first operation	Princeton	Los Alamos	Livermore and Berkeley	Oak Ridge	Other laboratories
1952		Perhapsatron			
1953	Model A		Table Top		
1954	Model B (renamed B-1)	Columbus I	Toy Top		
1955			Berkeley Linear Dynamic Pinch		
1956	Model B-2	Columbus II, Columbus S-4	Gamma		
1957		Perhapsatron S-3, Scylla		DCX (later DCX-1)	ZETA (UK), θ pinch (Naval Research Laboratory)
1958	Model B-3				
1959		Picket Fence			
1960			Levitron, Toy Top III (multistage compression), ALICE		
1961	Model C				
1962			2X	DCX-2	
1963		Scylla IV	Astron— accelerator		

Appendix B (continued)

Date of first operation	Princeton	Los Alamos	Livermore and Berkeley	Oak Ridge	Other laboratories
1964			Astron—E layer		
1965				INTEREM	Octupole (Wisconsin), Octopole (General Atomic)
1966		Fast linear z pinch	Baseball I	Elmo	
1967					
1968					
1969				IMP	Doublet-I (General Atomic)
1970	Symmetric Tokamak				
1971	FM-1 (multipole)	ZT-1 (toroidal z pinch), Scyllac toroidal sector	Baseball II, 2XII, Superconducting Levitron	Ormak	Doublet-II (General Atomic)
1972	Adiabatic Toroidal Compressor (ATC)	Scyllac linear experiment			Alcator (MIT; renamed Alcator A)
1973				Elmo Bumpy Torus	
1974		Scyllac—full torus			
1975	Princeton Large Torus (PLT)		2XIIB		

a. The chronology lists only experiments discussed in the text and gives the first year of operation.

Abbreviations in Notes

Berchtesgaden Conf.
Plasma Physics and Controlled Nuclear Fusion Research, 1976, Proceedings of the Sixth International Conference Held by the IAEA at Berchtesgaden, 6–13 October 1976, International Atomic Energy Agency, Vienna, 1977

Culham Conf.
Plasma Physics and Controlled Nuclear Fusion Research, Proceedings of a Conference Held by the IAEA at Culham, 6–10 September 1965. International Atomic Energy Agency, Vienna, 1966

DOE-HO
Archives of the Department of Energy, maintained in the Historian's Office

DOE-OMFE
Records under the jurisdiction of the Office of Magnetic Fusion Energy at the Department of Energy. These include records of magnetic fusion kept by the Division of Research (to 1971), the Division of Controlled Thermonuclear Research (1971–1976), and the Division of Magnetic Fusion Energy (1976–1977)

Geneva Conf.
Proceedings of the Second United Nations International Conference on the Peaceful Uses of Atomic Energy, Held in Geneva, 1 September–13 September 1958, United Nations, Geneva, 1958: Vol. 31, "Theoretical and Experimental Aspects of Controlled Nuclear Fusion"; Vol. 32, "Controlled Fusion Devices"

Innsbruck Conf.
Plasma Physics and Controlled Nuclear Fusion Research, 1978, Proceedings of the Seventh International Conference Held by the IAEA at Innsbruck, 23–30 August 1978, International Atomic Energy Agency, Vienna, 1979

JCAE
Joint Committee on Atomic Energy

JCAE, The Current Status of the Thermonuclear Research Program, 1971
Hearings . . . Subcommittee on Research, Development and Radiation of the Joint Committee on Atomic Energy, 92nd Congress, 1st session, The Current Status of the Thermonuclear Research Program, November 10, 11, 1971

LASL
Los Alamos Scientific Laboratory (presently Los Alamos National Laboratory)

LASL-RC
Records Center of the Los Alamos National Laboratory

LLNL
Lawrence Livermore National Laboratory (formerly Lawrence Livermore Laboratory)

Madison Conf.
Plasma Physics and Controlled Nuclear Fusion, 1971, Proceedings of the Fourth International Conference, Held by the IAEA at Madison, USA, 17–23 June 1971, International Atomic Energy Agency, Vienna, 1971

Novosibirsk Conf.
Plasma Physics and Controlled Nuclear Fusion Research, Proceedings of the Third International Conference held by the IAEA at Novosibirsk, 1–7 August 1968, International Atomic Energy Agency, Vienna, 1969

ORNL-CF
Central Files of the Oak Ridge National Laboratory

PPPL
Princeton Plasma Physics Laboratory

Salzburg Conf.
Proceedings of the Conference on Plasma Physics and Controlled Nuclear Fusion Research, 4–9 September 1961, Salzburg, Austria, (published as a 1962 supplement to *Nuclear Fusion*)

Sherwood Conf., 1952
"Classified Conference Held at Denver on June 28, 1952," WASH-115

Sherwood Conf., 1953
"Conference on Thermonuclear Reactions, Radiation Laboratory, University of California, Berkeley, California, April 7, 1953," WASH-146

Sherwood Conf., 1954
"Conference on Thermonuclear Reactions, Princeton University, October 26 and 27, 1954," WASH-184

Sherwood Conf., February 1955
"Conference on Thermonuclear Reactions, University of California Radiation Laboratory, Livermore, California, February 7, 8 and 9, 1955," WASH 289

Sherwood Conf., October 1955
"Conference on Controlled Thermonuclear Reactions, Held at Princeton University, October 17–30, 1955," TID-7502

Sherwood Conf., 1956
"Conference on Controlled Thermonuclear Reactions, June 4–7, 1956, Gatlinburg, Tennessee," TID-7520

Sherwood Conf., 1957
"Controlled Thermonuclear Reactions: A Conference Held at Berkeley, California, February 20–23, 1957," TID-7536

Sherwood Conf., 1958
"Papers Presented at the Controlled Thermonuclear Conference Held at Washington, D.C., February 3–5, 1958," TID-7558

Spitzer Papers
Project Matterhorn and Plasma Physics Laboratory Papers of Lyman Spitzer, Jr., in the Princeton University Archives

Strauss Papers
Papers of Admiral Lewis L. Strauss, presently in the possession of his family but to be located at the Herbert Hoover Presidential Library, West Branch, Iowa

Tokyo Conf.
Plasma Physics and Controlled Nuclear Fusion Research, Proceedings of the 5th International Conference, Held by the IAEA in Tokyo, 11–15 November 1974, International Atomic Energy Agency, Vienna 1975

Interviewees and Interview Dates in Notes

Don A. Baker	7/24/79	Gareth E. Guest	11/20/78
Thomas Batzer	6/28/78	Robert L. Hirsch	8/30/79
Lee A. Berry	2/07/79	Henry Hurwitz, Jr. with	
Edward Bettis	2/06/79	George Wise	4/06/78
Keith Boyer	7/79	Torkild Jensen	8/79
Norris E. Bradbury	7/25/79	John L. Johnson	11/30/78
Solomon J. Buchsbaum	9/20/79	Milton D. Johnson	6/21/79
James D. Callen	2/06/79	George Kelley	2/06/79
Gus Carlson	6/28/78	Edwin L. Kemp	7/12/79
John F. Clarke	12/13/79	Donald W. Kerst	10/25/78
Frederic H. Coensgen	6/28/78	Edwin E. Kintner	10/22/79
Stirling A. Colgate	7/23/79	Alan C. Kolb	8/02/79
Robert W. Conn	10/26/78	Robert Krakowski	7/13/79
Bruno Coppi	4/30/80	Gerald L. Kulcinski	10/27/78
Ray A. Dandl	8/03/79	Wolf Kunkel	11/13/78
N. Anne Davies	4/25/80	John D. Lawson	11/16/79
Stephen O. Dean	6/04/79	John Marshall, Jr.	7/16/79
Enzi De Renzis	8/29/80	Robert G. Mills	9/19/79
John M. Deutch	9/22/80	D. Bruce Montgomery	3/80
Julian L. Dunlap	2/06/79	O. B. Morgan	2/07/79
William R. Ellis	10/25/79	Tihiro Ohkawa	8/02/79
Spofford G. English	4/13/80	Ronald R. Parker	3/08/80
T. Kenneth Fowler	6/30/78	James A. Phillips	7/23/79
François N. Frenkiel	12/29/78	Richard F. Post	6/27 & 6/28/78
Harold P. Furth	12/01 & 12/15/78	Herman Postma	5/15/80
Melvin B. Gottlieb	11/30/78	James T. Ramey	9/15/80
William C. Gough	1/06/79	Fred L. Ribe	7/25/79
Roy W. Gould	6/14/79	William L. Rice with	
Harold Grad	4/09/79	George K. Hess, Jr.	11/30/77
Hans R. Griem	12/04/80	Werner B. Riesenfeld	7/16/79
Don J. Grove	12/01/78	Michael Roberts	3/05/79
		David J. Rose	3/05 & 3/07/80

Marshall N. Rosenbluth	9/21/79
Norman Rostoker	8/02/79
Arthur E. Ruark	2/28 & 3/01/79
Jorge Sabato	11/17/78
George A. Sawyer	7/12/79
F. Robert Scott with George K. Hess, Jr.	12/01/77
Robert C. Seamans, Jr.	3/07/80
Louis D. Smullin	3/06/79
Arthur H. Snell	2/07/79
Lyman Spitzer, Jr.	3/15/78
Julian C. Sprott	10/25/78
Don Steiner	2/05/79
Thomas H. Stix	12/01/78
Wolfgang Stodiek	12/01/78
Lewis H. Strauss	6/10/80
Richard F. Taschek	7/06/79
John M. Teem	1/06/80
James L. Tuck	7/27/79
Alan A. Ware	10/16/78
Richard W. Werner	4/17/79
Willem Westendorp	4/06/78
James M. Williams	6/18/79
Herbert F. York	8/20/78
Shoichi Yoshikawa	9/19/79
J. Ronald Young	6/21/79

Notes

Each citation to an unpublished item names the files or archival collection in which I found the item. I have also deposited xeroxes of documents that I judged hard to obtain at the Center for the History of Physics, American Institute of Physics, 335 East 45th Street, New York, New York. Tapes and notes of my interviews are included in this collection, as are the tapes of the May 14, 1981, meeting of fusion leaders cited in chapter 10. In general, neither reports issued by the laboratories nor items in the archives of the Historian's Office of the Department of Energy are included in the collection at the center. Access to the DOE archives can be arranged by contacting the department's archivist.

A few classes of documents are cited so frequently that I give their provenance here rather than in the notes. All minutes of the Steering Committee are taken from the Lyman Spitzer, Jr., papers at the Princeton University Archives. All minutes of the Standing Committee are taken from the files of the Magnetic Fusion Energy Division of Lawrence Livermore National Laboratory. All minutes of Atomic Energy Commission meetings are taken from the archives of the DOE's Historian's Office, as are all documents marked with an "AEC" followed by a number.

The two preceding lists give the abbreviations used in the notes and the names of those people whose interviews I cite, together with the date of our conversation. I extend here my deep thanks both to them and to the many others who are not cited, but whose remarks provided equally valuable information.

Chapter 1

1. See Steven Goldberg, "Controlling Basic Science: The Case of Nuclear Fusion," *Georgetown Law Journal* 68 (1980), 683–725, especially 702–704.

2. The program in inertial confinement reactors occupies a peripheral place in this book, as has been explained in the preface. For this reason, the bare noun "fusion" is often used where magnetic confinement fusion is meant.

3. Division of Magnetic Fusion Energy, "Fusion Power by Magnetic Confinement: Program Plan," Vol. I, "Summary," July 1976, ERDA-76/110/1; John M. Deutch to Mike McCormack, Sept. 20, 1979, DOE-OMFE.

4. G. L. Kulcinski, G. Kessler, J. Holdren, and W. Haefele, "Energy for the Long Run: Fission or Fusion," *American Scientist* 67 (1979), 78–89. My figure of one tenth is a great oversimplification. The comparison depends upon the length of time after plant shutdown and the hazard measured.

5. There is a large literature in the history of science and technology, the sociology of science, and the politics of science that discusses this directly or tangentially. See, for example, Spencer R. Weart, *Scientists in Power*, Cambridge, Mass., Harvard, 1979; R. Cargill Hall, *Lunar Impact*, Washington, D.C., NASA, 1977; Günther Küppers, "Fusionsforschung: Zur Zielorientierung im Bereich der Grundlagenforschung" in W. van den Daele, Wolfgang Krohn, Peter Weingart, eds., *Geplante Forschung*, Frankfurt, Suhrkamp, 1979, 287–327.

6. The question of how the citizenry can intervene to influence whether fusion should be pursued, and what final shape it will have is the central theme of S. Goldberg, "Controlling Basic Science."

7. The word "plasma" for a partially ionized gas had been coined at the General Electric Research Laboratory in the early 1920s by Irving Langmuir. Langmuir used the word to point out the similarity between the variety of particle types in a gas through which an electric discharge is occurring and the diversity of particle types carried by blood plasma. Memoir by Harold M. Mott-Smith, 1922, General Electric Research Laboratory Archives (courtesy of George Wise).

8. For example, Albert F. Siepert, "Creating the Management Climate for Effective Research in Government Laboratories," in Karl B. Hill, ed., *The Management of Scientists*, Boston, Beacon Press, 1963, 89; Adam Yarmolinsky, *The Military Establishment*, New York, Harper & Row, 1971, 378–394.

9. For references to sociological literature on competition among research teams, see Michael J. Mulkay, "Sociology of the Scientific Research Community," in Ina Spiegel-Rösing and Derek de S. Price, *Science, Technology and Society*, London, Sage Publications, 1977, 117–119. For the national laboratories as institutions, see Albert H. Teich, "Bureaucracy and Politics in Big Science: Relations between Headquarters and the National Laboratories in AEC and ERDA," U.S. House of Representatives, Committee on Science and Technology, *The Role of the National Energy Laboratories in ERDA and Department of Energy Operations: Retrospect and Prospect* (95 Congress, 2nd session), Jan. 1978, 353–392; Albert H. Teich and W. Henry Lambricht, "The Redirection of a Large National Laboratory," *Minerva 14* (1976–77), 447–474.

10. Another group of engineers was involved from the very start designing the hardware for the experiments. Their contribution to the program's achievements is substantial. They are hardly treated in this book, however, because they were by and large not strategy-makers, but fulfilled service roles under the direction of physicists.

11. See, for example, Richard G. Hewlett and Francis Duncan, *Nuclear Navy*, Chicago, University of Chicago Press, 1974; Harvey M. Sapolsky, *The Polaris System Development: Bureaucratic and Programmatic Success in Government*, Cambridge, Mass., Harvard, 1972; Hall, *Lunar Impact*; W. Henry Lambricht, *Governing Science and Technology*, New York, Oxford, 1976.

12. Deuterium can be separated from seawater but tritium must be manufactured from lithium. See Kulcinski et al., "Energy for the Long Run," 79.

13. Harold P. Furth and Richard F. Post, "Advanced Research in Controlled Fusion," University of California Radiation Laboratory Report UCRL-12234, December 10, 1964, 2.

14. Scientists at Princeton University and New York University's Courant Institute were in the forefront of this advance.

15. Furth and Post, "Advanced Research," 5.

16. Writing in 1979, British theorist John A. Wesson concluded that the role of theory in improving the tokamak over the decade of the seventies had been "quite small." J. A. Wesson, "The Theoretical Problem in Fusion Research," lecture to the Theoretical Plasma Physics Meeting of the Institute of Physics, April 4, 1979, preprint (courtesy of J. A. Wesson).

Chapter 2

1. *New York Times*, March 25, 1951, 1, and March 27, 8; "Abstract from Richter," filed with memorandum by A. S. Bishop, July 20, 1956, folder TR1956, LASL-RC; author's conversation with Jorge Sabato.

2. Oral history interview with Lyman Spitzer, Jr., by David DeVorkin, 1977, Sources for History of Modern Astrophysics, Center for History of Physics, American Institute of Physics.

3. Hannes Alfven, *Cosmical Electrodynamics*, Oxford, Clarendon, 1950; Lyman Spitzer Jr., "The Magnetic Field of the Galaxy," *Physical Review 70* (1945), 777–778; (with J. W. Tukey) "Interstellar Polarization, Galactic Magnetic Fields, and Ferromagnetism," *Science 109* (1949), 461–462; "On the Origin of Heavy Cosmic-Ray Particles," *Physical Review 76* (1949), 583; "Equations of Motion for the Ideal Plasma," *Astrophysical Journal 116* (1952), 299.

4. Richard G. Hewlett and Francis Duncan, *Atomic Shield*, University Park, PA, Pennsylvania State University Press, 1969, 535–537 and 539–545; John Major, *The Oppenheimer Hearing*, New York, Stein and Day, 1971, 91–146.

5. Carson Mark, "A Short Account of Los Alamos Theoretical Work on Thermonuclear Weapons, 1946–1950," 1975, LA-5647-MS; John A. Wheeler to Allen Shenstone, January 8, 1951, "Correspondence re Subcontract SC-B with Los Alamos Laboratory for Matterhorn B," Spitzer Papers.

6. Spitzer, oral history interview, 1977; Earl C. Tanner, *Project Matterhorn, 1951–1961*, revised 1979, Princeton University Plasma Physics Laboratory; author's interview with Lyman Spitzer.

7. Lyman Spitzer, Jr., "A Proposed Stellarator," July 23, 1951, USAEC NYO-993, 4.

8. "Project Matterhorn, Stellarator Division, Log Book," Spitzer Papers; Spitzer, "A Proposed Stellarator," 10. This was precisely the problem that had stymied the efforts of Fermi and others at Los Alamos to invent a fusion reactor in 1946. In April 1951, however, Spitzer knew nothing of these secret investigations.

9. Hydrogen exists in three isotopic forms with weights in the ratio 1 : 2 : 3. They are normal hydrogen, deuterium, and tritium. In the symbols, the subscripts are nuclear charge; the superscripts, weight.

10. Spitzer, "A Proposed Stellarator," 11–12. For a clear, elementary, technical history of fusion devices through 1957, see Amasa S. Bishop, *Project Sherwood*, Reading, Mass., Addison-Wesley, 1958.

11. "Project Matterhorn, Stellarator Division, Log Book," Spitzer Papers.

12. Spitzer, "A Proposed Stellarator"; Spitzer recalls that his own interest in the stellarator was predominantly the production of power for peaceful uses (author's interview with Spitzer).

13. J. L. Tuck "Controlled Thermonuclear Reactions," in Sherwood Conf., 1952; J. L. Tuck and S. Ulam, "The Possibility of Initiating a Thermonuclear Reaction by Non-nuclear Methods," abstract, July 23, 1946, folder TR 1956, LASL-RC; Edward Teller, "Preface," June 11, 1953, folder COTB-1, LASL-RC; Letter from Robert R. Wilson to the author, August 1980; author's interviews with Donald W. Kerst, and James L. Tuck.

14. G. P. Thomson and Moses Blackman, "Patent Specification"; A. A. Ware to M. Haines, December 1, 1976; A. A. Ware, "Controlled Thermonuclear Reactions," *Engineering*, November 15, 1957 (from personal files of A. A. Ware). Author's conversation with Alan A. Ware.

15. Interview with James L. Tuck by Neil S. Wolf, 1977 (from the files of N. S. Wolf); T. R. Kaiser and J. L. Tuck, "Air-Cored Synchrotron," *Nature 162*, (1948) 616–618.

16. J. L. Tuck, "Thermonuclear Reactions by Electrical Means," memorandum, April 8, 1948, UK AERE Atomic Energy Technical Committee; J. L. Tuck, untitled memorandum, no date; J. L. Tuck, "Memorandum to Fourth Meeting," January 1948 (from personal files of J. L. Tuck).

17. Interviews with Tuck by Wolf and the author. Tuck elected to insert a year in Chicago as a kind of cooling-off period between British and US government work.

18. Ford's paraphrase of Tuck in Kenneth W. Ford to Lyman Spitzer, Jr., May 3, 1951, "Scientific Correspondence, 1951–1956," Spitzer Papers.

19. Ford to Spitzer, and J. L. Tuck to Spitzer, May 3, 1951, "Scientific Correspondence, 1951–1956," Tuck to Spitzer, June 5, 1951, and Spitzer to Tuck, June 12, 1951, "Stellarator Computations Prior to AEC Contract . . . ," Spitzer Papers.

20. Interview with Tuck by Wolf; author's interview with James A. Phillips.

21. 582nd commission meeting, July 26, 1951, DOE-HO.

22. Since the bodies are charged, these are not billiard ball collisions but the so-called Coulomb collisions of electricity.

23. See, for example, L. Spitzer, Jr., R. S. Cohen, and P. McR. Routly, "The Electrical Conductivity of an Ionized Gas," *Physical Review 80* (1950), 230–238.

24. See Alfven, *Cosmical Electrodynamics*.

25. Spitzer to Tuck, June 12, 1951.

26. Their results were subsequently born out by experiments on the pinch. See chapter 4.

27. Lyman Spitzer, Jr., "The Princeton Stellarator Program," Sherwood Conf., 1952; Martin Kruskal and Martin Schwarzschild, "Some Instabilities of a Completely Ionized Plasma," *Proceedings of the Royal Society A223* (1954), 348–360; Schwarzschild, "Two Instabilities in a Magnetically Confined Plasma," Sherwood Conf., 1952; Spitzer, "A Proposed Stellarator."

28. George A. Kolstad to Lyman Spitzer, Jr., December 6, 1951, "Classification, Declassification, and Secrecy 1951–1959," Spitzer Papers.

29. "Research Program on Controlled Thermonuclear Processes," 1952, AEC 532, DOE-HO. Some of the division's consultants had urged that the program not be begun until an experienced experimentalist had been added to the Matterhorn group; Princeton met this condition by hiring the University of Iowa physicist James

A. Van Allen for 1953–1954. (AEC 532; Tanner, *Project Matterhorn*; author's interview with Spitzer.)

30. Spitzer, "The Princeton Stellarator Program," 22.

31. Lyman Spitzer, Jr., to Wolfgang K. Panofsky, October 23, 1951, "Scientific Correspondence, 1951–1956," Spitzer Papers.

32. Author's interviews with Tuck, Keith Boyer, and Phillips; interview with Tuck by Wolf.

33. Interview with Norris E. Bradbury by Arthur Norberg, 1976 (Bancroft Library, University of California).

34. James A. Phillips, George A. Sawyer, and Emory Stovall continued from the cross-section group. Louis C. Burkhardt and E. O. Swickard joined in 1952.

35. J. L. Tuck, "Controlled Thermonuclear Reactions," memorandum, January 5, 1952, folder TR 1956, LASL-RC; interviews with Tuck by Wolf and the author.

36. Tuck, "Controlled Thermonuclear Reactions," Sherwood Conf., 1952, 39–72. Schwartzschild, "Two Instabilities."

37. Author's interview with E. L. Kemp, and interview with Louis C. Burkhardt by Neil S. Wolf; Lyman Spitzer, Jr., to James A. Van Allen May 2, 1952, "Scientific Correspondence, 1951-1956," Spitzer Papers. Tuck, "Controlled Thermonuclear Reactions," in Sherwood Conf., 1952.

38. Herbert F. York, *The Advisors*, San Francisco, W. H. Freeman, 1976; Edward Teller and Allen Brown, *The Legacy of Hiroshima*, New York, Doubleday, 1962; Hewlett and Duncan, *Atomic Shield*; Major, *The Oppenheimer Hearing*, 140–142.

39. York, *The Advisors*; author's interview with Herbert York.

40. Author's interview with York.

41. Herbert F. York, memorandum to E. O. Lawrence, March 1952 (Ernest O. Lawrence Collection, Bancroft Library, University of California).

42. Author's interview with Richard F. Post; R. F. Post to Herbert York, April 22, 1952, "Possible Applications of Microwaves to the Generation of Thermonuclear Power"; R. F. Post, "Proposal for the Inauguration of a Research Group," September 1952 (from files of R. F. Post).

43. R. F. Post, "Denver Conference Report," Sherwood Conf., 1952, 81-94.

44. Author's interview with Spitzer.

45. Hewlett and Duncan, *Atomic Shield, passim*; AEC biographical files, DOE-HO.

46. Paul W. McDaniel to the author, March 2, 1979; AEC biographical files; 678th commission meeting, April 2, 1952, DOE-HO.

47. AEC 532 (Johnson stressed the "large number of uncertainties and complexities" that made his estimate an extremely precarious one); 678th commission meeting.

48. AEC 532; 678th commission meeting; Allan Shenstone to Lyman Spitzer, Jr., December 14, 1951, and J. C. Elgin to R. J. Woodrow, January 7, 1952, in "Correspondence and Memos re: Project Matterhorn, PPL. 1951-1961"; T. H. Johnson to Spitzer, December 21, 1951, in "Classification, Declassification and Secrecy, 1951-1959" (last three in Spitzer Papers); Spitzer, Oral history interview, 1977.

49. "Report on the Denver Meeting . . . by the Director of Research," 1952, folder R&D-7, CTR, DOE-HO; author's interviews with York and Spitzer; Sherwood Conf., 1952, *passim.*

50. "Report on the Denver Meeting."

51. Hewlett and Duncan, *Atomic Shield, passim.*

52. 900th commission meeting, July 31, 1953, DOE-HO.

53. AEC 532/5, DOE-HO, 2.

54. Allen L. Hammond, "Lithium: Will Short Supply Constrain Energy Technologies?," *Science 191* (1976), 1037–1038.

55. Tuck to Spitzer, May 3, 1951, cited in note 19.

56. Harold P. Furth, "High Magnetic Field Research," *Science 132* (1960), 388–393; author's interview with F. H. Coensgen; Spitzer, "The Princeton Stellarator Program."

57. "Report on the Denver Meeting."

58. Author's interview with Post.

59. AEC 532.

Chapter 3

1. Lewis L. Strauss, *Men and Decisions*, New York, Doubleday, 1962; Nuel Pharr Davis, *Lawrence and Oppenheimer*, New York, Simon and Shuster, 1968; *New York Times*, March 10, 1953, 13.

2. Strauss, *Men and Decisions*; Hewlett and Duncan, *Atomic Shield*; John Major, *The Oppenheimer Hearing*, chapter 4, "The Hydrogen Bomb."

3. L. L. Strauss, speech, October 29, 1957, "Strauss Speeches," DOE-HO; author's interview with Lewis H. Strauss; Paul W. McDaniel to author, March 2, 1979.

4. L. L. Strauss, "Draft," dated 5-1-56, "Sherwood," Strauss Papers; author's interviews with York, Spitzer, and Tuck. All but Glenn T. Seaborg opposed the bomb. Their reasons are given in Major, *Oppenheimer Hearing.*

5. Author's interview with Tuck.

6. For money and personnel, see *Hearings, Subcommittee on Research and Development, Joint Committee on Atomic Energy, 85th Congress, 2nd Session, on Physical Research Program . . . February 1958*, 784; Arthur E. Ruark to Steering Committee members, September 4, 1959, microfilm, GAMF 974, LASL-RC; "Fusion Power: An Assessment of Ultimate Potential and Research and Development Requirements," July 1972, "R&D-18, CTR, Vol. II, Bulky Package," DOE-HO, A33; 904th commission meeting, August 13, 1953.

7. Bishop, *Project Sherwood*, 75.

8. I owe this point to Robert B. Belfield.

9. Strauss, "Draft," 2.

10. Eugene M. Zuckert, "Memorandum for Chairman Strauss," August 25, 1953, "R&D-7, CTR," DOE-HO.

11. Thomas H. Johnson to John Von Neumann, Lyman Spitzer, James Tuck, Richard Post, Edward Teller, Ernest O. Lawrence, Robert R. Wilson, and Isador Rabi (Wilson did not attend), September 16, 1953, "R&D-7, CTR," DOE-HO.

12. 916th commission meeting, September 21, 1953. (My conclusions about the opinions expressed at this meeting differ from those of Bishop, *Project Sherwood*, 78.) 900th commission meeting, July 31, 1953.

13. 900th, 904th, 910th, and 916th commission meetings, July 31, August 13, September 3, and September 21, 1953. The CTR budget for FY 1955 was $4,300,000 as against $1,420,000 for FY 1954. A portion of the fiscal 1954 budget, however, represents additions made after the commission had assigned special priority to Sherwood. "Problems, Status and Outlook of the Sherwood Program, October 1955," 22, "Sherwood Classified," DOE-HO.

14. R. B. Belfield, draft, "The Rising Star."

15. Hewlett and Duncan, *Atomic Shield*.

16. Robert R. Wilson to the author, August 1980; R. R. Wilson to T. H. Johnson, August 17, 1953, "R&D-7, CTR," DOE-HO; 900th commission meeting, July 31, 1953; "Project Sherwood," January 15, 1954, AEC 532/15, DOE-HO; T. H. Johnson to M. W. Boyer, memorandum, August 25, 1953, "Director's Reading File, 1952-57," DOE-HO.

17. James G. Beckerley to Thomas H. Johnson, December 4, 1953, " R&D-7, CTR;" Beckerley testified to the Joint Committee on Atomic Energy, March 5, 1956; "Shortly after [the Denver Conference] the change in the Commission Chairmanship made the task of justifying declassification in this field essentially impossible" ("Statement of Dr. James G. Beckerley, . . . ," office files of R. F. Post, LLNL).

18. "Project Sherwood," January 15, 1954, AEC 532/15; T. H. Johnson to Oscar Smith, director of organization and personnel, "Nomination of A. S. Bishop for Arthur S. Fleming Award," September 20, 1955, "Director's Reading File," DOE-HO.

19. T. H. Johnson to William H. Brobeck, May 17, 1954, AEC 532/15; "Director's Reading File;" author's interviews with Spitzer and Tuck. Steering Committee minutes *passim* ("Steering Committee, Reports and Progress," Spitzer Papers.)

20. Compare the Steering Committee minutes with the Research Division staff papers for 1954 and 1955 in "R&D-7, CTR," DOE-HO. An important exception is the Research Division recommendation for detailed engineering design of the Model C, discussed in chapter 4.

21. Bishop, *Project Sherwood*, 196–197; interview with F. R. Scott by George K. Hess and the author; Steering Committee minutes for 1954 and 1955. Some of these information meetings were published in the form of "Sherwood Conference Reports," while others were not.

22. See E. O. Lawrence to Thomas H. Johnson, October 26, 1953 (office files of R. F. Post, LLNL); Thomas H. Johnson to H. D. Smyth, April 19, 1955, "Director's Reading File, CY 1952-57;" 1074th commission meeting, June 29, 1955; Amasa S. Bishop to T. H. Johnson, "Project Sherwood at Los Alamos," March 23, 1954, memorandum, folder TR 611, LASL-RC.

23. Author's interview with Tuck; Bishop to Johnson, "Project Sherwood at Los Alamos."

24. Norris E. Bradbury to T. H. Johnson, November 21, 1955, (E. O. Lawrence collection, Bancroft Library); Bishop to Johnson, "Project Sherwood at Los Alamos;" interview with N. E. Bradbury by Arthur L. Norberg, 1976 (Bancroft Library).

25. Author's interview with York; minutes of the Steering Committee of March 1954; E. O. Lawrence to McKay Donkin, April 20, 1954 (Lawrence collection, Bancroft Library); Davis, *Lawrence and Oppenheimer.*

26. Compare W. B. Reynolds to J. J. Flaherty, August 29, 1953 (office file of R. F. Post, LLNL), with T. H. Johnson to Francis McCarthy, November 3, 1953, "Director's Reading File," CY 1952-57. By July 1954, the larger model had been replanned as purely a research machine. ("UCRL Sherwood Program, Scheduling and Manpower Estimates for Fiscal Year 1955," office files of R. F. Post, LLNL.)

27. "Visit to Oak Ridge by Edward L. Heller, Staff Member, JCAE," January 11, 1954, 292/105, "Security S-1, Visitor Clearance and Control," DOE-HO. For more details of early Oak Ridge work, see Bishop, *Project Sherwood.*

28. Quotation from letter from K. E. Fields to E. O. Lawrence, October 19, 1953 (Lawrence collection, Bancroft Library); author's interview with Coensgen; Bishop, *Project Sherwood*, 68.

29. Johnson to Smyth, April 19, 1955, 1074th commission meeting, June 29, 1955; Bishop to Johnson, "Project Sherwood at Los Alamos"; JCAE *Hearings, February 1958*, cited in note 6, 784; general manager's report to commissioners for June 1954, DOE-HO. The several figures do not all harmonize. However, the general trends are clear.

30. Richard Courant had assembled the Institute of Mathematical Sciences in 1953, after he had succeeded in winning for his campus the surplus AEC computer UNIVAC-4, together with contracts to provide computer services for AEC laboratories (Constance Reid, *Courant in Goettingen and New York*, New York, Springer-Verlag, 1976); author's interview with Harold Grad.

31. Author's interview with Grad.

32. For example, R. L. Hirsch was to call him the "Jiminy Cricket" of fusion (Hirsch, "Diary," 1975–1976, DOE-HO).

33. Johnson to McCarthy, November 3, 1953. The table is from "Fusion Power, an Assessment of Ultimate Potential and Research and Development Requirements," July 1972. The NYU figures are from Edward Creutz, "Report on Project Sherwood," May 1956, TID 5447. The figures given in various sources are slightly discrepant. I have not attempted to reconcile them, as they all show the same overall behavior of funding.

34. Author's interview with Lewis H. Strauss; Stanislaw M. Ulam, *Adventures of a Mathematician*, New York, Scribner, 1976.

35. Author's interview with Spitzer; interview with Melvin B. Gottlieb by Neil S. Wolf.

36. Creutz, "Report on Project Sherwood," 61.

37. McDaniel to Author, March 1979. Johnson wrote Tuck on November 20, 1953, that they had begun to use the name ("Director's Reading File").

38. This point is adopted from a draft, "The Rising Star," by R. B. Belfield.

39. AEC 532/15.

40. Leonard M. Goldman and Lyman Spitzer, "Preliminary Experimental Results with the Model A Stellarator," Sherwood Conf., 1953, 28; Clodium H. Willis, "En-

gineering Aspects of Model B Stellarator," *ibid.*, 36; Tanner, *Project Matterhorn*. To place these hopes in perspective, it should be noted that these temperatures and densities were not achieved in the Princeton stellarators up to 1970, in which year the laboratory changed over from the stellarator to the variety of toroidal pinch called tokamak.

41. Lyman Spitzer to T. H. Johnson, December 9, 1953 ("Proposals to AEC, 1951–1953," Spitzer Papers).

42. Author's interview with Don J. Grove; Lyman Spitzer et al., "Problems of the Stellarator as a Useful Power Source," August 1975, NYO-6047; Steering Committee minutes of June 1954.

43. This could backfire, however. See chapter 6.

44. Lyman Spitzer to A. C. Monteith, Westinghouse Electric Corporation, December 16, 1953, "Subcontract Personnel, 1953–57," Spitzer Papers; Amasa S. Bishop to members of the Steering Committee, September 26, 1955, memorandum, "R&D-7, CTR," DOE-HO; there was some pulling and hauling on the question of how much Sherwood information could be relayed back to supervisors. The Division of Research did not want to give permission for any transfer of information. Westinghouse reacted strongly. The problem was solved by creating an Engineering Advisory Group of company supervisors. For Westinghouse W. E. Shoupp, Assistant Manager of the Atomic Power Division, served. ("Subcontract Personnel, 1953–1957," Spitzer Papers.)

45. Interview with Don J. Grove by Neil Wolf; author's interview with Willem Westendorp.

46. Spitzer et al., "Problems of the Stellarator," *passim*. The one issue conspicuously absent, from the present perspective, was environmental impact.

47. Spitzer et al., "Problems of the Stellarator," 260–261.

48. Author's interview with Thomas H. Stix.

49. Spitzer et al., "Problems of the Stellarator," introduction; Steering Committee minutes, June 1954.

50. J. A. Van Allen to Lyman Spitzer, Jr., July 6, 1954, "Model C Correspondence," Spitzer Papers.

51. D. J. Grove, "Princeton Model C Group," in Sherwood Conf., 1954; Tanner, *Project Matterhorn*.

Chapter 4

1. J. L. Tuck, "Controlled Thermonuclear Reactions," Sherwood Conf., 1952, 39–72; J. L. Tuck, "Report at the Berkeley Meeting," Sherwood Conf., 1953, 46–53; "Symposium on Plasma Instabilities," Sherwood Conf., February 1955, 356.

2. "UCRL Sherwood Program, Scheduling and Manpower Estimates for Fiscal Year 1955," July 6, 1954 (office files of R. F. Post, LLNL).

3. See L. M. Goldman, "Behavior of Current in the B-1′ Stellarator," Sherwood Conf., October 1955. Goldman was able to confirm the existence of the Kruskal limiting current in the B-1′. See Tanner, *Project Matterhorn*, for more information on the Model B.

4. See the Livermore reports in the October 1954, February 1955, and October 1955 Sherwood conferences. Most of my conclusions rely upon Post's summary reports: R. F. Post, "UCRL Sherwood Experimental Program-Progress Report," Sherwood Conf., 1954, 7–29; R. F. Post, "Outline of UCRL Sherwood Experimental Program and its Immediate Goals," Sherwood Conf., Feb. 1955, 127–134; R. F. Post, "Survey of the Magnetic Mirror Program," Sherwood Conf., October 1955, 88–94.

5. Lyman Spitzer, "Survey of the Stellarator Program," Sherwood Conf., February 1955, 258.

6. Post, "Outline of UCRL Sherwood," Sherwood Conf., February 1955, 133–134.

7. Edward Teller, "Comments on Plasma Stability . . . ," Sherwood Conf., October 1954, 6.

8. "Symposium on Plasma Instabilities," Sherwood Conf., February 1955, 342.

9. For light on Post's thinking, see R. F. Post to Amasa S. Bishop, November 3, 1954; R. F. Post to Hartland Snyder, November 3, 1954; R. F. Post to E. O. Lawrence, November 3, 1954, incorporating "an oddment I worked out on the train"; Lyman Spitzer to R. F. Post, November 8, 1954; Harold Grad to R. F. Post, November 12, 1954, and January 31, 1955 (office files of R. F. Post, LLNL).

10. E. Frieman, "The Energy Problem in Magnetohydrodynamic Stability Problems," in "Symposium on Plasma Instabilities," Sherwood Conf., February 1955, 343–350. The infinite configuration is easier because it is essentially two-dimensional, instead of three-dimensional.

11. "Symposium," 348–349.

12. "Symposium," 361.

13. Tuck, "Summary of the Review Given at Princeton, October 17, 1955," in Sherwood Conf., October 1955, 163–168. The Teller instability is also called the "flute" instability because in certain cases it corrogates the surface of a cylindrical plasma like the fluting on a Greek column, and it is sometimes called the "interchange instability" because the magnetic lines interchange their places.

14. R. F. Post, "Summary of UCRL Pyrotron (Mirror Machine) Program," Geneva Conf. 32, 245–265, quote on p. 259; See also Tuck, "Summary," Sherwood Conf., October 1955, and Edward Creutz, "Report on Project Sherwood," May 1, 1956, TID-5447, 28ff.

15. "Word of the Midwestern Theoretical Physicists on Sherwood, Summer 1955," folder TR 1956, LASL-RC.

16. Still another reason why mirrors might not sustain a flute instability was "line-tying." See chapter 7.

17. 916th commission meeting, September 21, 1953, "Mr. Lawrence said it seemed clear that ultimate success would come from broad basic research or from a purely empirical approach to the problem or from a combination of the two. He believed that at some time in the near future it would be profitable to undertake a certain amount of 'Edison-type experimentation'—large scale experiments which might provide direct and immediate solutions to many significant problems."

18. Post, "Survey," Sherwood Conf., October 1955, 89–90.

19. Author's interview with E. L. Kemp.

20. J. L. Tuck, "Summary," September 28, 1955, document #KT-11, LASL-RC; Steering Committee minutes, October 1954 and February 1955.

21. J. L. Tuck, "Picket Fence," Sherwood Conf., 1954, 77–86 (dated December 6, 1954).

22. Author's interview with Harold Grad; Harold Grad, "Sherwood Progress Report No. 1," October 1, 1957, NYO-7968; H. Grad to R. F. Post, January 31, 1955 (office files of R. F. Post, LLNL); H. Grad, "Theory of the Cuspidor," Sherwood Conf., October 1955, 319–325.

23. From a calculation by Hartland Snyder, quoted in Creutz, "Report," 23–24; Tuck, "Summary of the Review," Sherwood Conf., October 1955, 163.

24. J. Berkowitz, K. O. Friedrichs, H. Goertzel, H. Grad, J. Killeen and E. Rubin, "Cusped Geometries," Geneva Conf. *31*, 171ff.; Grad, "Sherwood Progress Report No. 1." NYU concluded that the losses were not incapacitating.

25. "Stellarator Program," March 18, 1955, AEC 532/18, DOE-HO.

26. Edward Teller to T. H. Johnson, February 19, 1955, filed with Steering Committee minutes, Spitzer papers.

27. Minutes of Steering Committee meeting, February 1955.

28. AEC 532/18.

29. Commission meeting of April 13, 1955 (incorrectly dated and numbered in the files), DOE-HO.

30. Minutes of the Matterhorn Review Committee, June 6, 1955, "Matterhorn (PPL) Review Committee," Spitzer Papers; R. Kulsrud, E. Frieman, J. Johnson, "On the Stabilization of a Confined Plasma," Sherwood Conf., October 1955, 232–235; J. L. Johnson et al., "Some Stable Hydromagnetic Equilibria," Geneva Conf. *31*, 199; Steering Committee minutes, June 1955; 1113rd commission meeting, July 28, 1955, DOE-HO.

31. Author's interview with Spitzer.

32. Minutes of the Matterhorn Review Committee, September 16, 1955, Spitzer Papers.

33. Author's interview with Robert G. Mills.

34. Lyman Spitzer, Jr. "Survey of Princeton Stellarator Program," Sherwood Conf., October 1955, 1–3.

35. Author's interviews with J. L. Johnson, and R. Mills.

36. Lyman Spitzer to Thomas H. Johnson, "Plans for the Model C Stellarator," January 21, 1957, Box, "Model C, Administrative," PPPL-Warehouse.

37. Bishop, *Project Sherwood*, chapter 9, "The Question of Stability," 85–89.

38. "Symposium," Sherwood Conf., February 1955, 357–358. See also Teller in June 1955 Steering Committee minutes, 1.

39. Representative Craig Hosmer, JCAE Hearings, November 1971, Vol. 1, 64–65.

40. Harold Grad, "Magnetic Confinement Fusion Energy Research," Peter D. Lax, ed., *Mathematical Aspects of Production and Distribution of Energy, Proceedings of Symposia in Applied Mathematics* 21 (1977), 10.

Chapter 5

1. General manager's monthly report to the commission for September 1955, DOE-HO. For the optimism which this result generated, also see Bishop, *Project Sherwood*, 66, and Creutz, "Report on Project Sherwood," 31. The scientists were theorist Albert Simon and experimentalist Roger V. Neidigh, in "Diffusion of Ions in a Plasma across a Magnetic Field," ORNL-1890, cited in Sherwood Conf., October 1955, 433.

2. R. M. Kulsrud et al., "On the Stabilization of a Confined Plasma," Sherwood Conf., October 1955, 232; J. L. Johnson et al., "Some Stable Hydromagnetic Equilibria," Geneva Conf., *31*, 198. Princeton's Koenig had already pointed out that spiral coils could overcome toroidal drift in early 1955, without noticing their other property of stabilizing the plasma.

3. *New York Times*, August 9, 1955, 1.

4. *New York Times*, August 9, 1955; 1113rd commission meeting, July 28, 1955; "Release of Information on Project Sherwood," August 1955; AEC 532/21. The location of the sites at which the research was carried on was revealed shortly after the conference.

5. Author's interview with Marshall N. Rosenbluth; A. Rosenbluth, M. Rosenbluth, and R. Garwin, "Infinite Conductivity Theory of the Pinch," Los Alamos Scientific Laboratory Report LA-1850.

6. Author's interviews with Phillips and with Stirling A. Colgate; Sherwood Confs., February 1955 and October 1955, *passim*.

7. James A. Tuck, "Summary of Review Given at Princeton, October 17, 1955," from Sherwood Conf., October 1955, 163 and 165.

8. Tuck, "Summary"; Bishop, *Project Sherwood*, 29ff and 90ff. Bishop expresses incredulity at Los Alamos's contemplation of such energies, which "scarcely . . . merit the term 'controlled' thermonuclear reactions!" (p. 32).

9. J. L. Tuck, "Sherwood Fiscal Forecast for 1956, 1957, 1958," February 16, 1956, LASL-RC.

10. For Washington opinions, see commission meetings, *passim*, Beckerley to Johnson, December 4, 1953, "R&D-7, CTR," DOE-HO, and Bishop, "Classification of Controlled Thermonuclear Program," draft, March 6, 1956, folder TR 611, LASL-RC. For the argument in favor of publication see 1113th commission meeting, July 28, 1955. For the project leaders' position, see Steering Committee minutes, *passim*.

11. See, for example, the June 1954 Steering Committee minutes: "It was generally agreed that progress in the Sherwood field had been impressive during the past six months and that the outlook for success . . . was at least sufficiently promising that a change in the general classification policy is not warranted." Again, according to the minutes of February 1955, a point in time when there seemed a good chance, on the basis of the M Theory, that fast nuclear pinches might give thermonuclear neutrons, the Steering Committee members were unenthusiastic about a recommendation sent by Los Alamos Director Norris E. Bradbury for a complete declassification on the occasion of the impending Geneva Conference of 1955. There were, however, substantial variations of opinion. Teller, for example, almost consistently favored complete declassification.

12. S. A. Colgate and H. P. Furth, "Stabilized Pinch and Controlled Fusion Power," *Science 128*, August 15, 1958, 337–342; author's interview with Colgate; W. R. Baker

and O. A. Anderson, "Linear Pinch Work in Berkeley," 134–143, and F. C. Gilbert, John Ise, Jr., Robert V. Pyle, and N. R. Stephen White, "Neutrons and Electrons from a Linear Deuterium Pinch," 144–147, Sherwood Conf., 1956. Colgate's apparatus, the Collapse, was one of the forerunners of the Scylla.

13. An English translation is in the British *Journal of Nuclear Energy 4* (1957), 193–202. Kurchatov's published talk was accompanied, in the Soviet journal *Atomnaya Energiya*, by a group of more technical accounts of the details of the Russian work.

14. Colgate and Furth, "Stabilized Pinch and Controlled Fusion Power"; author's interviews with Colgate and with Harold P. Furth; Bishop, *Project Sherwood*, chapter 10; Marshall Rosenbluth, "Theory of the Stability of a Pinch With a Longitudinal Magnetic Field and Conducting Walls," Sherwood Conf., 1956, 260–264.

15. James L. Tuck to Willard F. Libby, March 27, 1956. See also the addendum to the June 1956 Steering Committee minutes and J. L. Tuck to W. F. Libby, April 16, 1956.

16. J. L. Tuck to Edward Teller, February 16, 1956, "R&D-7, CTR," DOE-HO.

17. Creutz, "Report on Project Sherwood," 63.

18. J. L. Tuck, "Recent Developments in the Pinch Effect," Sherwood Conf., 1956, 22–27; S.A.Colgate, "The Stabilized Pinch," *ibid.*, 28–46; author's interview with Furth.

19. Edward R. Gardner, "Executive Director's Report on the Second International Conference on Peaceful Uses of Atomic Energy, Geneva, Switzerland, September 1–13, 1958," April 1959, DOE-HO; R. G. Hewlett and J. M. Holl, draft manuscript, volume 3 of the history of the Atomic Energy Commission. (I am indebted to the authors for making the manuscript available to me.)

20. "Planning for Second International Conference on Peaceful Uses of Atomic Energy," AEC 729/38, September 6, 1956, DOE-HO; 1223rd commission meeting, September 12, 1956; agenda planning session, September 21, 1956, "O&M Volume 2," DOE-HO; Gardner, "Executive Director's Report."

21. Lewis L. Strauss to Carl Hinshaw, February 13, 1956, "Congressional Files," DOE-HO; Lewis L. Strauss, "Draft," May 1, 1956, "Sherwood," Strauss Papers.

22. James G. Beckerley, statement to the Joint Committee on Atomic Energy, March 5, 1956 (office files of R. F. Post LLNL). See also Carl Hinshaw to Lewis L. Strauss, January 23, 1956, "Congressional Files," and "Notes on Informal Meeting of the Commissioners with Messrs. Tuck, Teller, and Spitzer—5:40 PM, Wednesday, May 23, 1956," Box 1296, file 3, DOE-HO.

23. Hinshaw to Strauss, January 23, 1956; letter exchange between L. H. Strauss and Clinton Anderson, January 1956, "Reactor Development-1, Project Sherwood, Vol. I," DOE-HO.

24. "Notes on Informal Meeting"; William P. Allis, "Summary of Academician Kurchatov's Paper . . . ," Sherwood Conf., 1956, 638; *New York Times*, April 26, 1:6 and April 28, 9:1.

25. Steering Committee minutes, June 1954 and February 1955; Division of Research staff papers AEC 532, March 1952, AEC 532/15, January 1954, AEC 532/19, March 1955, AEC 532/22, September 1955; 678th and 939th commission meetings, April

2, 1952, and December 11, 1953; Bishop, "Classification of the Controlled Thermonuclear Program," cited in note 10.

26. AEC 532/27.

27. ZETA is an acronym for Zero Energy Thermonuclear Assembly. It was an apt name; in its first runs it produced about 10^{-12} (one million millionth) of the energy that it used. Author's interview with John D. Lawson (the date of 1954 is taken from that interview).

28. "Classification of the Controlled Thermonuclear Program," AEC 532/28, May 1, 1956 (submitted jointly with the Division of Classification); Warren C. Johnson to Willard F. Libby, April 26, 1956. "Reactor Development I, Project Sherwood, CTR," DOE-HO.

29. "Notes on Informal Meeting," May 23, 1956 and 1197th commission meeting, May 2, 1956. The fifth commissioner, John von Neumann, did not participate.

30. "Classification of the Controlled Thermonuclear Program," AEC 532/34, September 17, 1956; 1228th commission meeting, September 25, 1956.

31. General manager's monthly report to the commission, October 1956, DOE-HO.

32. Lewis H. Strauss, "Status of the SHERWOOD Program as of Gatlinburg Meeting . . . ," memorandum, June 13, 1956, "Sherwood," Strauss Papers. Lewis H. Strauss had been trained in physics.

33. Gardner, "Report;" R. W. Cook to Paul F. Foster, "1958 International Conference Exhibit," "O&M 12-5," DOE-HO.

34. Author's interview with Gottlieb and telephone interview with Wolf Kunkel.

35. The men were in agreement that the United States program was "currently better" and, of course, considerably more diversified. Steering Committee minutes, November 19, 1956, and Gardner, "Report."

36. T. H. Johnson to files, "Notes on Princeton Visit," January 2, 1957, "Director's Reading File," DOE-HO; "Construction of Model C Stellarator at Princeton," AEC 532/37, February 13, 1957; 1269th and 1272nd commission meetings, February 27, 1957, and March 20, 1957.

37. Author's interview with Colgate; Colgate and Furth, "Stabilized Pinch." The British had already done some independent thinking along these lines.

38. *New York Times*, July 26, 1957, 2:4, and July 28, 1957, IV, 9.

39. *New York Times*, September 17, 1957, 130:5. In their published paper, ZETA scientists made the more sober claim of an ion temperature of at least 300 eV, based on data yielding figures between about 0.6 and 5.5 keV (P. C. Thonemann et al., "Production of High Temperatures and Nuclear Reactions in a Gas Discharge," *Nature* 181 (1958), 220).

40. R. M. Heckstall-Smith, "The Zeta Episode" (B.Sc. Thesis, Polytechnic of the South Bank, 1975), 87.

41. "Discussion of Classification Policy for Project Sherwood," appendix C, AEC 532/43; 1312th commission meeting, October 30, 1957.

42. Late examples of British comments are reported in the *New York Times* of January 13 and 17, 1958; Steering Committee minutes, October, 1957; Heckstall-Smith, "Zeta Episode," 88.

43. Lewis Strauss to W. Libby and H. Vance, March 5, 1957, memorandum, "Congressional Files, Sherwood," DOE-HO; J. L. Tuck "Interim Report on Recent Columbus II Results . . . ," August 26, 1957, document PT-26, LASL-RC.

44. Tuck, "Interim Report"; Bishop, *Project Sherwood*, 98, 99. The new switch was largely the work of Hagerman, and divided the voltage over many stages.

45. "U.S. Technical Exhibit for Second International Conference . . . ," written September 12, circulated October 7, 1957, AEC 930/8.

46. Paul McDaniel to the author; author's interviews with Post and Tuck; Steering Committee minutes for April 1957; "Magnetic Fusion Energy Program Funding Outlays" (office files of J. Ronald Young, DOE-OMFE).

47. Bibliographic files, DOE-HO; *New York Times*, 1957, *passim*.

48. "U.S. Technical Exhibit," AEC 930/8. (Graham did not comment and there is no record of Vance's reaction.)

49. Steering Committee minutes, October 1957; Spofford G. English to Herbert F. York, October 25, 1957 (E. O. Lawrence Collection, the Bancroft Library, University of California).

50. Quotations from letter of Paul McDaniel to the author, March 2, 1979; "List of People Attending Sherwood Meeting on Saturday, October 19, 1957 at 4:00 P.M. Chairman's Office," "Sherwood," Strauss Papers; author's interview with York.

51. English to York, October 25, 1957.

52. "Accelerated Sherwood Activities," AEC 532/44, November 7, 1957; English to York, October 25, 1957.

53. Creutz, "Report on Project Sherwood," 62; J. R. McNally, "Survey of ORNL Sherwood Program," September 13, 1956, ORNL-CF, 56-9-145; "Oak Ridge National Laboratory Thermonuclear Project for Fiscal Year 1958," ORNL-CF, 57-4-45.

54. John S. Luce, "Proposed Sherwood Experiment," April 11, 1955, ORNL-CF, 55-4-73; John S. Luce, "Molecular Ion Breakup; Preliminary Note," July 26, 1956, ORNL-CF, 56-7-119; J. S. Luce, "A Flexible Facility for Sherwood Experiments," August 28, 1956, ORNL-CF, 56-8-171; Bishop, *Project Sherwood*, chapter 13, "The Molecular Ion Ignition Program."

55. E. D. Shipley and A. Simon, "Survey of the Oak Ridge Sherwood Program," Sherwood Conf., 1957, 42; P. R. Bell, "Status of Sherwood at the Oak Ridge National Laboratory," Sherwood Conf., 1958, 6; McNally, "Survey of ORNL Sherwood Program." (McNally argued for retention of a "broad basic program, unfettered from any commitment to a particular device.")

56. Author's interview with Arthur E. Ruark; "Dr. Arthur E. Ruark named Chief of Controlled Thermonuclear Branch," January 28, 1957, press release, "Reactor Development-1, Sherwood," DOE-HO.

57. E. D. Shipley to S. G. English, "ORNL Sherwood Program," November 15, 1957, ORNL-CF, 57-11-59; J. A. Swartout to H. M. Roth, "ORNL Program Plans for Project Sherwood," November 15, 1957, ORNL-CF, 57-11-58; "Accelerated Sherwood Activities," AEC 532/44; P. R. Bell, "Status of Sherwood at Oak Ridge National Laboratory," Sherwood Conf., 1958, 6–7; Steering Committee minutes for February 6, 1958 (quotation is from here).

58. "Visit to Harwell by Representatives of Sherwood . . . ," December 20, 1957, AEC 532/47, "R&D-1, Project Sherwood (CTR), Vol. 2," DOE-HO.

59. "Visit to Harwell," AEC 532/47; *Hearings, Joint Committee on Atomic Energy, Physical Research Program, February 1958, 85th Congress, Second Session*, testimony of J. L. Tuck, 426–427.

60. Lyman Spitzer, Jr., "Cooperative Phenomena in Hot Plasmas," *Nature 181*, January 25, 1958, 221–222. The experiment was done by Thomas H. Stix.

61. S. A. Colgate, John Ferguson, and H. P. Furth, "Nonthermonuclear Neutrons from the Stabilized Pinch," Sherwood Conf., 1958, 85–97; author's interviews with Colgate and Furth. In the event, they were both right and wrong. The ZETA neutrons were not thermonuclear, and the well-known theories of conductivity were inapplicable because of a hitherto unknown disturbance of plasma conductivity by instabilities.

62. Heckstall-Smith, "The Zeta Episode"; *New York Times*, January 25, 1958, 9:8; USAEC Press Conference, January 23, 1958, Strauss biographical file, DOE-HO. Sir John Cockcroft, "The Next Stages with Zeta," *The New Scientist*, January 30, 1958.

63. 1336th commission meeting, February 21, 1958; "O&M 7, Vol. 3," DOE-HO, *passim*.

64. Author's interviews with Keith Boyer, and Tuck; F. R. Scott, " 'Totempole', The Shock Preheated Linear Pinch System," Sherwood Conf., 1957, 432–438; W. C. Elmore, E. M. Little, W. E. Quinn, and K. Boyer, "Neutrons from Plasma Compressed by an Axial Magnetic Field (Scylla)," Geneva Conf. *32*, 337–342; *Hearings . . . JCAE*, February 1958, testimony of Tuck and Phillip, 425–430 and 442–451.

65. Alan C. Kolb, "Magnetic Compression of Shock Preheated Deuterium," Geneva Conf. *31*, 328–340. For a discussion of the θ pinch, see Glasstone and Lovberg, *Controlled Thermonuclear Reactions*, Princeton, Van Nostrand, 1960, 414–421.

66. *Hearings . . . JCAE, February 1958*, 433–435.

67. John D. Lawson, "On the Power Efficiency of a Pulsed DD Fusion Reactor," December 31, 1954 (files of J. D. Lawson); J. D. Lawson, "Some Criteria for a Power Producing Thermonuclear Reactor," *Proceedings of the Physical Society B 70* (1957), 2–10.

68. B. Rose, A. E. Taylor, and E. Wood, "Measurement of the Neutron Spectrum from ZETA," *Nature*, June 14, 1958, 1630–1632; Thomas F. Stratton, "Report on ZETA," contained in a letter to Arthur E. Ruark, April 3, 1958, folder TR 611, LASL-RC; author's interview with John D. Lawson.

69. Gardner, "Report," cited in note 20, 41–42.

70. Other parts of the US exhibit were two operating reactors, a computer facility, a radioisotope laboratory, a hydrogen bubble chamber, and a whole-body radiation counter. "Atomic Industrial Progress and Second World Conference," *25th Semi-Annual Report of the AEC, July–December 1958*.

71. *25th Semi-Annual AEC Report*, 154–161.

72. Author's interviews with Torkild Jensen and Furth.

73. Colgate and Furth, "Stabilized Pinch," 342.

Chapter 6

1. AEC 146/10.

2. Spitzer, "Princeton Stellarator Program," Sherwood Conf., February 1957, part 2, 35. Don J. Grove headed the Princeton vacuum group. See Tanner, *Project Matterhorn*, 14–16.

3. Their coefficient, however, was two orders of magnitude greater than the classical diffusion coefficient, for fields from 3 to 14 kilogauss. R. V. Neidigh, "Pressure vs. Ion Density Experiments," Sherwood Conf., October 1955, 433.

4. See chapter 8 for additional discussion of pumpout.

5. Lyman Spitzer, Jr., to Arthur E. Ruark, August 8, 1958, "Outgoing Correspondence 1958," "E. A. Frieman, General files," in PPPL warehouse.

6. Author's interviews with Stix, Kunkel, and Gottlieb.

7. See the address of L. Biermann, "Recent Work on Controlled Thermonuclear Fusion in Germany (Federal Republic)" in Vol. 31 of the conference *Proceedings* and Küppers, "Fusionsforschung," cited in chapter 1, note 5. The *Proceedings* contain 111 papers, 19 from countries other than the USA, the UK and the USSR. Of these, 11 are from Germany, France, and Sweden (with the German paper a review over all their work).

8. Author's interview with Colgate; Martin Kruskal to Dr. and Mrs. Frieman, June 16, 1957, "Incoming Correspondence, 1956–57," "E. A. Frieman, General files," PPPL warehouse. The meeting was the Third International Conference on Ionization Phenomena in Gases, June 11–15, 1957.

9. Geneva Conf. *31*, 10, 20.

10. They calculated that power radiated increases with the square of temperature. The mechanism is called synchrotron radiation. Classical electromagnetic theory predicts that when charged particles change direction, as electrons do when they circle magnetic lines, they lose energy by radiation. Trubnikov and Kondryatsev, in M. A. Leontovich, ed., *Plasma Physics and the Problem of Controlled Thermonuclear Reactions*, J. Turkevitch, trans. ed., Oxford, Pergamon Press, 1959–60, Vol. III, 141; *New York Times*, October 30, 1958, 11:1.

11. "Controlled Nuclear Fusion Research, September 1961: Review of Experimental Results," Salzburg Conf. *1*, 15–16. This state of affairs is vividly reflected in the book, *Controlled Thermonuclear Reactions* (Princeton, NJ, Van Nostrand, 1960), which the AEC arranged to have written by Samuel Glasstone and Ralph Lovberg, as a teaching aid and record of Sherwood results. Originally conceived as a volume for presentation at Geneva under the authorship of Glasstone and R. F. Post, it was delayed because of the demands made by the crash program upon Post's time, and ultimately the coauthor was also changed.

12. For example, Artsimovich, Geneva Conf. *31*, 20; Biermann, *31*, 25; Teller, *31*, 27; Alfven, *32*, 463.

13. For example, Biermann and Teller, Geneva Conf. *31*, 25 and 27.

14. *New York Times*, October 30, 1958, 11:1; "John McCone: New AEC Chairman," *Bulletin of the Atomic Scientists 14* (1958), 334–335; George B. Kistiakowsky, *A Scientist at the White House*, Cambridge, Harvard Univ., 1976, *passim*; Irving C. Bupp, *Priorities*

in Nuclear Technology: Program Prosperity and Decay in the USAEC 1956–1971, Harvard PhD thesis, 1971; author's interview with York, and interview with Dwight Ink by Robert B. Belfield, April 18 and 30, 1980; J. A. McCone to the general manager, memorandum, July 24, 1959, DOE-HO.

15. P. W. McDaniel to the author, March 2, 1979. Williams was promoted into Libby's place as the commission's scientific member in August 1959, and the Division of Research reverted to the stewardship of Paul McDaniel.

16. John H. Williams to N. E. Bradbury, September 22, 1958, and A. E. Ruark to Steering Committee members, September 4, 1959, microfilm, GAMF 974, LASL-RC.

17. McCone to the general manager, July 24, 1959; author's interview with A. E. Ruark; interview with Dwight Ink by Belfield.

18. AEC 532/44, cited in chapter 5, note 55.

19. 1533rd commission meeting, April 8, 1959. The inclination to cut back the number of approaches conflicts with McCone's recognition that fusion was in the research stage, and may be a manifestation of McCone's lack of interest in and sympathy for, research. See Kistiakowsky, *A Scientist at the White House.*

20. See the report on the meeting in "Recommendation of Ruark to Steering Committee Members, September 4, 1959," AEC 532/54, GAMF 974, LASL-RC.

21. "Recommendation of Ruark," AEC 532/54.

22. V. S. Emelyanov, "Toward Close International Cooperation in Atomic Research," translation in *Bulletin of Atomic Scientists 16*, October 1960, 322–325; *New York Times*, September 16, 1959, 1:5; September 25, 1959, 1:6; September 26, 1959, 13:5, October 1, 1959, 17:1.

23. Thomas H. Stix, "Report on Visit to the USSR July 4–21, 1960," GAMF 975, LASL-RC. Besides Ruark and Stix, the American delegation included Tuck from Los Alamos, Van Atta from Livermore, and Persa R. Bell from Oak Ridge.

24. Author's interviews with David J. Rose and William C. Gough; JCAE, *Hearings*, *passim.*

25. Author's interview with Kerst.

26. Author's interview with Gottlieb; Division of Plasma Physics, folders "First Division Meeting, Monterey, California, December 1959," "New York, January 27–30, 1960," "1959," "Montreal, Canada, June 15–17, 1960" (office files of William Grossman, the Courant Institute, NYU). The division's first executive committee was constituted entirely of people involved in the fusion aspects of plasma physics; they were F. Coensgen, M. Gottlieb, A. Kolb, W. Kunkel, J. Phillips, M. Rosenbluth, A. Simon, and H. Hurwitz.

27. There were fewer interactions with gas discharge researchers at this stage. (See division folders, and author's interview with Rose.)

28. *The Physics of Fluids* started in January 1958, "with financial underwriting from the Air Force Office of Scientific Research and the Office of Naval Research." *Physics Today*, June 1959, 19; interview with François N. Frenkiel.

29. Matterhorn Review Committee minutes, meeting, January 14, 1963, 2; "Matterhorn (PPL) Review Committee, 1955–65," Spitzer Papers.

30. Thomas H. Stix, "Education in Plasma Physics at Princeton University," December 6, 1963 (office files of T. H. Stix, PPPL); "Matterhorn Review Committee Minutes," March 9, 1959, Spitzer Papers; Matterhorn Review Committee minutes, 1959, *passim*. In 1962, a summer school was added, to train university professors, industry personnel, and others in the new field (Raymond J. Woodrow to James Bethel, August 8, 1961, "Plasma Physics Program—NSF," Spitzer Papers). In 1962 also, a new Department of Astrophysical Sciences was created to supplement the interdepartmental committee and to serve as a joint sponsor of courses with the original departments (Stix, "Education in Plasma Physics").

31. Even in the early sixties, when Kerst moved to the University of Wisconsin, he was unsure whether fusion was at a state to admit of acceptable experimental theses (author's interview with Kerst).

32. Matterhorn Review Committee minutes, meeting of September 28, 1960.

33. Tanner, *Project Matterhorn*, 77.

34. J. A. Stratton, "R.L.E.—The Beginning of an Idea," *R.L.E.: 1946 + 20*, Research Laboratory of Electronics, MIT, Cambridge, Mass., May 1966, 1–6; William P. Allis, "Along the Road From Electrical Discharges to Plasmas," *ibid.*, 30–32; Sanborn C. Brown, "A Brief History of the Plasma Physics Group," *ibid.*, 33–35.

35. Author's interview with Rose.

36. Author's interviews with Rose and Louis D. Smullin.

37. M. N. Rosenbluth et al., "Plasma Physics," report of the Plasma Physics Subcommittee of the Physics Survey Committee of the National Academy of Science, 1966 (the Pake committee); see also course listings described in Stix, "Education in Plasma Physics at Princeton." For an intimation of the distribution of plasma scientists among fields other than fusion research, see the panel report.

38. Rosenbluth, "Plasma Physics"; author's interview with Gottlieb.

39. JCAE, *Hearings before the Subcommittee on Research and Development, February 10, 1958*, 399–401.

40. See Bruno Coppi and Harold Grad, JCAE, *The Current Status of the Thermonuclear Research Program, 1971*, part 2, 639–642. The Pake committee found that "the cost per Ph.D. physicist in a government laboratory was more than twice the cost in a university laboratory," $83,000/year against $36,000. Some of the difference presumably is in equipment. "Outlook for U.S. Physics," *Physics Today* 19, 23–36.

41. T. H. Johnson to Ivan A. Getting of Raytheon, February 20, 1956, "Subcontract Personnel, 1953–1957," Spitzer Papers.

42. Francis Bello, "Fusion Power: The Trail Gets Hotter," *Fortune*, July 1957, 135ff.

43. Kenneth H. Kingdon to C. Guy Suits, April 21, 1955; K. H. Kingdon, "Project 43-561—Fusion Group," memorandum, May 4, 1955; "Project 43-561 . . . Minutes of Meeting, May 20, 1955" (General Electric Research Laboratory Archives; I am indebted to George Wise for this and the other GE documents).

44. Press release of June 17, 1957 (GE Archives).

45. Henry Hurwitz, Jr., "Electrical Power Problems in Fusion Research," August 20, 1958, draft MS (GE Archives). See Hewlett and Duncan, *Nuclear Navy*, on General Electric's role in the forties and fifties.

46. Hurwitz, "Electrical Power Problems in Fusion Research;" interview with Henry Hurwitz, Jr., and Herbert C. Pollack by the author and George Wise.

47. Author's interviews with Grove and Johnson.

48. "Model C Meeting," April 7, 1955, and April 22, 1955, "AC Correspondence 5/12/55 to 6/30/57," Box 174, #32, Spitzer Papers; "Minutes of the First Meeting of the Model C Contractor Selection Board," March 15, *ibid.*, 1957.

49. Leonard J. Linde, "Thermonuclear Fusion Research for Power," *American Power Conference, Proceedings, March 31, April 1–2, 1959*, Vol. XXI, 51–59. Forty of the 60 were technical personnel.

50. Robert Sheehan, "General Dynamics vs. the U.S.S.R.," *Fortune*, February 1959, 87ff; author's interview with Norman Rostoker and Alan C. Kolb.

51. Author's interview with Rostoker; telephone interview with Howard R. Drew by Robert B. Belfield. There were also tax advantages in putting money into research.

52. Author's interview with Kerst; high-intensity accelerators also involve scientists in plasma physics because the particle density is so high that cooperative effects modify the behavior of the charges, and the analysis by single-particle motion in magnetic fields no longer applies. "Biographical Data" (office files of D. W. Kerst); author's interview with Jensen; interview with F. Robert Scott by George K. Hess and the author; Donald W. Kerst, "Joint General Atomic-TAERF Fusion Program," Geneva Conf. *32*, 106–110.

53. *Hearings, JCAE, 86th Congress, 2nd Session, Development, Growth and State of the Atomic Energy Industry, 1960*, Statement of Frederic de Hoffman, 193.

54. The figures represent budget outlays. "Magnetic Fusion Energy Program Funding," June 8, 1979 (office files of J. Ronald Young).

Chapter 7

1. The most definitive papers were I. Bernstein, E. Frieman, M. Kruskal, and R. Kulsrud, "An Energy Principle for Hydromagnetic Stability Problems," *Proceedings of the Royal Society (London) A244* (1958), 17–40; J. Berkowitz, H. Grad, and H. Rubin, "Magnetohydrodynamic Stability," Geneva Conf. *31*, 177–189.

2. R. F. Post, R. E. Ellis, F. C. Ford, and M. N. Rosenbluth, "Stable Confinement of a High-Temperature Plasma," *Physical Review Letters 4* (1960), 166–170.

3. They did not have good data on ion temperature.

4. This physical explanation of the flute instability, which greatly facilitated theorizing, was worked out by Conrad Longmire.

5. Post et al., "Stable Confinement," 168–169.

6. F. H. Coensgen, "Experiment Sequence Toy Top → 2XIIB → Future," August 23, 1977, MFE/CP/77-164 (office files of Dr. Coensgen).

7. F. H. Coensgen, W. F. Cummins, and A. E. Sherman, "Multistage Magnetic Compression of Highly Ionized Plasma," *Physics of Fluids 2* (1959), 351.

8. In ordinary fluids composed of several substances, for example, air, the different components come to a common, equilibrium temperature. In plasmas confined for small fractions of a second, on the other hand, there is often insufficient time for

the electrons and ions to interact to the point at which their temperatures are equalized.

9. A. E. Ruark to P. W. McDaniel, October 28, 1960, "A Recent Advance in Mirror-Machine Energies at Livermore," memorandum, "Reactor Development 1, Sherwood, Vol. 1," DOE-HO; F. H. Coensgen, W. F. Cummins, W. E. Nexsen, Jr., and A. E. Sherman, "Evidence of Containment of a 3-kev Deuterium Plasma," *Physical Review Letters 5* (1960), 459–461.

10. The problem is analogous to the one that Spitzer had to overcome in 1951. In a completely symmetrical field, the electrons drift by rotating in a closed curve around the central axis and the ions rotate in the opposite direction, so that there is no drift out of the vessel and no separation of charge.

11. R. F. Post, "Present Status and Future Prospects of Pyrotron (Mirror Machine) Program," "Reactor Development 1, Sherwood, Vol. 1," DOE-HO. Post understood as well as anyone in the program how little was yet known of the theory and practice of plasma physics; this grasp coexisted as another trait alongside his unvarying optimism. For a thorough survey of known and unknown territory at the time, see R. F. Post, "High-Temperature Plasma Research and Controlled Fusion," *Annual Review of Nuclear Science 9* (1959), 367–435.

12. "Some Aspects of the Economics of Fusion Reactors," Dansk Atomenergikom-missionen, *International Summer Course in Plasma Physics 1960*, 367–429. He also postulated large mirror ratios, from 3 to 10. Subsequently, it was realized that economic constraints set an upper limit on mirror ratios. Post knew of the ambipolar potential, but undervalued it, as well as the cooling effect the electrons would exercise upon the ions (author's interview with T. Kenneth Fowler, June 30, 1978).

13. Post, "Present Status and Future Prospects of Pyrotron."

14. *New York Times*, April 25, 1961, 11.

15. Acting Chairman John Graham, *Hearings before the Joint Committee on Atomic Energy . . . Pursuant to Section 202 . . . February 21, 23, 24, 27, 28, March 1, 2, and 3, 1961,"* 50–53; Commissioner Loren K. Olson, speech, December 7, 1960, AEC biographical file, DOE-HO; "Summary Notes on Review of Controlled Thermonuclear Research Program, Tuesday, January 10, 1961," "RD-1, CTR, Vol. 1," DOE-HO.

16. Author's interviews with Coensgen and Furth.

17. A. E. Ruark to L. L. Strauss, December 11, 1961, "Ruark," Strauss Papers.

18. *Salzburg Conf. 1*, 210. R. F. Post, "Critical Conditions for Self-Sustaining Reactions in the Mirror Machine," *ibid.*, 99–123; F. H. Coensgen, W. F. Cummins, W. E. Nexsen, Jr., and A. E. Sherman, "Production and Containment of Hot Deuterium Plasmas in Multistage Magnetic Compression Experiments," *ibid.*, 125–133; M. S. Ioffe and E. E. Yushmanov, "Experimental Results on Plasma Instabilities in Traps with Magnetic Stoppers" (in Russian), *ibid.*, 177–182; "Discussion of Session II," *ibid.*, 207–211.

19. See L. A. Artsimovich, *Controlled Thermonuclear Reactions*, New York, Gordon & Breach, 1964, trans. P. Kelly and A. Peiperl, eds. A. C. Kolb and R. S. Pease, 377–379.

20. This holds if the rate at which the instability would grow (if unhampered) is sufficiently small, compared to the ion's period for one complete gyration, and the ion's gyration is sufficiently large, compared to the wavelength of the instability. M. N. Rosenbluth, N. A. Krall, and N. Rostoker, "Finite Larmor Radius Stabilization of 'Weakly' Unstable Confined Plasmas," *Salzburg Conf. 1*, 143–144.

21. Author's interview with Post.

22. F. H. Coensgen, "Review of Multistage Magnetic Compression Experiments," October 17, 1961 (office files of R. F. Post, LLNL).

23. Coensgen, "Experiment Sequence Toy Top → 2XIIB → Future." Whether or not to rely upon line-tying for stabilization was, at the time, a disputed point.

24. W. A. Perkins and R. F. Post, "Observations of Plasma Instability with Rotational Effects in a Mirror Machine," *Physics of Fluids 6* (1963), 1537–1555. Perkins and Post also supplemented their work with hot-electron plasmas by studying a hot-ion plasma (ions hotter than electrons).

25. They brought the vacuum down to a pressure of a few times 10^{-10} millimeters of mercury, and they brought the beam current up from 2 mA to 40 mA (C. C. Damm, memoir, September 1977, courtesy of Dr. Coensgen).

26. C. C. Damm, J. H. Foote, A. H. Futch, and R. F. Post, "Cooperative Oscillations in a High-Temperature Plasma Formed by Neutral Atom Injections," *Physical Review Letters 10* (1963), 323–326; R. F. Post, "The LRL Mirror Machine Program—FY '64 and '65," Oct. 28, 1963, "RD-1, Sherwood, Vol. 2," DOE-HO.

27. R. F. Post, "Self-Sustaining Reactions in Mirror Machines," Salzburg Conf. *1,* 115. R. F. Post to Lyman Spitzer, Jr., August 28, 1962, "General Correspondence" (office files of R. F. Post). The point was that they knew, in principle, how to overcome the macroscopic flute instability, whereas they did not have the same hold on the microinstabilities.

28. Post, "The LRL Mirror Machine Program . . ."; author's interview with Furth.

29. McDaniel remembers that he "probably so requested" (McDaniel to author, March 2, 1979).

30. "Report of the Ad Hoc Sherwood Subcommittee," August 7, 1962, "O&M 8, GAC Reports, BP," DOE-HO, quotations on 11 and 12.

31. Manson Benedict to Glenn T. Seaborg, January 11, 1963, and excerpts on Sherwood from the General Advisory Committee minutes of January 7–9, 1963 (courtesy of the Chief Historian, DOE).

32. "Meeting with Princeton University to Discuss Recommendations of GAC, January 18, 1963" (office files of W. C. Gough, DOE); author's interview with Gottlieb; 1899th commission meeting, February 26, 1963; Paul W. McDaniel to A. R. Luedecke, "Comments on the 1962 GAC Report of the Sherwood Program," November 12, 1962, "RD-1, Sherwood, Vol. 2," DOE-HO.

33. A pinch with a metal core down its throat had already been invented by Oscar Anderson and William Baker, the Berkeley members of the California pinch group, and called "Triax." Furth's machine differed from the Berkeley device in one crucial respect; it had a longitudinal magnetic field outside the plasma, while the Berkeley pinch had an azimuthal field there. The hydromagnetic theory then in use predicted stability for the former, but not for the latter, magnetic field configuration, and this made the Livermore hard-core pinch suited to test stability theories.

34. Author's interview with Furth; H. P. Furth, "Mirror Machine with Absolute Hydromagnetic Stability," manuscript, June 8, 1963 (files of Dr. Furth); "Existence of Mirror Machines Stable against Interchange Modes," *Physical Review Letters 11*

(1963), 308–310. Similar calculations were made, independently, in Europe, by J. Andreoletti.

35. Post, "The LRL Mirror Machine Program. . . ." The magnetic well was widely regarded in the fusion community as a kind of "overkill" to make possible the exploration of microinstability.

36. Harold P. Furth and Richard F. Post, "Advanced Research in Controlled Fusion," December 10, 1964, URCL-12234 (TID-4500), 4.

37. The promise of a breakthrough caused a flurry of anger and annoyance among the other CTR project leaders. See James A. Phillips " 'Flap' in Washington," and J. M. B. Kellogg, "Sherwood Meeting, Washington, November 18 and 19," microfilm GAMF 974, LASL-RC. For the budget cut, see file "R&D 7, CTR," Box 1306, *passim*, DOE-HO.

38. Richard J. Barber, *The Politics of Research*, Washington, Public Affairs Press, 1966; Daniel S. Greenberg, *The Politics of Pure Science*, New York, New American Library, 1967.

39. *Hearings, JCAE, 88th Congress, 2nd Session, AEC Authorizing Legislation, Fiscal Year 1965*, part 3, 1377, 1378.

40. Harold Orlans, *Contracting For Atoms*, Washington, D.C., Brookings Institute, 1967, 78ff.

41. A. W. Trivelpiece to Senator Thomas H. Kuchel, October 9, 1963, "R&D, CTR, Vol. II," DOE-HO.

42. *Hearings, JCAE, Fiscal Year 1965*, part 3, 1378.

43. Cited in Paul McDaniel to General Manager A. R. Luedecke, memorandum, December 13, 1963, "Reactor Development I, Vol. II (CTR)," DOE-HO.

44. *Hearings, JCAE, Fiscal Year 1965*, part 3, 1427, 1456.

45. McDaniel to Luedecke, December 13, 1963; "Report of the Ad Hoc Sherwood Subcommittee," cited in note 30.

46. Thus, when Christofilos in 1950 had laid out the new principle of accelerator beam focusing by means of alternating gradients, little attention had been paid to him. It was only in 1952, when the same focusing principle had been independently invented at Brookhaven National Laboratory, that Christofilos's work was recognized in the United States. The American inventors had been Ernest D. Courant, M. Stanley Livingstone, and Hartland Snyder: press release, University of California, Lawrence Radiation Laboratory, June 11, 1964; *Time*, March 20, 1959; *Washington Star*, April 29, 1959 (all courtesy of NASA historical office).

47. James G. Beckerley to Robert L. Thornton, March 2, 1953 (E. O. Lawrence papers, Bancroft Library).

48. N. C. Christofilos, "Astron Thermonuclear Reactor," Geneva Conf. *32*, 279–290.

49. Hilliard Roderick to Files, "The Astron," August 28, 1959, "LLL Astron 1960–1972," Job 6868, Box 2775, DOE-OMFE.

50. George A. Kolstad to R. L. Thornton, March 13, 1953 (E. O. Lawrence Papers, Bancroft Library); Steering Committee minutes, February 1955; Christofilos, "Astron," Geneva Conf.; "Source Material for History from Retired Files," "LLL Astron, 1960–1972," DOE-OMFE.

51. T. Kenneth Fowler, "A Tribute to Nicholas C. Christofilos," *Energy and Technology*, Lawrence Livermore Laboratory, January 26–27, 1981. Over the 2-year period spanning fiscal years 1961 and 1962, ARPA was asked to furnish nearly $4,000,000 for the Astron accelerator ("Source Material for History"). Christofilos had worked with the Department of Defense and ARPA in 1957–1959 when he had proposed "Project Argus" to explode bombs above the atmosphere and study the way in which the relativistic electrons emitted were captured in the mirror field created by the earth's magnetism. See, for example, "Top Scientists Find *Argus* Opening New Avenue of Experimentation," *Missiles and Rockets*, May 18, 1959, 37ff.

52. John S. Foster, T. Kenneth Fowler, and Frederick E. Mills, "Nicholas C. Christofilos," obituary, *Physics Today*, January 1973, 109ff.

53. *Times*, March 20, 1959.

54. H. P. Furth to N. C. Christofilos, "Remarks on the Astron Project," March 26, 1963, "LRL Astron, 1963–65," DOE-OMFE; "Report of Ad Hoc Panel on Astron Program (Preliminary Version)," January 1968, "Ad Hoc Panel on Astron," DOE-OMFE; author's interview with Smullin. Christofilos's scientific style was controversial, and Furth was one of his severer critics. Theorists who held more favorable views include T. K. Fowler and Gareth Guest.

55. AEC biographical files, DOE-HO; author's interview with Ramey.

56. John S. Foster, Jr., to Leland Haworth, February 6, 1953, "Lawrence Livermore Lab., 1963–64," DOE-OMFE.

57. W. B. McCool to file, January 15, 1964, "LRL (Astron), 1963–65," DOE-OMFE: Paul W. McDaniel to A. R. Luedecke, Dec. 20, 1963, "Sherwood Program—FY 1964," "R&D CTR, Vol. II," DOE-HO.

58. Foster to Haworth, February 6, 1953, Chester M. Van Atta to Arthur E. Ruark, January 22, 1964, "LRL 1963–64," DOE-OMFE; testimony by N. Christofilos, *Hearings, JCAE, 88th Congress, 2nd Session, AEC Authorizing Legislation, Fiscal Year 1965*, part 3, February and March 1964, 1456.

59. McDaniel to Luedecke, December 13, 1963, cited in note 43; McDaniel to Luedecke, December 20, 1963, cited in note 57. The laboratories at Berkeley and Livermore, however, had done unusually well in fiscal year 1963. Their actual operating expenses had exceeded those budgeted by $700,000 while Princeton and Oak Ridge spent about $250,000 and $200,000 more, respectively, and Los Alamos actually spent more than $300,000 less.

60. Chester Van Atta to Arthur E. Ruark, December 18, 1963, "LRL 1963–64," DOE-OMFE.

61. W. A. Perkins and W. L. Barr, abstract 05, *Bulletin of the American Physical Society* 9 (1964), 328. Perkins had begun this work in the summer of 1963 (John R. Hiskes to R. Hobart Ellis, November 21, 1966, "LRL, 1963–66," DOE-OMFE).

62. C. C. Damm, J. H. Foote, A. H. Futch, A. L. Gardner, and R. F. Post, "Suppression of Convective Losses from a Steady-State Plasma by a Positive-Gradient Field," *Physical Review Letters 13*, 464–467.

63. A. H. Futch, Jr., et al., "Stability of a Magnetically Confined 20-kev Steady-State Plasma," Culham Conf. *II*, 3–21; W. A. Perkins and W. L. Barr, "Observations of a Velocity-Distribution Instability," *ibid.*, 115–134.

64. Hiskes to Ellis, November 21, 1966; C. C. Damm, memoir, September 1977 (office files of F. H. Coensgen); author's interview with Furth.

65. "Lorentz dissociation"; the affected particles are those which enter the confinement chamber in an excited state.

66. Author's interview with Julian L. Dunlap; C. F. Barnett, P. R. Bell, J. S. Luce, E. D. Shipley, and A. Simon, "The Oak Ridge Thermonuclear Experiment," Geneva Conf. *31*, 298–304; J. L. Dunlap, C. F. Barnett, R. A. Dandl, and H. Postma, "Radiation and Ion Energy Distributions of the DCX-1 Plasma," Salzburg Conf. *1*, 233–237; J. L. Dunlap, memoir, May 10, 1976 (office files of J. L. Dunlap); A. H. Snell et al., "Planning Document: Controlled Fusion Research at Oak Ridge National Laboratory," December 1, 1961, ORNL-CF, 61-12-8; A. H. Snell, "Planning Document: Controlled Fusion Research at Oak Ridge Laboratory," January 20, 1964, ORNL-CF, 64-1-31.

67. J. L. Dunlap, H. Postma, G. R. Haste, and L. H. Reber, "Severe Microinstability-Driven Losses in an Energetic Plasma," Culham Conf. *II*, 67–75.

68. The other design was John Luce's. A vigorous contention between these two forceful men seems to have been resolved on the ground that Bell's machine could be put into operation more quickly (author's interview with Dunlap).

69. This was the work of William Gauster. Another technological achievement was the refinement of the accelerator by Bell and Kelley, to the point where it could produce currents of several hundred milliamperes, making it one of the foremost particle accelerators of the early sixties (author's interview with George G. Kelley).

70. Dunlap, memoir; reports of the Thermonuclear Advisory Committee to Alvin M. Weinberg, for 1961, 1962, 1963, 1964, (office files of J. Rand McNally); P. R. Bell, R. A. Gibbons, G. G. Kelley, N. H. Lazar, J. F. Lyon, and R. F. Stratton, "Oak Ridge Multiple-Pass Injection Experiment, DCX-2," Colham Conf. *II*, 77–91.

71. Lyman Spitzer, Jr., "Controlled Nuclear Fusion Research, September 1965: Review of Experimental Results," Culham Conf. *1*, 3.

72. R. F. Post, "Controlled Fusion Research and High-Temperature Plasmas", *Annual Review of Nuclear Science 20*, (1970), 509–588, quotation on 557.

73. R. F. Post and M. N. Rosenbluth, "Electrostatic Instabilities in Finite Mirror-Confined Plasmas," *Physics of Fluids 9* (1966), 730–749.

74. James L. Tuck to Stephen O. Dean, June 2, 1965, "Review, June–July 1965," DOE-OMFE.

Chapter 8

1. This section makes use of Earl C. Tanner, *The Model C Decade, 1961–1969*, Princeton University Plasma Physics Laboratory, 1980. I thank Dr. Tanner for making a draft of this monograph available to me.

2. Tanner, *Model C Decade*; D. J. Grove, R. M. Sinclair, W. Stodiek, W. L. Harries, and L. P. Goldberg, "Some Preliminary Results on Containment Time in the Model C Stellarator," Salzburg Conf. *1*, 203–206.

3. R. A. Ellis, Jr., L. P. Goldberg, and J. G. Gorman, "Loss of Charged Particles in a Stellarator During Ohmic Heating," *Physics of Fluids 3* (1960) 468–473. (In this early paper, they got a pumpout dependence of $1/\sqrt{B}$); W. Stodiek, R. A. Ellis, Jr., and

J. G. Gorman, "Loss of Charged Particles in a Stellarator," Salzburg Conf. *1*, 193–198; author's interview with Stodiek, December 1, 1978.

4. Tanner, *Model C Decade*, 27.

5. "Report of the Ad Hoc Sherwood Sub-Committee," "O&M 8, GAC Reports," DOE-HO. This recommendation may have reflected Commissioner Wilson's hesitations. "In the mind of Commissioner Wilson, the Princeton Plasma Physics Laboratory should produce something with the Model C. He wants a moment of truth, not a long-range endless program." "Meeting with Princeton University to Discuss Recommendations of GAC, January 18, 1963," 10 (office files of W. C. Gough).

6. "Meeting . . . January 18, 1963." Paul W. McDaniel to A. R. Luedecke, November 16, 1972, "Comments on the GAC Report of the Sherwood Program."

7. Author's interviews with Ruark and Gough.

8. "Meeting . . . January 18, 1963." They projected a reactor β of 10%.

9. David J. Rose and Melville Clark, Jr., *Plasmas and Controlled Fusion*, MIT Press and John Wiley, 1961, 13.3.

10. J. E. Kunzler, E. Buehler, F. S. L. Hsu, and J. H. Wernick, "Superconductivity in Nb_3Sn at High Current Density in a Magnetic Field of 88 Kilogauss," *Physical Review Letters 6* (1961), 89–91.

11. Lyman Spitzer, Jr., "Particle Diffusion across a Magnetic Field," *Physics of Fluids 3* (1960), 659–661. Also see Ira B. Bernstein, Edward A. Frieman, Russell M. Kulsrud, and Marshall N. Rosenbluth, "Ion Wave Instabilities," *Physics of Fluids 3*, 136–137.

12. Stodiek, Ellis, Gorman, "Loss of Charged Particles in a Stellarator," cited in note 3; R. W. Motley, "Diffusion of Plasma from the Stellarator," Salzburg Conf. *1*, 199–201.

13. Author's interview with Furth and John L. Johnson; H. P. Furth, J. Killeen, and M. N. Rosenbluth, "Finite-Resistivity Instabilities of a Sheet Pinch," *Physics of Fluids 6* (1963), 459–484.

14. W. Stodiek, D. J. Grove, and J. O. Kessler, "Plasma Confinement in Low-Density C Stellarator Discharges," Culham Conf. *2*, 687–703, quotation on 689.

15. Lyman Spitzer, Jr., "Controlled Nuclear Fusion Research, September 1965: Review of Experimental Results," Culham Conf. *1*, 3–11.

16. E. P. Butt, H. C. Cole, A. N. Dellis, A. Gibson, M. Rusbridge, and D. Wort, "Conditions for Improved Stability in Zeta," Culham Conf. *2*, 751–764.

17. J. L. Tuck, "Los Alamos Research on Controlled Thermonuclear Reactions," *Proceedings of the Fourth International Conference on Ionization Phenomena in Gases, Uppsala, 17–21 August 1969*, Vol. II, 920–928, North Holland, 1960; Samuel Glasstone, ed., "Quarterly Status Report of the LASL Controlled Thermonuclear Research Program for Period Ending May 20, 1960," LASL Report LAMS-2444.

18. Author's interviews with Kerst and Tihiro Ohkawa.

19. L. A. Artsimovich, *Controlled Thermonuclear Reactions*, 1964, chapter 6. A. L. Bezbachenko et al. in *Plasma Physics and the Problem of Controlled Thermonuclear Reactions 5*, 135ff. G. G. Dolgov-Saveliev et al., "Investigations of the Stability and Heating of Plasma in Toroidal Chambers" *Proceedings, Geneva, 1958*, Vol. 32, 82–91. L. A. Artsimovich, "Soviet Research on Controlled Fusion," *New Scientist 279* (1962), 702–704.

In 1962 Artsimovich and a collaborator discovered that an additional, vertical field would permit higher temperatures. L. A. Artsimovich and K. B. Karteshev, "Effect of a Transverse Magnetic field on a Toroidal Discharge," *Soviet Physics—Doklady* 7, 919.

20. Author's interview with Gottlieb.

21. Spitzer, "Review of Experimental Results," cited in note 15, 9.

22. M. N. Rosenbluth to A. E. Ruark, May 18, 1964, "General Atomic, 1960–67," DOE-OMFE. See also the Pake committee report by the Plasma Physics Subcommittee, to which Rosenbluth is directing his remarks (M. N. Rosenbluth et al., "Plasma Physics," office files of G. K. Hess).

23. See the Standing Committee minutes for August 1966, when the members were obliged to discuss how to give CTR a better image in the eyes of the physics profession.

24. Author's interviews with Gough, Gottlieb, and Ramey; Michael D. Stiefel, "Government Commercialization of Large Scale Technology: The United States Breeders Reactor Program 1964–1976," Thesis, Department of Nuclear Engineering, MIT, 1981.

25. Interview with Hurwitz and Pollack by George Wise and the author.

26. George Wise to the author, June 3, 1981, with enclosures, "GE Research and Development Center," *Research Laboratory Bulletin*, Summer 1965, 3–4 (quotation by board chairman Gerald L. Phillippe, 3), and Leslie G. Cook, "How to Make R&D More Productive," *Harvard Business Review* 44, (1966), 145–151.

27. Irvin C. Bupp and Jean-Claude Derian, *Light Water*, New York, Basic Books, 1978, chapter 2.

28. L. G. Cook, H. C. Pollack, W. L. Robb, and J. E. Burke, "The Thermonuclear Fusion Reactor as a Source of Central Station Electricity," General Electric Research Laboratory, Report No. 65-RL-4012X, July 1965.

29. Cook et al., "The Thermonuclear Fusion Reactor," 7, and "Summary and Conclusions"; Henry Hurwitz, Jr., to George Wise, "Comments on Bromberg Draft," June 8, 1981 (courtesy of G. Wise).

30. H. Hurwitz, Jr., to W. C. Gough, July 19, 1965, "Review, June–July 1965," DOE-OMFE; "GE 1966–1972," *passim*, DOE-OMFE. The Cook report was not widely distributed, in part due to AEC intervention. (See file "GE 1966–1972," DOE-OMFE; author's interview with Hurwitz and Pollack; P. W. McDaniel to A. E. Ruark, August 12, 1965, "Review August–December 1965," DOE-OMFE.)

31. A. E. Ruark to J. M. B. Kellogg, December 16, 1964, "Review to May 1965," DOE-OMFE.

32. The others were Thomas H. Johnson; Solomon J. Buchsbaum, head of Bell Laboratories Solid State and Plasma Research Division; David D. Jacobus, a highly regarded mechanical engineer with long experience with accelerators and nuclear reactors; Gordon S. Brown, electrical engineer and dean of the MIT School of Engineering; Eugene N. Parker, an astrophysicist from the University of Chicago; and Peter L. Auer, a physicist and deputy director for ballistic missile defense of the Defense Department's Advanced Research Projects Agency. Biographical summaries, "Review to May 1965," DOE-OMFE.

33. Author's interview with Buchsbaum; "Review to May 1965," *passim*, DOE-OMFE.

34. Allison to McDaniel, August 4, 1965; W. C. Gough, "Informal Rough Notes on the Meeting of the Controlled Thermonuclear Research Review Panel. . . July . . . 1965" and "Rough Notes on the Meeting between the CTR Review Panel and the Commissioners . . . December 2, 1965. . . ," "Review, August–December 1965," DOE-OMFE; author's interview with Buchsbaum. It would be worthwhile to look more fully into the committee's deliberations because of the contrast between the sharpness of some of the criticisms recorded in committee meetings and the mildness of the final report.

35. 2176th commission meeting, February 8, 1966; Daniel J. Kevles, *The Physicists*, New York, Knopf, 1978; "Report of Review Panel." In particular, as a corrective to what they saw as a parochial, and laboratory-centered rather than program-centered attitude, and a stagnation of ideas, the committee advocated the establishment of a national fusion laboratory. This was also strongly supported by members of the CTR branch, but, predictably, was opposed by the laboratories: "Summary of Meeting on Controlled Thermonuclear Research Policy Report, Jan. 20–21, 1966" (office files of W. C. Gough). The Bureau of the Budget also opposed it because of its expense, and it eventually was eliminated from the policy document that the commissioners ratified.

36. Author's interview with Dean.

37. P. W. McDaniel to the author, March 2, 1979; author's interviews with Young and Gough; Norris E. Bradbury in "Summaries of Talks to the Panel at Los Alamos— July 12–13, 1965," "Review . . . June, July 1965," GAC minutes, April 1966, DOE-HO.

38. A. E. Ruark, "Material for the McKinney Report," December 31, 1959, microfilm GAMF 974, LASL-RC.

39. Author's interview with Dean.

40. Author's interviews with Furth and Dunlap.

41. GAC minutes, April 1966, DOE-HO.

42. "Amasa Bishop assesses U.S. effort in controlled fusion," *Physics Today*, May 1966, 74–78, quotation on 74.

43. A. S. Bishop to R. G. Herb, February 18, 1966, "Policy Report," DOE-HO.

44. P. W. McDaniel to R. E. Hollingsworth, "Controlled Thermonuclear Research Report," memorandum February 3, 1966, "Reactor Development II Project Sherwood," DOE-HO.

45. 2176th commission meeting, February 8, 1966.

46. Adam Yarmolinsky, *The Military Establishment* (1971), chapter 16, "Military Spending in the Economy."

47. Charles L. Schultze to G. T. Seaborg, February 16, 1966, and Donald F. Hornig, director of the Office of Science and Technology, to G. T. Seaborg, February 11, 1966, "RD-1, CTR," DOE-HO.

48. *Hearings, Joint Committee on Atomic Energy, 89th Congress, 2nd Session, AEC Authorizing Legislation, Fiscal Year 1967*, part 3 (February 2, March 8, 9, 10, 11 and 15, 1966), 1292–1303, with quotations on 1294 and 1292.

49. A. S. Bishop to CTR project directors, "Meeting with Presidents' Science Advisory Committee," March 25, 1966, "Policy Report," DOE-OMFE.

50. L. R. Hafstad to Glenn T. Seaborg, April 30, 1966, DOE-HO.

51. 2205th commission meeting, June 21, 1966; A. S. Bishop to G. F. Tape, June 13, 1966, "Policy Report," DOE-OMFE.

52. "AEC Policy and Action Paper on Controlled Thermonuclear Research," AEC 532/70, June 17, 1966; P. W. McDaniel to W. B. McCool, June 27, 1966, "Policy Report."

53. "Informal Rough Notes on the Meeting of the Controlled Thermonuclear Research Review Panel . . . July . . . 1965," "Review, Aug.–Dec. 1965," DOE-OMFE; Minutes of the General Advisory Committee, April 1966, DOE-HO.

54. Standing Committee minutes, May 1968, 4 (office files of W. C. Gough); author's interview with Taschek.

55. Author's interview with Grad. These paragraphs elaborate a point made to me by him.

56. Author's interviews with Ribe and Taschek.

57. E. M. Little, W. E. Quinn, and G. A. Sawyer, "Plasma End Losses and Heating in the 'Low-Pressure' Regime of a Theta Pinch," *Physics of Fluids 8* (1965), 1168–175; "Review of Controlled Thermonuclear Research at Los Alamos, 1965," LA-3253-MS (includes LA-3289-MS); author's interview with Ribe.

58. David Rose of MIT argued that this placement would be unworkable. The mechanical stresses on the magnet could not be supported in a satisfactory way, and the degree to which the magnet would obstruct the passage of neutrons from the containment vessel to the blanket would be unacceptable (author's interview with Rose).

59. "Review of Controlled Thermonuclear Research at Los Alamos, 1965," *passim*.

60. Author's interview with Riesenfield; "Review of Controlled Thermonuclear Research at Los Alamos, 1965."

61. H. P. Furth, " 'Dynamic Stabilization' with Oscillating Electric Fields Parallel to a Static Magnetic Field," August 3, 1966, "Ad Hoc Panel—Scyllac," DOE-OMFE.

62. J. L. Tuck to J. M. B. Kellogg, "A plan for Sherwood and Some Remarks on Planning," February 27, 1961 (office files of J. A. Phillips); author's interview with John Marshall, Jr.

63. J. A. Phillips, " 'Flap' in Washington," and J. M. B. Kellogg, "Sherwood Meeting, Washington, November 18 and 19 [1963]," microfilm GAMF 974, LASL-RC. It is possible, according to Phillips and Kellogg, that this was only being considered as a political gambit, that is, McDaniel felt that LASL's Sherwood had a sufficiently excellent reputation, and sufficient support from Senator Clinton Anderson, JCAE member from New Mexico, to ensure that such a cut would be restored.

64. Testimony to the Joint Committee on Atomic Energy in March 1964, reprinted in "Review of Controlled Thermonuclear Research at Los Alamos, 1965," 38.

65. Author's interviews with Norris E. Bradbury, Reisenfield, and Taschek.

66. "Report of the Ad Hoc Panel on Scyllac," *passim*, quotation on 26–27, "Ad Hoc panel on Scyllac," DOE-OMFE; author's interview with Hans Griem.

67. Harold P. Furth to Hans Griem, August 19, 1966, microfilm GAMF 974, LASL-RC.

68. Standing Committee minutes, August and September 1966. It is important to note, however, that the plans for the 15-meter pinch were never completely carried out (see chapter 12).

69. D. W. Kerst to A. E. Ruark, January 27, 1961, "General Atomics 1960–67," DOE-OMFE; author's interview with Kerst.

70. Author's interviews with Kerst and Ohkawa.

71. James L. Tuck, "A New Plasma Confinement Geometry," *Nature 187* (1960), 863–864. See chapter 4 for the picket-fence concept.

72. T. Ohkawa and D. W. Kerst, "Multipole Magnetic Field Configurations for Stable Plasma Confinement," *Il Nuovo Cimento 22* (1961), 784–799 (picture on 786, quotation on 785). D. W. Kerst to fusion staff, "Elements of the Thermonuclear Problem," May 8, 1961 (office files of T. Ohkawa).

73. R. A. Dory, D. W. Kerst, D. M. Meade, W. E. Wilson, and C. W. Erickson, "Plasma Motion and Confinement in a Toroidal Octupole Magnetic Field," *Physics of Fluids 9* (1966), 997.

74. Author's interview with Ohkawa.

75. T. Ohkawa, A. A. Schupp, M. Yoshikawa, and H. G. Voorhies, "Toroidal Multipole Confinement Experiment," Culham Conf. 2, 531–543; Dory et al., "Plasma Motion and Confinement in a Toroidal Octupole"; "Further General Atomic Experiments with Multipoles," included with letter from Bernard B. Smyth to A. S. Bishop, March 30, 1967, "General Atomic 1960–1967," DOE-OMFE.

76. D. M. Meade, "Stability of a Collisionless Plasma Confined by a Toroidal Octupole Magnetic Field," abstract 6C-8, *Bulletin of the American Physical Society*, 1967, 790. In mid-1967, however, the evidence was not yet extensive enough to merit the general conclusion that multipoles had evaded Bohm diffusion generally. See the report of the Ad Hoc Panel on Low Beta Toroidal Research, 30 ("Ad hoc Panel—Low Beta Toroidal Research," DOE-OMFE).

77. Author's interview with Gough. Dory et al., "Plasma Motion and Confinement in a Toroidal Octupole"; Kerst deliberately worked with cool plasmas in order that every available diagnostic instrument, including material probes which would vaporize if the plasma is too hot, could be utilized (author's interview with J. C. Sprott). Ohkawa's temperatures were about twice as high.

78. Interview with H. R. Drew by Robert B. Belfield; author's interview with Ohkawa. Ohkawa believes that the nationwide disaffection of industry with fusion research also came into play here. A. E. Ruark to P. W. McDaniel, "Dame Rumor Says. . . ," "General Atomic 1960–1967," DOE-OMFE; Bernard B. Smyth to Bishop, September 1, 1966, "General Atomic 1960–1967," DOE-OMFE.

79. Author's interview with Coppi; file, "General Atomic 1960–1967," DOE-OMFE, *passim.* Standing Committee minutes for November 1966.

80. Standing Committee minutes, December 14–15, 1967.

Chapter 9

1. L. A. Artsimovich et al., "Experiments in Tokamak Devices," *Nuclear Fusion, Special Supplement 1969, English Translations of the Russian Papers . . . At Novosibirsk . . .* , 17–24;

L. A. Artsimovich, "Survey on Closed Plasma Systems," Novosibirsk Conf. *1*, 11–17; Harold P. Furth, "Progress Toward a Tokamak Fusion Reactor," *Scientific American*, August 1979, 55.

2. Artsimovich, "Survey on Closed Plasma Systems"; D. J. Grove et al., "Confinement of Gun-Injected Hydrogen Plasmas and Microwave-Produced Xenon Plasmas in the C-Stellarator," Novosibirsk Conf. *1*, 479–496. German investigators were also getting confinement times many multiples of Bohm times in low-density stellarator experiments (Artsimovich, "Survey").

3. PPPL Staff, "The Proposed Princeton Quadrupole," April 1967 (Box "M. B. Gottlieb, 1952–1971," PPPL warehouse); "Rough Notes" on the December 12–13, 1968, Standing Committee meeting (filed with Standing Committee minutes), author's interview with Shoichi Yoshihawa, September 19, 1979.

4. Author's interviews with Furth and Bruno Coppi.

5. Artsimovich et al., "Experiments on Tokamak Devices."

6. Author's interviews with Furth and Gottlieb.

7. Standing Committee minutes for May 1968, December 1968, and March 1969.

8. "Rough Notes," Standing Committee, December 1968.

9. This comment, the notes record, was followed by "commotion."

10. See chapter 7.

11. Harold Grad, David J. Rose, and Warren Heckrotte, "Report of the 1965 Outside Committee on the Thermonuclear Division to the Director," February 1, 1966 (office files of J. Rand McNally, ORNL).

12. Alvin M. Weinberg to Drs. H. Grad, Waren Heckrotte, and D. J. Rose, March 30, 1966 (office files of J. Rand McNally).

13. "Report of the 1965 Outside Committee."

14. Weinberg to Grad, Heckrotte, and Rose (the committee had recommended speed); Standing Committee minutes, August 1967. Superconducting magnets were called for because "to get a magnetic well, you must have two sets of magnetic fields that buck each other, in part at least, leaving a difference field that still has to be strong enough to contain the plasma. This means that you have to pour much more magnet power into a given volume, you have to supply much more cooling, and much more mechanical strength to prevent the coils from twisting. . . . These problems forced us at Oak Ridge to go superconducting" (Arthur H. Snell to the author, October 17, 1980.)

15. Author's interviews with Ray A. Dandl and Arthur H. Snell; Arthur H. Snell et al., "Controlled Thermonuclear Research at Oak Ridge National Laboratory," February 1965; W. B. Ard et al., "Use of a Hot-Electron Target Plasma for Accumulation of Energetic Ions in Stabilized Magnetic Mirror Traps," Madison Conf. *1*, 619–629.

16. "Special Analytic Study No. 67-6 CTR Research Program," AEC 532/72, April 10, 1967; Snell to the author, October 17, 1980; A. H. Snell, "ORNL Statement on the Alice II - DCX 4 Situation" and memorandum to members of the Standing Committee, April 14, 1967 (office files of R. F. Post); author's interview with Snell; minutes of the Standing Committee, February, 1967.

17. Author's interviews with Snell, Rose, Dandl, Herman Postma, and Gareth E. Guest.

18. Report of the Outside Advisory Committee of 1965: "The main question is whether the present plasma [of DCX-2] is a fluke, peculiar to the exact combination of highly directional injection, long uniform field, small auxiliary mirror field, and specific operating parameters. If so, it certainly does not pay to spend more time studying it." Author's interview with John F. Clarke.

19. Author's interviews with Clarke and George G. Kelley.

20. Author's interviews with Clarke and Michael Roberts. The six delegates were Postma, Clarke, Roberts, Norman Lazar, Gareth Guest, and Roger Neidigh (M. Roberts, "The Birth of Ormak," *Oak Ridge National Laboratory Review*, winter 1974, 12–17).

21. Author's interviews with Postma and Guest.

22. They spoke with Artsimovich.

23. Author's interview with Clarke.

24. J. F. Clarke and M. Roberts, "Report of Foreign Travel . . . Dubna, September 29–October 3, 1969," 69-11-50 ORNL-CF.

25. Author's interview with Roberts.

26. G. G. Kelley, M. Roberts, and J. F. Clarke, "Conceptual Design of Ormak Facility," December 3, 1969, ORNL-TM-2821; author's interviews with Roberts and Clarke; Roberts, "The Birth of Ormak."

27. Artsimovich, "Experiments in Tokamak Devices," 24.

28. Estimate in G. G. Kelley, O. B. Morgan, L. D. Stewart, and L. Stirling, "Neutral Beam Injection Heating of Toroidal Plasmas for Fusion Research," *Nuclear Fusion 12* (1972), 171.

29. Author's interview with O. B. Morgan and Lee Berry, February 7, 1979.

30. Author's interviews with Smullin and Coppi; L. A. Artsimovich, "Researches in High-Temperature Plasma Physics in the U.S.S.R." lecture, and "Acad. L. A. Artsimovich-Itinerary" (office files of Bruno Coppi).

31. "Biographical Sketch, David J. Rose" (office files of D. J. Rose); author's interviews with Postma, Smullin, and Coppi.

32. Author's interviews with Ohkawa, Smullin, and Robert L. Hirsch.

33. The reader may recall that the ions can lose their charge to neutral atoms desorbing from the walls, and, when they do, the magnetic field no longer contains them. They are "charge-exchange neutrals," and a measurement of their energy gives the energy distribution of the ion population.

34. "Scientific Committee Report on Recent Advances of Importance to Controlled Thermonuclear Research Planning," AEC 532/76, May 15, 1969.

35. Author's interviews with Roberts and Coppi; Roberts, "The Birth of Ormak."

36. Bruno Coppi and Jan Rem, "The Tokamak Approach in Fusion Research," *Scientific American 227* (1972), 65–75; JCAE, *The Current Status of the Thermonuclear Research Program in the United States, 1971*, testimony and statements of William E. Drummond.

37. Author's interviews with Smullin, Coppi, Ronald R. Parker, and D. Bruce Montgomery.

38. Interviews with Montgomery, Parker, and Coppi; Coppi and Rem, "The Tokamak Approach in Fusion Research," *Scientific American* 227 (1972), 65–75; Ronald R. Parker, "Ohmic Heating in Tokamaks," draft, March 27, 1979, MIT Plasma Fusion Center; "High Field Tokamak Experiment at MIT," AEC 532/82, January 7, 1970. ("The design target for plasma density, temperature, confinement times and magnetic fields are those typical of a steady state fusion reactor.") Coppi and Montgomery contemplated fields of 130 kilogauss, about four times the maximum fields in T-3.

39. Coppi and Rem, "The Tokamak Approach," 75. Coppi sketched a program for Alcator weighted toward basic studies of both fusion and astrophysical interest ("An Advanced Experimental Facility, ALCATOR . . . ," Proposal, September 1969, office files of B. Coppi).

40. Amasa S. Bishop, "Recent Developments in the CTR Program," May 2, 1969 (office files of W. C. Gough).

41. Author's interview with Ohkawa; R. L. Hirsch to Paul W. McDaniel, January 21, 1970, "Doublet II CTR Experiment at Gulf General Atomic," ("R&D 18, CTR-Vol. 1, 1966–1970," DOE-HO); Testimony by Tihiro Ohkawa in JCAE, *The Current Status of the Thermonuclear Research Program, 1971.* Current through a double-lobed plasma emulates current through two separate conductors.

42. Author's interview with Solomon J. Buchsbaum.

43. By removing the straight sections and joining the curved ends into a torus.

44. F. L. Ribe, "Comments on the Tokamak Experiments," April 25, 1969, microfilm GAMF-2405, 1C-1666, LASL-RC.

45. Author's interviews with Buchsbaum and James D. Callen.

46. Interview with Gottlieb by Neil S. Wolf, March 1977.

47. *Standing Committee minutes, June 1969.*

48. N. J. Peacock, D. C. Robinson, M. J. Forrest, P. D. Wilcock, and V. V. Sannikov, "Measurement of the Electron Temperature by Thomson Scattering in Tokamak T3," *Nature*, November 1, 1969, 488–490; Amasa S. Bishop, "Confirmation of the Soviet Tokamak Data," August 13, 1969, AEC 532/78.

49. Author's interview with Furth; V. V. Sannikov and Yu. A. Sokolov, "Escaping Electrons in Tokamak TM-3," *Fizika Plazmy (USSR)* 2 (1976), 204–211.

50. Minutes of the June and October 1969 Standing Committee meetings; author's interview with Postma.

51. October 1969 Standing Committee minutes.

52. *Hearings, JCAE, Authorizing Legislation, FY 1971*, March 3, 1970, 563.

53. Richard F. Taschek to Roy W. Gould, May 14, 1970 (see also Harry Dreicer to R. F. Taschek, May 13, 1970, both in microfilm GAMF 975, LASL-RC).

54. G. G. Kelley, O. B. Morgan, H. Postma, and H. K. Forsen, "Neutral Beam Ion Heating in Ormak II," July 1970, ORNL-TM-3076.

55. M. B. Gottlieb to Roy W. Gould, October 23, 1970, "Princeton Plasma Physics Laboratory 1969–1970," DOE-OMFE; Roy W. Gould to Paul W. McDaniel, "Ten Year Plan for the Princeton Plasma Physics Laboratory," April 28, 1971, "Princeton Plasma Physics Laboratory, 1971," DOE-OMFE.

56. Harold P. Furth to the author, November 10, 1981.

57. Roberts, "The Birth of Ormak," 14.

58. Author's interviews with Kelley, Roberts, and Clarke; S. G. English to Edward J. Bauser, February 13, 1970, reprinted in JCAE, *Hearings . . . (91/2), AEC Authorizing Legislation, FY 1971* part 2, 561–562.

59. Author's interview with Postma. Author's interviews with Morgan and Kelley; Herman Postma to G. G. Kelley, memorandum, "Ormak Future Plans," June 1, 1970 ("O. B. Morgan retired files 404.1.1," Fusion Energy Division Storage Room, ORNL).

60. Kelley, Morgan, Postma, and Forsen, "Neutral Beam Ion Heating in Ormak II."

61. This is because very intense beams of ions tend to disperse from mutual repulsion as they travel down the long distance of the magnetic lens. Plasma instabilities also occur (author's interview with Morgan and Berry).

Chapter 10

1. John Quarles, *Cleaning Up America*, Boston, Houghton-Mifflin, 1976.

2. Philip M. Boffey, "Energy Crisis: Environmental Issue Exacerbates Power Supply Problem," *Science 168* (1970), 1554–1559.

3. See, for example, James Everett Katz, *Presidential Politics and Science Policy*, New York, Praeger 1978, 188–191; Philip M. Boffey, "Energy Crisis;" Richard M. Nixon, energy message of June 4, 1971, reprinted in *Hearing, JCAE, 93rd Congress, 1st Session . . . Atomic Energy Commission [Dec. 1, 1973] Report (on) . . . Energy Research and Development*, Dec. 11, 1973; "Remarks Presented by Dr. Edward E. David Jr., Science Advisor to the President, at National Academy of Sciences Symposium . . . April 26, 1971 . . . ," "PSAC-OST 1971," DOE-OMFE.

4. *New York Times*, May 26, 1971, 84; Katz, *Presidential Politics*.

5. "Impact of Recent Developments on the U.S. CTR Program," May 12, 1969, AEC 531/76; "Invigoration of the US CTR Program," August 1969, AEC 532/77; A. S. Bishop, "Conference Summary," *Nuclear Fusion Reactors, Proceedings of the. . . . Conference, Culham Laboratory, September 1969*, 576–577.

6. B. I. Eastlund and W. C. Gough, "The Fusion Torch," *Nuclear Fusion Reactors*, 421. Laboratory leaders, like Gottlieb of Princeton and Rose of MIT, pointed out that there was no adequate scientific and technological basis for touting the fusion torch. See, for example, the testimony of David J. Rose, *House Committee on Science and Astronautics, Task Force on Energy, Hearings . . . November 1971*.

7. *New York Times*, March 8, 1971, 1:5; March 11, 1971, 30; July 7, 1971, 1:7; July 20, 1971 (editorial), 32:2.

8. Issue of March 9, 1971, reproduced in the *Hearings, JCAE, 92nd Congress, 1st Session, AEC Authorizing Legislation, Fiscal Year 1972*, part 3, March 9, 16, 17, 1971, 1225–1226.

9. *Cleveland Plain Dealer*, April 1, 1971, "PPPL-1971," DOE-OMFE.

10. Mike Gravel to Hannes Olof Alfven, March 16, 1971, "R&D-18, CTR Vol. II," DOE-HO; Egan O'Connor to Chester Van Atta, June 9, 1970, Livermore Magnetic Fusion Energy Division, General Correspondence Files.

11. *New York Times*, April 27, 1971, 40:4; see also Standing Committee minutes for January 1971.

12. Many appear in the AEC controlled thermonuclear research files for 1970, 1971, 1972, either sent directly to the commissioners or transmitted by Congress, DOE-HO.

13. *Hearings, JCAE, AEC Authorizing Legislation, Fiscal Year 1972, 92nd Congress, 1st Session*, part 4, May 13, 1971, 2486. Holifield was a breeder advocate whose "main goal" before he "hangs up his gloves" was to get the breeder going, according to W. Henry Lambricht (*Governing Science and Technology*, New York, Oxford, 1976). Similar remarks about constituent pressure by other congressmen can be found throughout the JCAE 1971 hearings, for example, by John O. Pastore, in *The Current Status of The Thermonuclear Research Program*, November 1971, part I, 12.

14. This analysis omits the further step between reactor studies and actual conditions: "How difficult and how expensive it will be to keep track of tens of kilograms of tritium to an accuracy of one part in 10^6 per day in the real world of leaky valves, faulty seals, scratched diffusion barriers, and so on, will not really be known until we have tried it." W. Haefele, J. P. Holdren, C. Kessler, "Normal Operating Losses and Exposures," in W. Haefele, J. P. Holdren, G. Kessler, and G. L. Kulcinski, *Fusion and Fast Breeder Reactors*, 1977, 271–272.

15. D. J. Rose, "On the Engineering Problems of Controlled Fusion, 6/65," "Review: June–July 1965," DOE-OMFE; author's interview with Rose; D. J. Rose, "Engineering Feasibility of Controlled Fusion," *Nuclear Fusion 9*, (1969), 183–203.

16. See chapter 7.

17. D. J. Rose, "Engineering Feasibility of Controlled Fusion," especially section II, "Summary of Work Hitherto." T. K. Fowler and M. Rankin, "Plasma Potential and Energy Distribution in High-Energy Injected Machines," *Plasma Physics 4*, (1962), 311–320.

18. H. P. Furth and R. F. Post "Advanced Research in Controlled Fusion," December 10, 1964, UCRL-12234.

19. Author's interview with Ruark.

20. Hilliard Roderick and Arthur E. Ruark, "Thermonuclear Power," *International Science and Technology*, September 1965.

21. A. E. Ruark to P. W. McDaniel, June 25, 1965, "Review, June–July 1965," DOE-OMFE; A. E. Ruark to R. Herb, October 5, 1965, "Review August–December 1965," DOE-OMFE; author's interviews with Gough, Rose, Ruark, and Buchsbaum.

22. Author's interviews with Rose, Postma, and Don Steiner.

23. Amasa S. Bishop to CTR Project Heads, "The Question of Convening a Study Project on the Technological Aspects of CTR," January 17, 1968, microfilm GAMF-2063 LASL-RC; author's interview with Gough.

24. Author's interview with Gough.

25. Memorandum from D. J. Rose to Herman Postma, "Controlled Fusion Feasibility Plans," February 13, 1968 (office files of Don Steiner, ORNL).

26. W. C. Gough to Roy W. Gould, "Fusion Reactor Technology Program," July 19, 1972 (files of W. C. Gough); W. C. Gough to A. S. Bishop, "Meeting on a Proposed

Study Project on the Technological Aspects of CTR," March 5, 1968, microfilm GAMF-2063 LASL-RC.

27. Author's interviews with Werner, Post, and Fowler.

28. R. W. Moir, W. L. Barr, R. P. Freis, and R. F. Post, "Experimental and Computational Investigations of the Direct Conversion of Plasma Energy to Electricity," Madison Conf. *3*, 316; author's interviews with Post and Werner. Post's scheme was based upon a kind of reversal of the mirror effect. In mirrors, particles are kept within the plasma by the circumstance that as they move toward the ends, they see an increasing magnetic field and respond by converting the longitudinal component of their velocity into the transverse component until those particles with sufficiently small initial longitudinal velocities lose all endward movement and are stopped. Post proposed that the ions driven out through the mirrors be made to see a gently decreasing magnetic field outside the fusion region. Such particles would then go through the inverse process of converting their transverse velocity into a streaming, longitudinal motion. The electrons accompanying them could be separated off, and the ions could be gradually decelerated by a series of electrodes until they lost all forward motion and fell into a collector. The currents from these collectors would be combined and sent out directly over transmission lines.

29. *Nuclear Fusion Reactors, Proceedings of the British Nuclear Energy Society Conference . . . Held at the UKAEA Culham Laboratory, September 17–19, 1969.* Only 2 papers, by the Russians Golovin, Dnestrovsky, and Kotomarov and by the Frenchman Hubert, were contributed by scientists outside the United Kingdom or United States. The other 39 were approximately evenly divided between the British and the Americans. The conference was followed by workshop sessions extending over several weeks.

30. Gough to Gould, "Fusion Reactor Technology Program."

31. Author's interview with Mills; Mills did not believe that enough was known about plasmas in 1971 to make intelligent assumptions about reactors, but he was interested in reactor studies and was spurred on by the fact that Princeton had a reactor design tradition and, above all, by pressure from Washington that was transmitted through laboratory director Gottlieb.

32. Author's interviews with Fred L. Ribe and Robert Krakowski.

33. Author's interview with Gough and Buchsbaum.

34. Author's interviews with Gerald L. Kilcinski, Robert W. Conn, Mills and Krakowski.

35. Commissioner Theos J. Thompson, speech at the Division of Plasma Physics session at the meeting of the American Physical Society, November 5, 1970, reprinted in *JCAE, Current Status of the Thermonuclear Research Program, November 1971, 1,* 7.

36. Author's interviews with Kulcinski and Conn. It was also important to these men and women that the tokamak was working so much better than other machines had by 1971. This meant that fusion also had begun to seem a much more realizable aim.

37. C. M. Van Atta to James Tuck, April 10, 1964 and H. W. Van Ness and T. H. Batzer to E. L. Kemp, November 17, 1964, microfilm GAMF-974, LASL-RC; author's interviews with Kemp, Thomas Batzer, and Edward Bettis. See Tanner, *Project Matterhorn,* for some material on Princeton hardware engineers.

38. By June 1974, it was to have 446 members, a 73% increase over 1973. Minutes of the Technical Group for Controlled Nuclear Fusion, Executive Committee, meeting, June 23, 1974; author's interview with Werner.

39. Vugraphs for "Review, Univ. of Wisconsin Fusion Engineering Program," April 25, 1979; author's interview with Conn.

40. Robert L. Hirsch, "Lasers and Intense Relativistic Electron Beams," *JCAE, The Current Status of the Thermonuclear Research Program, November 1971*, part 2, 385–400.

41. Hirsch, "Lasers."

42. Robert Gillette, "Laser Fusion: An Energy Option, but Weapons Simulation Is First," *Science 188* (1975), 30–34.

43. Hirsch, "Lasers."

44. Author's interview with Roy W. Gould.

45. Gillette, "Laser Fusion."

46. Gillette, "Laser Fusion"; author's interview with Boyer.

47. Author's interview with Edwin E. Kintner.

48. K. M. Siegel to Glenn T. Seaborg, January 16, 1970, and November 25, 1969, "R&D-18, CTR Vol. I," DOE-HO; author's interviews with Gould and Boyer.

49. Chet Holifield to G. T. Seaborg, May 27, 1970, "R&D-18, CTR, Vol. I," DOE-HO.

50. "R&D-18, CTR," *passim*.

51. Author's interviews with Gould, Rose, and Boyer.

52. J. L. Tuck to G. T. Seaborg, August 14, 1970, "R&D-18, CTR Vol. I," DOE-HO.

53. *Hearings, JCAE, AEC Authorizing Legislation, FY 1972* (92/1), part 4, May 13, 1971, 2486.

54. J. T. Ramey to Howard R. Drew, November 25, 1970, "R&D-18, CTR Vol. I."

55. Blake Myers to T. K. Fowler, January 17, 1971, "Draft Proposal for Fusion Reactor Design Program," files of Magnetic Fusion Energy Division, LLNL; Daniel Axelrod, "Draft," "Project Heads—1965–1973," DOE-OMFE; Standing Committee minutes, "Rough Notes," December 12, 1968.

56. Standing Committee minutes, January 1971, 4.

57. Standing Committee minutes, July 1970, 3, and June 1972, 4; author's interviews with Gould and Rose.

58. General manager to Hubert H. Humphrey, April 7, 1971; general manager to George McGovern, February 12, 1971, "R&D-18, CTR Vol. II."

59. Testimony of Commissioner C. E. Larson, *JCAE, The Current Status of the Thermonuclear Research Program, November 1971*, part 1, 8.

60. Author's interview with Hirsch; "Miami Meeting, Introductory Notes," for the Standing Committee meeting at Key Biscayne, Florida, July 1973 (filed with Standing Committee minutes); interview with William L. R. Rice by George K. Hess and the author.

61. Author's interviews with Gould and Hirsch.

62. Author's interviews with Gould and J. Ronald Young.

63. This section draws heavily on research and interviews conducted by Robert B. Belfield.

64. Roy W. Gould to Paul W. McDaniel, memorandum, "Utility Industry Interest in the CTR Program," March 17, 1971, SECY-1190, DOE-HO; minutes of the July 1971 Standing Committee.

65. Robert A. Baker to James T. Ramey, August 27, 1970 (files of W. C. Gough). For background, see Arturo Gandara, "Electric Utility Decisionmaking and the Nuclear Option," Rand Report R-2148-NSF, June 1977.

66. R. L. Hirsch to files, "Utility Support for CTR (Revised)," September 7, 1971, "Electric Utilities #1 (EEI)," DOE-OMFE.

67. "Electric Utilities Industry Research and Development Goals through the Year 2000," Report of the R&D Goals Task Force to the Electric Research Council, June 1971, abridged reprint in *JCAE, The Current Status of the Thermonuclear Research Program, November 1971, 1,* 207.

68. Howard Drew to Roy Gould, January 11, 1971, "Electric Utilities #1," DOE-OMFE.

69. "IEEE Fellow Grade Nomination Form for R. A. Huse" (courtesy of Dr. Betty K. Jensen); "Breeder is Riding High, Fusion Outlook Appears Hopeful," *Nuclear Industry 22* (1975), 23–26; interview with Howard R. Drew by R. B. Belfield; H. R. Drew to R. B. Belfield, February 21, 1980.

70. James A. Phillips to R. F. Taschek, "Some Comments on LASL Meeting with New Jersey Utilities, March 26, 1971," March 29, 1971, microfilm GAMF 3073 LASL-RC; author's interview with Taschek.

71. R. L. Hirsch to R. W. Gould, no date, "Electric Utilities #1"; author's interview with Hirsch. For a time, the fusion report was handled by PSE&G research engineer Daniel Axelrod, but he was retired for enthusiasm unbridled by realism.

72. *Hearings, JCAE, 92nd Congress, 1st Session, AEC Authorizing Legislation Fiscal Year 1972,* part 3, March 9, 16, and 17, 1971, 1234–1235; Standing Committee minutes, July 1970, 3; author's interview with Hirsch.

73. Author's interviews with Hirsch and Gould; Edward E. David, Jr. to G. T. Seaborg, April 2, 1971; E. E. David to G. T. Seaborg, May 6, 1971, "OST-PSAC 1971," DOE-OMFE.

74. Roy W. Gould to P. W. McDaniel, May 11, 1971, "OST-PSAC 1971."

75. Much of the following is inferred from tapes of the May 14 meeting, which have kindly been lent to me by W. C. Gough.

76. PLT was ultimately cut to 1.2 million amperes current in the design, for engineering reasons, and was further cut to 0.6 million amperes to keep down the costs (author's interview with Yoshihawa); *Hearings, JCAE, 92nd Congress, 2nd Session, AEC Authorizing Legislation, FY 1973, Jan. 26, Feb. 3 & 17, 1972,* part 3, appendix 19; Roy Bickerton, "Review of Experimental Work," Madison Conf. *3,* 619–625. The temperature given is without any auxiliary to the ohmic heating.

77. Tapes of May 14 meeting, side 1. It was probably Gottlieb, serving here as devil's advocate.

78. "Definition of the Scientific Feasibility of Controlled Fusion," document "sent to O[ffice of] M[anagement and] B[udget] on 7/28/(71?)" (office files of W. C. Gough).

79. "Proposed Reply to Dr. David's May 6 Request for Analysis of Approaches to CTR Development," May 25, 1971, SECY 1590, "R&D-18, CTR," DOE-HO.

80. Standing Committee minutes, April 1971.

81. *JCAE, The Current Status of the Thermonuclear Research Program in the United States,* Nov. 1971, 161–162; author's interviews with Gould and Hirsch.

82. Letter to R. A. Huse, Apr. 13, 1971, "Electric Utilities #1."

83. *JCAE, Current Status, November 1971,* part 1, 160–162.

84. *JCAE, Current Status, November 1971,* part 1, 230.

85. File, "National Science Foundation, 1965–1971", DOE-OMFE, *passim.*

86. *New York Times,* February 7, 1971, 61; Roy W. Gould to Fred Ribe, Telecon, Box, "Ribe correspondence," LASL-RC.

87. John P. Abbadessa to Mike Gravel, December 2, 1971, "R&D-18, CTR."

88. Author's interview with Ben Loeb; *New York Times,* July 22, 1971, 1, 18; *ibid.,* August 20, 1971, 10, *ibid.,* November 26, 1972, 1; Robert Gillette, "Schlesinger and the AEC: New Sources of Energy," *Science 175* (1972), 147–151.

89. USAEC Announcement No. 231, December 14, 1971, DOE-HO; author's interviews with Spofford G. English and Enzi DeRenzis.

90. Author's interviews with Hirsch, Gould, English, DeRenzis, and Stephen O. Dean.

Chapter 11

1. Robert Gillette, "Schlesinger and the AEC," *Science 175* (1972), 147–151.

2. "America's Fusion Director," *The New Scientist,* April 12, 1973, 88.

3. Guenter Kueppers, "Fusionsforschung-Zur Zielorientierung im Bereich der Grund-lagenforschung," in W. v. d. Daele, W. Krohn, and P. Weingart, eds., *Geplante Forschung* (1972).

4. R. L. Hirsch, "Organizational Developments in the Controlled Thermonuclear Research Program," December 7, 1972, memorandum; John C. Ryan, "Organizational Developments in the Controlled Thermonuclear Research Program," December 2, 1972, SECY 2833; "R&D-18, CTR, Vol. 2," DOE-HO.

5. CTR organization charts, June 1972–June 1977 (office files of Roberta Cunningham).

6. Author's interview with Fowler; minutes of the Standing Committee of October 10–12, 1972; T. K. Fowler to R. L. Hirsch, September 29, 1972, "Lawrence Livermore Laboratory 1972," DOE-OMFE; R. L. Hirsch to CTR Standing Committee, December 12, 1972, "December 20–21 Meeting at Princeton," memorandum with marginal notes, file 17-4, Magnetic Fusion Division, LLNL; interview with William L. Rice by G. K. Hess and the author.

7. "Fusion Power: An Assessment of Ultimate Potential and Research and Development Requirements," July 1972, "R&D 18 (CTR)—Bulky Package," DOE-HO.

8. R. L. Hirsch to the commissioners, October 24, 1972, "The Next Major Goal for the Controlled Thermonuclear Research Program," "R&D-18 CTR," DOE-HO; minutes of the Standing Committee for October 1972.

9. Author's interviews with Fowler and Dean; Standing Committee minutes, October 1972, 15; Bennett Miller to R. L. Hirsch, "Comments on the Superconducting Levitron and its Relationship to FM-1," September 13, 1972, "Lawrence Livermore Laboratory 1972," DOE-OMFE.

10. Keith Brueckner, preliminary draft of the report of the ad hoc panel on Astron, p. 26, "Ad Hoc Panel on Astron Program," DOE-OMFE.

11. "Report of the Review Panel on Controlled Thermonuclear Research," 4th draft, November 2, 1965, "Review August–December 1965," DOE-OMFE.

12. Roy W. Gould, "Report on the Livermore Astron Program," June 13, 1972, "LLL-Astron 1960-72," DOE-OMFE.

13. L. D. Smullin, "Comments on THE ASTRON PROGRAM," December 6, 1967, "Ad Hoc Panel on Astron Program," DOE-OMFE.

14. Author's interview with Dean; Standing Committee minutes, 1967–1971, *passim*.

15. J. J. Bzura, T. J. Fessenden, H. H. Fleischmann, D. A. Phelps, A. C. Smith, and D. M. Woodall, "Generation of Field-Reversing Rings of Relativistic Electrons Using Injection of High-Current Electron Beams," *Fifth European Conference on Controlled Fusion and Plasma Physics, Grenoble, France . . . 1972, 1*, 96; Hans H. Fleischmann, "Use of Strong Relativistic Electron Rings for the Confinement of Thermonuclear Plasmas," *IEEE Transactions, Nuclear Science*, NS-20, #3, 966–969. These articles cite M. L. Andrews et al., *Physical Review Letters 27* (1971), 1428, and J. J. Bzura et al., *Physical Review Letters 29* (1972), 256.

16. L. D. Smullin, chairman, "Report of Astron Review Committee," March 8, 1972 (office files of G. Lubkin), 17.

17. Smullin, "Report"; "Minutes of the Meeting [of the Astron Review Panel] of November 5–6, 1971," microfilm GAMF-975, LASL-RC.

18. Author's interview with Dean; R. W. Gould to S. G. English, January 10, 1972, and T. K. Fowler to R. W. Gould, April 14, 1972, "Lawrence Livermore Lab. 1972," Division; Standing Committee minutes, March 1972; R. W. Gould to S. G. English, "Report on the Livermore Astron Program," June 13, 1972, "LLL Astron, 1960–1972," DOE-OMFE.

19. Because of the combination of closed magnetic lines and absolute minimum-B (author's interview with Fowler).

20. Roger E. Batzel to R. L. Hirsch, September 28, 1972, "Lawrence Livermore Lab. 1972," DOE-OMFE; "Organizational Developments in the Controlled Thermonuclear Research Program," December 7, 1972, "R&D-18 CTR, Vol. 2," DOE-HO; the quotation is by S. O. Dean, from his interview (italics added).

21. Author's interview with Hirsch; R. L. Hirsch to Floyd Culler, April 24, 1973, "CTR 38 ORNL, January–June 1973," DOE-OMFE.

22. Author's interview with Hirsch.

23. Robert L. Hirsch, "Reflections on DT Tokamaks," memorandum, December 26, 1973, "CTR 38 PPPL, TFTR 1973," DOE-OMFE.

24. Author's interview with Dean; William D. Metz, "Nuclear Fusion: The Next Big Step Will Be A Tokamak," *Science 189* (1975), 421–423.

25. "Status and Objectives of Tokamak Systems for Fusi.... Research," WASH 1295, 55–57.

26. George K. Hess, Jr., "History of the Magnetic Fusion Energy Program," manuscript, 1977, 19 and 24 (office files of G. K. Hess, DOE-OMFE); K. Bol et al., "Adiabatic Compression of the Tokamak Discharge," *Physical Review Letters 29*, (1972), 1495–1498;

K. Bol et al., "Neutral Beam Heating in the Adiabatic Toroidal Compressor," *Physical Review Letters 32*, (1974), 661–664.

27. Author's interview with Callen; S. O. Dean, J. D. Callen, J. F. Clarke, H. P. Furth, T. Ohkawa, and P. H. Rutherford, "Status and Objectives of Tokamak Systems for Fusion Research," WASH 1295; M. N. Rosenbluth to R. L. Hirsch, June 29, 1973 (office files of W. C. Gough).

28. See WASH 1295, cited in note 27; B. B. Kadomtsev and O. P. Pogutse, "Trapped Particles In Toroidal Magnetic Systems," *Nuclear Fusion 11* (1971), 67–92. As temperatures rise, plasma particles bump into each other less and less frequently until the plasma enters a "collisionless regime." At this point, the inhomogeneity of toroidal magnetic fields introduces into the tokamak some of the properties of a mirror machine. Recall that the geometry of the torus causes the magnetic field to be stronger on the inside of a tokamak than on the outside, and recall further that the magnetic lines in the whole family of tokamak- and stellaratorlike devices are engineered so that they spiral around the axis of the doughnut, passing alternately near the inside and the outside. A charged particle being guided along such a line will alternately see regions of strong and weak magnetic fields. This is just what happens to particles in a mirror, and like a mirror particle an ion or electron in a toroidal field may be reflected back from the strong field. Such a particle will fail to complete a full circuit of the torus and will instead be "trapped." Bouncing back and forth between two regions of strong field, the trapped particles trace an orbit that is called a "banana" orbit because of its characteristic shape.

29. Rosenbluth to Hirsch, June 29, 1973.

30. Alvin M. Weinberg to H. P. Furth, D. W. Kerst, and M. N. Rosenbluth, June 9, 1972 (office files of J. Rand McNally, ORNL).

31. Author's interview with Clarke.

32. ORNL, "A Plan to Demonstrate the Feasibility of Controlled Fusion," March 1972, revised May 1972, 1–4 (office files of M. Roberts, DOE-OMFE). Author's interview with Guest.

33. Author's interviews with Morgan and Clarke. "Ormak Data Review Panel Report," February 28, 1973 (retired file 110.2, Thermonuclear Division, ORNL).

34. H. P. Furth, D. W. Kerst, and M. N. Rosenbluth, "1972 Report of the ORNL Review Committee on Thermonuclear Research," February 1, 1973 (files of J. Rand McNally, ORNL).

35. Author's interviews with Postma, Guest, Callen, and Dandl.

36. "Summary of Discussions on DT Burning Tokamak Experiments," December 19–20, 1973, "CTR 38 PPPL TFTR 1973," DOE-OMFE.

37. Author's interview with Dean.

38. "Furth views PPPL as solely a plasma physics laboratory. They are not interested in DT burning or reactor engineering. Bussard got the impression that Princeton is worried about not getting a bigger physics machine." Paraphrase of Bussard's trip report in "Minutes of Director's Meeting with Assistant Directors, June 14, 1973," "CTR Staff Meeting Minutes" (office files of W. C. Gough).

39. Author's interview with Gould.

40. Rosenbluth to Hirsch, June 29, 1973, cited in note 27.

41. "Status and Objectives of Tokamak Systems for Fusion," cited in note 27. The work started shortly after March 1973 and the report was published in 1974.

42. Arthur M. Sleeper, "Scaling Laws in Operating Tokamaks," June 3, 1975; "CTR 13, 1975," DOE-OMFE.

43. The other members were Dean; James D. Callen, who, although he was also from Oak Ridge, served as an arbitrator; Tihiro Ohkawa from General Atomic's Doublet tokamak program; and Paul Rutherford of Princeton.

44. Author's interview with Clarke; Metz, "Nuclear Fusion: The Next Big Step."

45. See, for example, *Newsweek*, January 22, 1973; Robert Gillette's articles in *Science 181* (1973), *passim*; CTR Staff minutes, April 20, 1973 (files of W. C. Gough); presidential message, and White House fact sheet, reproduced in *Hearing . . . Joint Committee on Atomic Energy . . . 93rd Congress, 1st Session, Atomic Energy Commission Report . . . for . . . Energy Research and Development Program, Dec. 11, 1973,* appendix 12. Congress was pressing even more vigorously for energy research dollars. Indeed, much of the $100 million that Nixon announced as a fiscal year 1974 add-on was merely a ratification by the administration of additions that Congress was making to Nixon's January budget submission. See the press conference of October 11, 1973, in the JCAE reference above, in which Nixon's proposal, then up to $115 million, is estimated as $20 to $30 million above congressional add-ons already made.

46. *New York Times*, February 7, 1973, 18; author's interviews with Teem and Hirsch; Dixy Lee Ray to Melvin Price, July 24, 1973, and Dixy Lee Ray to Mike McCormack, July 5, 1973, "R&D-18, CTR, Vol. 2," DOE-HO.

47. R. L. Hirsch, "Thinking Paper," June 1973 (file 17-8, Magnetic Fusion Energy Division, LLNL); Standing Committee meeting, July 1973, with marginal notes (file 17-8, Magnetic Fusion Energy Division, LLNL).

48. Author's interview with Hirsch. Stiefel, "The United States Breeder Reactor Program," cited in note 54.

49. Present were the four project leaders—F. L. Ribe, T. K. Fowler, M. B. Gottlieb, and H. Postma—two outside members—S. J. Buchsbaum and E. Creutz—and Hirsch and his three assistant directors: S. O. Dean, Robert Bussard, and A. W. Trivelpiece (author's interviews with Ribe, Buchsbaum, and Postma). Some of Fowler's and Ribe's objections may be gleaned from the minutes of the executive session of the FPCC, July 10-11, 1974, "OM7 Committees 1974," DOE-OMFE.

50. Hirsch, "Thinking Paper": "With respect to mirror research, it is recognized that the available margin for a practical economical power reactor is theoretically relatively small." See also "Position Paper on Controlled Thermonuclear Research," July 12, 1973, included in R. L. Hirsch to Edward J. Bauser, July 13, 1973, "R&D-18 CTR II," DOE-HO.

51. Author's interview with James M. Williams; report on AEC Budget Review Committee, minutes of director and assistant director meeting, February 12, 1972 (filed with CTR Staff Meeting minutes, files of W. C. Gough).

52. "Postion Paper on Controlled Thermonuclear Research."

53. Hirsch to Bauser, July 13, 1973; Ray to Price, July 13, 1973; "Implementation . . . would represent an acceleration of the program . . . a shortening of the

overall time-scale to a demonstration power plant" ("Position Paper on Controlled Thermonuclear Research").

54. Michael Stiefel, "Government Commercialization of Large Scale Technology: The United States Breeder Reactor Program 1964-1976," MIT thesis, June 1981; author's interviews with Hirsch and Teem; CTR staff meeting minutes for 1972 and 1973, *passim.*

55. Mike McCormack to D. L. Ray, June 30, 1973, and "Press Release by Congressman Mike McCormack," "R&D-18, CTR, Vol. 2," DOE-HO. Mike McCormack to Roy L. Ash, Sept. 21, 1973, *ibid.*

56. "Ormak Data Review Panel Report," cited in note 34; Hirsch, "Thinking Paper"; author's interviews with Furth, Dean, and Callen.

57. In the event, the real breakthrough in temperature for Ormak did not come until 1976. For Washington's disaffection see file, "CTR 38 ORNL HIGH FIELD ORMAK 1974," DOE-OMFE.

58. K. Bol et al., "Neutral Beam Heating in the Adiabatic Toroidal Compressor," cited in note 27.

59. John F. Clarke to Robert L. Hirsch, April 18, 1975 (retired file 110.1.1, Thermonuclear Division, ORNL).

60. "Summary of Discussions on DT Burning Tokamak Experiments, December 19-20, 1973," "CTR 38 PPPL TFTR 1973," DOE-OMFE.

61. Author's interviews with Dean, Furth, and N. Anne Davies; "Summary of Discussions."

62. J. M. Dawson, H. P. Furth, and F. H. Tenney, "Production of Thermonuclear Power by Non-Maxwellian Ions in a Closed Magnetic Field Configuration," *Physical Review Letters 26* (1971), 1156-1160. Tokamak reactors in 1971 could not be run to provide less than 1,000 megawatts.

63. Author's interviews with Dean, Callen, and Furth.

64. Author's interview with Furth.

65. Author's interviews with Dean and Callen.

66. Author's interview with Clarke; R. L. Hirsch, "Background for the July 9-11 Meeting of the Fusion Power Coordinating Committee," June 27, 1974, with added marginal notes (file 17-11, Magnetic Fusion Energy Division, LLNL).

67. R. L. Hirsch, "Vugraphs, DT Tokamak Meeting, December 19-20, 1973" (office files of W. C. Gough).

68. Metz, "Nuclear Fusion: The Next Big Step"; minutes of the CTR Staff meetings for 1974; author's interviews with Clarke and Callen. The opposing views of the two laboratories sound more clear-cut than they were. Fuzziness was introduced by the ever present element of speculations about possible future innovations. Perhaps, for example, two-component reactors were not a dead end. Indeed, the knowledge that one was about to be built gave rise immediately to new schemes about how it might be put to uses other than mere demonstration. Hopes for future and for unexpected inventions, however vague, were a significant part of the considerations. See Daniel L. Jassby, "Neutral-Beam-Driven Tokamak Fusion Reactors," *Nuclear Fusion 17* (1977), 309ff, for a discussion of possibilities, and for a bibliography of schemes.

69. Author's interview with Hirsch; "Background for the July 9–11 Meeting of the Fusion Power Coordinating Committee," cited in note 66.

70. Dixy Lee Ray, "The Nation's Energy Future," reprinted in *Hearing JCAE, 93rd Congress, 1st Session, . . . A.E.C. . . . Energy Research and Development Program, Dec. 11, 1973*, 76 and 189–190; R.L. Hirsch to files, December 7, 1972, "Princeton Plasma Physics Laboratory, 1972," DOE-OMFE; Richard L. Garwin to Melvin Price, February 14, 1973, "R&D-18, CTR, Vol. I," DOE-HO.

71. CTR Staff meeting minutes, March to July 1974, *passim*.

72. Paul W. McDaniel to the author, March 2, 1979: "The Division of Research had a policy from its inception of allowing the scientists involved to manage their own research."

73. The writing of plans also had political aspects. Since fusion energy was a technology that had not yet been proved possible, it was important to exhibit something specific out there in the long term, a demonstration reactor and a timetable to reach it. In the short term, it was important to show that the program was sufficiently under control for milestones to be set and met (interviews with Teem and Hirsch).

74. Computed by subtracting the percentage for reactor oriented development and technology: $186 million divided by the 5-year total of $1,200 million. The figures are taken from the February 1974 program plan.

75. Author's interviews with Don A. Baker and Grove.

Chapter 12

1. R. F. Post and M. N. Rosenbluth, "Electrostatic Instabilities in Finite Mirror-Confined Plasmas," *Physics of Fluids 9*, (1966), 730–749. F. H. Coensgen, "Experiment Sequence Toy Top → 2XIIB → Future," August 23, 1977, MFE/CP/77-164 (office files of F.H. Coensgen, LLNL); "Planning Recommendations of the Ad Hoc Executive Panel on Open Magnetic Confinement Systems," November 17, 1972, WASH-1299, DOE-OMFE.

2. C. C. Damm, untitled memoir, September 1977.

3. Minutes of the CTR staff meetings, fall 1972 (office files of W. C. Gough); author's interviews with Dean and Callen; "Planning Recommendations of the Ad Hoc Executive Panel on Open Magnetic Confinement Systems"; Stephen O. Dean, "Future Directions in Open Systems Research," October 31, 1972, and S. O. Dean, "Proposed Office Position Regarding Open Systems Research," n.d. (included in "CTR Staff Meeting Minutes," Gough files).

4. Author's interviews with Fowler and Post; file "CTR 38 LLL 1973," *passim*, DOE-OMFE, F. H. Coensgen et al., "2XII Program Plan" February 28, 1973, LLL Prop.-102; F. H. Coensgen to the author, December 29, 1981.

5. Author's interview with Fowler; "CTR 38 LLL 1973," *passim*.

6. Don Steiner, "The Technological Requirements for Power by Fusion," *Nuclear Science & Engineering 58* (1975), 124–126.

7. Author's interview with Werner. R. W. Moir, ed., "Standard Mirror Fusion Reactor Design Study," UCID-17655. This paper, dated January 30, 1978, refers to work done in fiscal year 1976.

8. R. L. Hirsch, "Thinking Paper," June 25, 1973 (file 17-8, Magnetic Fusion Energy Division, LLNL).

9. "CTR Program Review," October 23, 1974 (file 14-5, Magnetic Fusion Energy Division files, LLNL); author's interview with Milton Johnson, June 21, 1979.

10. F. L. Ribe to H. Grad, July 11, 1968, "Ribe Correspondence," LASL-RC; author's interview with Grad; A. A. Blank, H. Grad, and H. Weitzner, "Toroidal High-β Equilibria," Novosibirsk Conf. 607–617; H. Grad to F. L. Ribe, January 16, 1968, F. L. Ribe to H. Grad, January 22, 1968, and H. Grad to F. L. Ribe, July 8, 1968, all in "Ribe Correspondence," LASL-RC.

11. Author's interview with F. L. Ribe; Fred L. Ribe to Ewald Fuenfer, November 5, 1968; F. L. Ribe to John L. Johnson, February 28, 1969, F. L. Ribe to Harold Grad, December 31, 1968, and January 2, 1969, all in "Ribe Correspondence," LASL-RC; author's interview with Werner B. Riesenfeld, July 16, 1979; "Discussion Paper, DCTR and LASL Meeting, December 9–10, 1974" (office files of George A. Sawyer, LASL).

12. W. R. Ellis, "History of Planned Scyllac Feedback Experiments," December 9, 1974 (office files of W. B. Riesenfeld). Ellis uses τ for the sensor reaction time. I have used T to avoid confusion with the confinement time.

13. R. F. Taschek to Amasa S. Bishop, May 8, 1969, "Ribe Correspondence," LASL-RC.

14. R. F. Taschek, "Status Report on Scyllac," July 20, 1979, "LASL-1970," DOE-OMFE. This paragraph also draws upon the author's interviews with Taschek and Ribe.

15. "Discussion Paper, DCTR and LASL Meeting, December 9–10, 1974"; F. L. Ribe to R. F. Taschek, "Discussion Concerning Scyllac . . . June 26, 1970," "Ribe Correspondence, 1970," LASL-RC.

16. Author's interview with Ribe.

17. Ellis, "History of Planned Scyllac Feedback Experiments;" author's interviews with Riesenfeld and Ribe; R. L. Hirsch to files, November 9, 1972, and Sibley C. Burnett to Robert L. Hirsch, November 27, 1972 ("Reference Material for the History of the Scyllac Program, DCTR and LASL Meeting December 9 and 10, 1974," office files of G. A. Sawyer).

18. Ellis, "History."

19. Feedback experiments had been started on an 8-meter sector of this machine, but, again, were not completed. They were carried to a point where it was seen that feedback was not working, but the sector was then absorbed into the full torus (Ellis, "History").

20. Ellis, "History;" Richard E. Siemon, "A Summary of Scyllac Results," April 1978, LASL-7125-MS.

21. Minutes of the executive session of FPCC, July 10–11, 1974, "OM7 Committees, 1974, Fusion Power Coordinating Committee," DOE-OMFE.

22. Interview with Robert Krakowski; Don Steiner, "The Technological Requirements for Fusion," *Nuclear Science & Engineering* 58 (1975), 126–129. The Los Alamos fusion reactor study group had collaborated with engineers at Westinghouse to work out

a solution for the energy storage and transfer problem. It was a mechanical system, a "homopolar generator" in which the energy is passed to, and taken from, a rotating flywheel. The group had less success, however, in designing a vacuum wall capable of withstanding the predicted thermal stress.

23. R. L. Hirsch, "Diary," entries for May 20, June 18, June 24, June 25, and October 21, 1975 (office files of Maurice J. Katz). Author's interviews with Williams, Ribe, and George A. Sawyer, July 12, 1979, and Krakowski and Rose.

24. Hewlett and Duncan, *Nuclear Navy*. Author's interviews with Hirsch and Kintner.

25. Author's interview with Robert C. Seamans, Jr.; articles by Philip M. Boffey and William D. Metz, *Science 189* (1975), *passim*.

26. Author's interview with Milton D. Johnson.

27. The particles that enter the plasma vessel are neutral. The beam strength, however, is measured in terms of the total current that would be carried by an equivalent number of positive ions.

28. F. H. Coensgen et al., "2XIIB Plasma Confinement Experiments," IAEA, Berchtesgaden Conf. *3*, 135–143; author's interview with Milton D. Johnson; file "CTR 38 LLL 2XIIB 1975," *passim*, DOE-OMFE; Coensgen, "Toy Top → 2XIIB → Future."

29. Hirsch, "Diary," October 16, 1975.

30. Warren E. Quinn and George A. Sawyer, "Meeting at ERDA Headquarters on LASL CTR Program," memorandum to files, June 19, 1975 (office files of G. A. Sawyer, LASL). Quinn and Sawyer remark, "It was clear that DCTR put reactor design considerations first, before plasma physics," 12; Hirsch, "Diary," June 18, 1975; author's interview with Williams.

31. J. A. Phillips, "Rebirth of the Z-pinch," October 7, 1969 (office files of J. A. Phillips, LASL); author's interview with Phillips; J. A. Phillips, "Proposal for a Shock Heated Toroidal Z-Pinch Experiment," December 1, 1970, "Sherwood Reports," LASL-RC; author's interview with Don A. Baker; D. A. Baker and J. N. Di Marco, "The LASL Reversed-Field Pinch Program," December 1975, LA-6177-MS; USDOE, "Report on the Concept Review Committee Recommendations for Proof-of-Principle Alternate Concept Programs," March 1979 (James F. Decker, chairman), DOE/ET-0085.

32. Author's interview with Ribe; F. L. Ribe to the author, January 9, 1982.

33. Author's interviews with Parker, Coppi, and Montgomery.

34. B. Coppi, M. Oomens, R. Parker, L. Pieroni, C. Schuller, S. Segre, and R. Taylor, "Electron Slide-away Regime in High Temperature Plasmas," MIT Plasma Research Report #PRR-7417, Research Laboratory of Electronics, December 1974; author's interview with Coppi.

35. B. Coppi, G. Lampis, F. Pegoraro, L. Pieroni, and S. Segre, "High Temperature Regimes in Magnetically Confined Plasmas," MIT Research Laboratory of Electronics, Report #PRR-7524, October 1975, E 18. The analogy was with the density dependence of confinement time as calculated from the "neo-classical" theory. The confinement time of the energy is the parameter under discussion. The factor of proportionality between confinement time and density was not constant, but varied with magnetic field strength and current. Most of the work was done by a member of a group

om Frascati, Italy. The Frascati team was one of a number of European groups
hat Coppi had enlisted to bring to the Alcator needed practical expertise.

6. Hirsch, "Diary," October 27 and November 10, 1975.

7. Author's interview with Parker.

8. The French also achieved this milestone at the same time on their TFR tokamak.
. A. Berry et al., "Confinement and Neutral Beam Injection Studies in Ormak,"
erchtesgaden Conf. *1* 49–68; author's interviews with Berry, Morgan, Clarke, and
allen.

9. Hirsch, "Diary," September 26, 1975.

0. Hirsch, "Diary," December 19, 1975.

1. Hirsch, "Diary," *passim*; interviews with Hirsch, Kintner, Teem, and William R.
llis.

2. Hirsch, "Diary," December 10, 1975.

3. "An Assessment of the Role of Magnetic Mirror Devices in Fusion Power De-
elopment," April 1976, ERDA 76–74; author's interview with Clarke.

4. Author's interview with Kintner.

5. William R. Ellis, chairman, "Report of the Open Systems Technical Review
anel," appendix C, 84, September 1976, ERDA - 76/140.

5. Author's interview with Werner.

7. Author's interviews with Sawyer and Ribe; R. E. Siemon, "A Summary of Scyllac
esults," cited in note 20.

8. Interviews with Kintner and Ellis.

9. E. E. Kintner to R. F. Taschek, December 21, 1976 (office files of George A.
awyer, LASL).

0. Author's interview with Kintner.

1. *House of Representatives, Subcommittee on Fossil and Nuclear Energy Research of the
ommittee on Science and Technology, 95th Congress, 1st Session Hearings, 1978 ERDA
uthorizing Legislation*, part 3, February 24, 1977.

2. "Fusion Power By Magnetic Confinement Program Plan, Vol. I, Summary," July
976, ERDA-76/110/1.

3. R. L. Hirsch to the author, February 12, 1981.

4. *New York Times*, February 23, 1977, 1:5, February 25, 1977, 34:2, January 24,
978, 17:5; *Energy Daily*, January and February 1977, *passim*.

5. Author's interviews with Kintner, Hirsch, and John M. Deutch.

6. William D. Metz, "MIT Chemist, Schlesinger Ally, Assumes Energy Research
ost," *Science 198* (1977), 1125–1126.

7. Author's interviews with Kintner and Deutch.

8. Author's interviews with Buchsbaum and Deutch.

9. Particle fusion is a species of inertial confinement fusion that uses particle beams
stead of laser light beams to irradiate pellets of fusion fuel.

0. In addition to Foster (of TRW, Inc., with a background as director of Livermore
aboratory and of the Defense Research and Engineering Program in the Department

of Defense), there was Solomon J. Buchsbaum of Bell Laboratories; plasma theorist John M. Dawson of the University of California; accelerator physicist and particle fusion specialist Burton Richter of Stanford; Michael M. May, director of Livermore Laboratory; Edward T. Gerry, laser fusion expert, from W. J. Shaffer Assoc.; Thomas M. O'Neil, University of California, San Diego; and John W. Simpson from the Office of Energy Research, DOE.

61. Author's interviews with Deutch, Kintner, and Clarke. "Final Report of the Ad Hoc Experts Group on Fusion," June 7, 1978, DOE-OMFE.

62. For a fine review of all the studies through 1974 see Don Steiner,"The Technological Requirements for Power by Fusion," *Nuclear Science & Engineering 58* (1975), 107–165.

63. The University of Wisconsin Fusion Feasibility Group, "UWMAK-I," November 20, 1973 (revised March 15, 1974), UWFDM-68, 1, DOE-OMFE.

64. R. G. Mills, ed., *A Fusion Power Plant*, August 1974, MATT-1050.

65. Author's interview with Kulcinski.

66. Interview with Harold R. Drew by Robert B. Belfield; John P. Holdren, "Fusion Energy in Context: Its Fitness for the Long Term," *Science 200* (1978), 175.

67. The pages on utilities are largely based upon materials collected by Robert B. Belfield. EPRI absorbed the Electric Research Council as well as the Edison Electric Institute's Fusion Committee.

68. Helium-3 itself would have to be produced in a remote plant. The shielding would be less than half by Ashworth's computations, however, since the DD reaction producing He^3 yields 2.5-MeV neutrons, as against the DT reaction's 14-MeV neutrons.

69. $1,000–$3,000 per kilowatt in 1974–1975 dollars. These numbers, however, were speculative in the extreme, since as UWMAK leaders Robert Conn and Gerald Kulcinski pointed out, "Five or six years is only enough time to scratch the surface of an area like fusion reactor technology, and it should be clear that our most promising engineering discoveries still lie in the future." Letter in *Science 193*, 630–161.

70. Clinton P. Ashworth, "A Utility View of Fusion," June 27, 1977, typescript.

71. William D. Metz, "Fusion Research (III): New Interest in Fusion-Assisted Breeders," *Science 193*, (1976), 307–309.

72. John M. Dawson and Burton D. Fried to Edwin E. Kintner, April 29, 1977, "ONM 7-19-77," DOE-OMFE; W. C. Wolkenhauer, R. A. Huse, and Betty Jensen, "The Hybrid Reactor, What is the Next Step?" US/USSR Fusion Fission Hybrid Symposium, January 22–26, 1979, typescript; Michael Lotker, "Commercializing Fusion," talk prepared for the American Nuclear Society, November 1976; Raymond A. Huse, James M. Burger, and Michael Lotker, "Fusion-Fission Energy Systems, Some Utility Perspectives," December 1974, talk prepared for AEC Meeting on Fusion Driven Fission Systems, typescript. Huse was not against the tokamak, as might be anticipated from his ties to Princeton.

73. W. D. Metz, "Fusion Research (I): What is the Program Buying the Country?," *Science 192* (1976), 1320–1323; "II: Detailed Reactor Studies Identify More Problems," *ibid. 193*, (1976), 38–40; "III: New Interest in Fusion-Assisted Breeders," *ibid.* (1976), 307–309. See also the editorial "Glamorous Nuclear Fusion" in the July 23, 1976, issue by Philip H. Abelson.

74. Foster committee, "Final Report," 1 and 3; interview with Deutch.

75. Foster committee, "Final Report," 12. See also Dawson and Fried to Kintner, April 29, 1977.

76. Foster committee, "Final Report," 5–6: "Tandem mirror machines promise better efficiency," *Physics Today 31* (1978), 18–19.

77. Foster committee, "Final Report," 6.

78. John M. Deutch, "The Department of Energy Policy for Fusion Energy," September 1978, DOE/ER-0018, 6 and 18–19. See also J. M. Deutch to secretary, deputy secretary, and acting under secretary, "Information Memorandum, FY 1981 R&D Strategy for Fusion," June 22, 1979, DOE-OMFE.

79. James F. Decker et al., "An Evaluation of Alternate Magnetic Fusion Concepts," May 1978, DOE/ET-0047.

80. R. A. Dandl et al., "The Elmo Bumpy Torus Experiment," November 1971, ORNL-TM-3694; R. A. Dandl et al., "Plasma Confinement and Heating in the Elmo Bumpy Torus (EBT)," IAEA, Tokyo Conf., 141–149; R. A. Dandl et al., Executive Summary of "The EBT-II Conceptual Design Study," September 1978, ORNL/TM-5955, 1–3; author's interviews with Dandl and Guest.

81. Kintner had inserted a Proto-EPR into the 1976 plan, but it had not found favor with the Foster committee.

82. Author's interview with Kintner.

83. Foster committee, "Final Report," 5; author's interview with Clarke.

84. J. M. Deutch, "FY 1981 R&D Strategy for Fusion," 35–36.

85. Author's interview with Deutch; file "ONM9-1977: Management Planning," *passim*, DOE-OMFE.

86. E. E. Kintner to M. B. Gottlieb, March 23, 1977 ("ENP 13-5 PPPL/PLT, 1977," DOE-OMFE). D. Ignat to S. O. Dean, February 1, 1977, and S. O. Dean to Harold Eubank, February 11, 1977, *ibid.*

87. L. R. Grisham, "Neutral Beam Heating in the Princeton Large Torus," *Science 207* (1980), 1301–1309. It is worth remarking that by this time the fusion community had begun to arrive at a general appreciation that the formula that had been incorporated into the tokamak Bible was erroneously pessimistic. Linear theory had been used to derive it, whereas nonlinear theory was necessary. In addition, the linear theory used had been unacceptably rough (author's interview with Coppi). See Walter L. Sadowski, "Trapped Particle Instabilities Workshop," May 1979, DOE/ET-0097.

88. This ignores the very large losses in the coils that produce the ohmic heating and magnetic fields. In a reactor, however, these coils would not use much power because they would be superconducting. For this, and also a discussion of the power actually lost in the neutral injectors, see Grisham, "Neutral Beam Heating."

89. D. L. Jassby, letter, *Science 202* (1978) 370; W. D. Metz, rejoinder.

90. The PLT results did cause a contretemps in the department, however, because they coincided with the determination of budget figures for fiscal year 1980. Kintner had wanted to release the results contemporaneously with Princeton's presentation of them at the International Atomic Energy Agency's upcoming fusion conference, scheduled for August in Innsbruck, Austria. His plan was frustrated, however, when

the Fusion Energy Foundation, an arm of the political cult group, the U.S. Labor Party, leaked the story to the *Miami Herald* on August 11. In consequence, the good news received an initially angry reception in Schlesinger's office (author's interviews with Kintner and Clarke; William D. Metz, "Report of Fusion Breakthrough Proves to Be a Media Event," *Science 201* (1980), 792–794; on the U.S. Labor Party, see the *New York Times*, October 9, 1979, 1).

Chapter 13

1. General Atomic got approximately $18.4 million and $22.1 million in fiscal years 1980 and 1981, respectively, as against Los Alamos's $16.5 and $18.9 millions. The figures for Oak Ridge, Princeton, and Livermore range from $63 to $100 million (Michael Roberts, Director, OFE Division of Planning and Projects, to the author, October 1981).

2. E. Michael Blake, "Fusion in the United States: An Overview," *Nuclear News 23* (1980), 67.

3. For examples, see "Historical Notes on Important Tubes and Semiconductor Electronic Devices," *IEEE Transactions on Electron Devices, ED 23,* 7 (1976). See also Ernest Braun and Stuart MacDonald, *Revolution in Miniature: The History and Impact of Semiconductor Electronics*, Cambridge University Press, 1978, and Robert W. Wilson, Peter K. Ashton, and Thomas P. Egan, *Innovation, Competition, and Government Policy in the Semiconductor Industry*, Lexington Mass., Lexington, Books, 1981.

4. Author's interview with Westendorp.

5. John M. Logsdon, *The Decision to Go to the Moon: Project Apollo and the National Interest*, Cambridge, Mass., MIT Press, 1970; Jonathan F. Galloway, *The Politics and Technology of Satellite Communications*, Lexington, Mass., Lexington Books, 1972.

6. Harold P. Green and Alan Rosenthal, *Government of the Atom*, Atherton Press, 1963.

7. R. Jeffrey Smith, "Legislators Accept Fast-Paced Fusion Program," *Science 210* (1980), 290–291; *Subcommittee on Energy Research and Development of the Senate Committee on Energy and Natural Resources, Hearings, July 28 and August 5, 1980, to Consider the Magnetic Fusion Energy Engineering Act*; Public Law 96-386, in *U.S. Code, 96th Congress, 2nd Session, 1980.*

8. Daniel S. Greenberg, "200 BEV; Harmony Prevails as Physicists Close Ranks," *Science 155* (1967), 983–985; Daniel S. Greenberg, *The Politics of Pure Science*, New York, New American Library, 1967.

9. Testimony by Roy W. Gould, *JCAE, The Current Status of the Thermonuclear Program, 1971,* 2, 285.

10. David O. Edge and Michael J. Mulkay, *Astronomy Transformed: The Emergence of Radio Astronomy in Britain*, New York, John Wiley, 1976, chapter 7, "Discovery, Competition and Cooperation."

11. Author's interview with Yoshikawa.

12. A survey of plasma heating methods and plasma diagnostics, together with references to more extensive sources, is given in chapters 14 and 15 of Kenro Miyamoto, *Plasma Physics for Nuclear Fusion*, Cambridge, Mass., MIT Press, 1980 (original Japanese edition 1976). Two older books that remain standard sources on all aspects

of fusion science and technology are David J. Rose and Melville Clark, Jr., *Plasmas and Controlled Fusion* (New York: Wiley, 1961) and Samuel Glasstone and Ralph H. Lovberg, *Controlled Thermonuclear Reactions* (Princeton, NJ: Van Nostrand, 1960).

13. Aluminum alloys were brought forward by Commissioner Willard Libby in *Bulletin of the Atomic Scientists 14* (1958), 191. Compare "Radioactive Inventories of Reactor Economies," G. Kessler and G. L. Kulcinski, in W. Haefele, J. P. Holdren, G. Kessler, and G. L. Kulcinski, *Fusion and Fast Breeder Reactors*, 224–227.

14. Bruno Coppi comments that the situation is no different from other fields like astrophysics or high-energy physics. A new regime, in his view, always brings new phenomena. It is only after the new phenomena have been explored experimentally and the theory reworked according to the data uncovered that the revised and completed theory becomes available to guide engineering design.

15. J. L. Tuck, "Controlled Thermonuclear Reactions," Sherwood Conf., 1952, 44.

Index

Lightning Source UK Ltd.
Milton Keynes UK
UKOW04f2258300414

230890UK00001B/65/P